Knowledge-Intensive Economies and Opportunities for Social, Organizational, and Technological Growth

Miltiadis D. Lytras
Effat University, Saudi Arabia

Linda Daniela
University of Latvia, Latvia

Anna Visvizi
Effat University, Saudi Arabia

A volume in the Advances in Knowledge Acquisition, Transfer, and Management (AKATM) Book Series

IGI Global
DISSEMINATOR OF KNOWLEDGE

Published in the United States of America by
IGI Global
Information Science Reference (an imprint of IGI Global)
701 E. Chocolate Avenue
Hershey PA, USA 17033
Tel: 717-533-8845
Fax: 717-533-8661
E-mail: cust@igi-global.com
Web site: http://www.igi-global.com

Library of Congress Cataloging-in-Publication Data

Names: Lytras, Miltiadis D., 1973- editor. | Daniela, Linda, 1967- editor. | Visvizi, Anna, 1976- editor.
Title: Knowledge-intensive economies and opportunities for social, organizational, and technological growth / Miltiadis D. Lytras, Linda Daniela, and Anna Visvizi, editors.
Description: Hershey, PA : Information Science Reference, [2019]
Identifiers: LCCN 2018024575| ISBN 9781522573470 (h/c) | ISBN 9781522573487 (eISBN)
Subjects: LCSH: Technological innovations--Economic aspects. | Technological innovations--Social aspects. | Knowledge management. | Continuing education.
Classification: LCC HC79.T4 K5999 2019 | DDC 303.48/34--dc23 LC record available at https://lccn.loc.gov/2018024575

This book is published in the IGI Global book series Advances in Knowledge Acquisition, Transfer, and Management (AKATM) (ISSN: 2326-7607; eISSN: 2326-7615)

British Cataloguing in Publication Data
A Cataloguing in Publication record for this book is available from the British Library.

Advances in Knowledge Acquisition, Transfer, and Management (AKATM) Book Series

Murray E. Jennex
San Diego State University, USA

ISSN:2326-7607
EISSN:2326-7615

Mission

Organizations and businesses continue to utilize knowledge management practices in order to streamline processes and procedures. The emergence of web technologies has provided new methods of information usage and knowledge sharing.

The **Advances in Knowledge Acquisition, Transfer, and Management (AKATM) Book Series** brings together research on emerging technologies and their effect on information systems as well as the knowledge society.**AKATM** will provide researchers, students, practitioners, and industry leaders with research highlights surrounding the knowledge management discipline, including technology support issues and knowledge representation.

Coverage

- Cognitive Theories
- Cultural Impacts
- Information and Communication Systems
- Knowledge acquisition and transfer processes
- Knowledge management strategy
- Knowledge Sharing
- Organizational Learning
- Organizational Memory
- Small and Medium Enterprises
- Virtual communities

IGI Global is currently accepting manuscripts for publication within this series. To submit a proposal for a volume in this series, please contact our Acquisition Editors at Acquisitions@igi-global.com or visit: http://www.igi-global.com/publish/.

The Advances in Knowledge Acquisition, Transfer, and Management (AKATM) Book Series (ISSN 2326-7607) is published by IGI Global, 701 E. Chocolate Avenue, Hershey, PA 17033-1240, USA, www.igi-global.com. This series is composed of titles available for purchase individually; each title is edited to be contextually exclusive from any other title within the series. For pricing and ordering information please visit http://www.igi-global.com/book-series/advances-knowledge-acquisition-transfer-management/37159. Postmaster: Send all address changes to above address.

Titles in this Series

For a list of additional titles in this series, please visit: www.igi-global.com/book-series

Global Information Diffusion and Management in Contemporary Society
Zuopeng (Justin) Zhang (State University of New York at Plattsburgh, USA)
Information Science Reference • copyright 2019 • 331pp • H/C (ISBN: 9781522553939) • US $195.00 (our price)

Contemporary Knowledge and Systems Science
W. B. Lee (Hong Kong Polytechnic University, China) and Farzad Sabetzadeh (Hong Kong Polytechnic University, China)
Information Science Reference • copyright 2018 • 340pp • H/C (ISBN: 9781522556558) • US $185.00 (our price)

Enhancing Knowledge Discovery and Innovation in the Digital Era
Miltiadis D. Lytras (The American College of Greece, Greece) Linda Daniela (University of Latvia, Latvia) and Anna Visvizi (Effat University, Saudi Arabia)
Information Science Reference • copyright 2018 • 363pp • H/C (ISBN: 9781522541912) • US $215.00 (our price)

N-ary Relations for Logical Analysis of Data and Knowledge
Boris Kulik (Russian Academy of Science, Russia) and Alexander Fridman (Russian Academy of Science, Russia)
Information Science Reference • copyright 2018 • 297pp • H/C (ISBN: 9781522527824) • US $225.00 (our price)

Radical Reorganization of Existing Work Structures Through Digitalization
Punita Duhan (Meera Bai Institute of Technology, India) Komal Singh (Meera Bai Institute of Technology, India) and Rahul Verma (Bhai Parmanand Institute of Business Studies, India)
Business Science Reference • copyright 2018 • 287pp • H/C (ISBN: 9781522531913) • US $235.00 (our price)

Maximizing Social Science Research Through Publicly Accessible Data Sets
S. Marshall Perry (Saint Mary's College of California, USA)
Information Science Reference • copyright 2018 • 328pp • H/C (ISBN: 9781522536161) • US $195.00 (our price)

Promoting Interdisciplinarity in Knowledge Generation and Problem Solving
Mohammed Nasser Al-Suqri (Sultan Qaboos University, Oman) Abdullah Khamis Al-Kindi (Sultan Qaboos University, Oman) Salim Said AlKindi (Sultan Qaboos University, Oman) and Naifa Eid Saleem (Sultan Qaboos University, Oman)
Information Science Reference • copyright 2018 • 324pp • H/C (ISBN: 9781522538783) • US $195.00 (our price)

IGI Global
DISSEMINATOR OF KNOWLEDGE

701 East Chocolate Avenue, Hershey, PA 17033, USA
Tel: 717-533-8845 x100 • Fax: 717-533-8661
E-Mail: cust@igi-global.com • www.igi-global.com

Table of Contents

Detailed Table of Contents

This chapter summarizes the development and transformation of the International Journal of Knowledge Society Research (IJKSR) into a new form called the International Journal of Smart Education and Urban Society (IJSEUS). In the chapter, there is used literature review methodology to analyze the 199 articles published in 32 issues from 2010 to 2017. The obtained results allow us to ascertain what were the emerging topics published in the journal, which types of research article are published more often, and which countries are represented. The aggregated data also allow us to identify future directions in developing a journal and promoting its visibility and citation. The descriptive analysis method was adopted for data analysis, the quantitative data were compiled using Excel software, and the content analysis was done descriptively by defining the research categories.

As technologies become more exciting, interactive, and reachable, various technological solutions are used in higher education. On the one hand, there is the conviction that technologies are indispensable, both for improving learning and for making learning process more effective, both in terms of learning outcomes and in terms of costs. Additionally, technology and technological solutions can provide sustainability of knowledge because students develop the competences that they will need in their future professional work. This chapter summarises the systematic literature review (SLR) carried out by the authors in analysing research that has been done on the impact of technology-enhanced learning on learning outcomes in order to understand what emerging research has been done, as the authors published a similar study one year ago. The SLR was conducted for the period of 2010 to 2018 by selecting peer-reviewed articles on specified terms. The selected articles were then analysed following sub-purposes. The descriptive analysis method was adopted for the data analysis.

While online education continues to grow, virtual instructors face certain asynchronous uncertainties when it comes to knowledge exchange with students. These challenges are especially prevalent in the cybersecurity and programming domains. To counteract such uncertainties and minimize teaching deficiencies expected to occur in asynchronous learning environments, this chapter assesses knowledge factors that impact virtual knowledge transfer and absorption processes. Synchronicity framework is proposed to integrate knowledge-seeking behavior, knowledge properties, knowledge domains, knowledge types, knowledge tools, and technology synchronicity. A real-life case is provided to integrate the framework in practice.

This chapter examines critical thinking as a veritable tool in the usage of values, norms, and beliefs for sustainable development in contemporary societies. The problem statement raises the need for knowledge of what critical thinking, social norms, and values are, as well as the competence in how to employ and apply its tools in social policy designed for development. Illustrating from an extensive contemporary literature on social norms and values, this chapter examines how social norms, values, and beliefs can enhance the development and transformation of the society. Findings hypothesize that critical thinking, social norms, and values, based on thoughts and experience, are capable of guiding human behavior towards truth, development, and functional transformation toward solution to many societal problems. The study suggests collaborative measures to encourage inculcating social norms and values in the society.

This chapter investigates seven Gulf cities (Kuwait City, Manama, Doha, Abu Dhabi, Dubai, Sharjah, and Muscat) that have grown rich due to large reserves of oil and natural gas. Now, with the threat of ending resources, governments focus on the development towards a knowledge society with knowledge-based

Volunteer geographic information and user-generated content represents a source of updated information about what people perceive from their environment. Its analysis generates the opportunity to develop processes to study and solve social problems that affect the people's lives, merging technology and real data. One of the problems in urban areas is the traffic. Every day at big cities people lose time, money, and life quality when they get stuck in traffic jams; another urban problem derived from traffic is air pollution. In the present approach, a traffic event classification methodology is implemented to analyze VGI and internet information related to traffic events with a view to identify the main traffic problems in a city and to visualize the congested roads. The methodology uses different computing tools and algorithms to achieve the goal. To obtain the data, a social media and RSS channels are consulted. The extracted data texts are classified into seven possible traffic events, and geolocalized. In the classification, a machine learning algorithm is applied.

E-services are becoming increasingly important for government and for business services. They provide benefits to all parties concerned, including service providers, users, and the society at large. This chapter presents a framework of key issues that any e-service should enjoy. The framework is based on three main domains: the e-service strategy, its architecture, and its performance. Each of these domains is structured in three groups of issues, which are associated with three main dimensions: management, targeted features, and the people concerned. Per this structure, the framework considers "45" key e-services issues, and these issues are derived from the available literature and previous experience. Awareness of the importance of these issues among staff working in the field is investigated by means of a survey designed for this purpose. The outcome shows that there is a clear lack of awareness with regards to various key issues, and this problem would demote e-services. In response, recommendations are derived for solving the problem and toward enhanced future development.

In the concept of internet of things (IOT), physical position of smart object is very useful for relevant function over sensor networks. However, the invalid information of indoor geo-localization systems relative to these wireless sensor compromises the intelligence of IOT network. Therefore, this chapter produces the recent progress in the indoor geo-localization systems and the IOTs area. It defines the best indoor geo-localization technologies that meet their needs while respecting the constraints related to sensor networks. This framework combines between simplicity of Bluetooth low energy (BLE), popular wi-fi infrastructure, and the k-nearest neighbor (KNN) algorithm (in order to filter the initial fingerprint dataset). This new conception increases real-time detection accuracy and guarantees the low energy consumption.

Cloud computing is considered a great paradigm that enables access to large scale, on demand, flexible computing infrastructure. Lower infrastructure cost and better application performance are some of the benefits of utilizing a cloud computing model. With the appearance of the new generation of multi-cloud computing, clients can benefit from a diversity of services. However, data security has become an important requirement for clients when dealing with clouds that may fail due to faults in the software or hardware, or attacks from malicious insiders. Hence, building a highly dependable and reliable cloud system has become a critical research problem. To that end, this chapter investigates and presents the results in relation to that how Byzantine fault tolerance (BFT) and secure cloud data storage and sharing techniques can be deployed to manage multi-cloud data dependability faults.

With the great mass of the pages managed through the world, and especially with the advent of the web, it has become more difficult to find the relevant pages after an interrogation. Furthermore, the manual filtering of the indexed web pages is a laborious task. A new filtering method of the annotated web pages (by a semantic annotation process) and the non-annotated web pages (retrieved from search engine Google) is then necessary to group the relevant web pages for the user. In this chapter, the authors first synthesize their previous work of the semantic annotation of web pages. Then, they define a new filtering method based on three activities. They also present their querying and filtering component of web pages; their purpose is to demonstrate the feasibility of our filtering method. Finally, the authors present an evaluation of this component, which has proved its performance for multiple domains, and they discuss the use of the extended Boolean retrieval method in the new filtering method.

The size of semi-structured data is increasing continuously. Handling semi-structured data efficiently is a challenging task. Keyword search is an important task, and required information can be retrieved without having knowledge of data storage hierarchy. There are several challenges in handling XML data. This chapter discusses various challenges in terms of lowest common ancestor (LCA) semantics, processing of queries efficiently, retrieving top-k results for user needed data. The existing approach is defined under many classes based on how the problem and solution are tackled. Analysis of keyword search and ranking techniques for retrieving desired information are discussed in detail.

Section 4
New Skills and New Mental Models for the New Era of Education

Today, with increased internet access, users are often interested in new content-based multimedia applications of high added value such as interactive TV, video on demand (VoD), and catch-up TV services such as YouTube or Dailymotion frameworks. Despite the easy and rapid access to media information of these services, they present the risk of the wide propagation of fake news. As a solution, the authors propose that the input for these services must be from a trustworthy traditional media, precisely TV program content. So, the automatic process of TV program identification and their internal segmentation facilitate the availability of these programs. In this chapter, the major originality of the authors' approach is the use of contextual and operational characteristics of TV production rules as prior knowledge that captures the structure for recurrent TV news program content. The authors validate their approach by experiments conducted using the TRECVID dataset that demonstrate its robustness.

In specific fields, such as e-training in nursing, involving computer-intensive training scenarios, there is an increased demand to deliver training services to a larger number of learners, and with it, the need for cloud services. However, to date there has been a lack of a formalized framework relating to the use of cloud computing for on-demand interactive e-training resources in nursing education. To this end, this chapter formalizes a conceptual framework for a cloud-based e-training system in nursing education. The conceptualization takes into consideration nursing e-training system requirements, with a focus on applying cloud computing technologies to ensure the dynamic scalability of virtual distributed services and computing power while maintaining QoS and security.

The value of education is a dynamic and multi-dimensional concept that refers not only to the educational model but also to the institutional mission and its goals, as well as to the specific standards of the system, facility, program, or event. In today's competitive scenario, the academic institutions need to focus explicitly on providing effective and quality education to the students with the help of experienced academicians. The education quality would increase when the faculty members are having a better quality of work life,

Preface

The quest for and the intention to contribute to sustainable and inclusive socio-economic growth and development constitute the key rationale behind this edited volume. Our day-to-day reality is filled with problems, challenges and concerns that to a large extent remain unresolved. It is imperative that members of the academia take a resolute stance towards these issues. Academic research should and can represent, both, a bold response to problems our societies are exposed to today and a source of alternative solutions to those problems. Researchers can generate novel creative ideas and responses. The twist of the matter is how to create incentives to do that and how to convince the otherwise disinterested stakeholders that it is their business too to get involved.

From this perspective, the new summation volume of IJSEUS, *Knowledge-Intensive Economies and Opportunities for Social, Organizational, and Technological Growth*, provides a fresh approach to the critical challenges of our times. It claims that smart education and smart urban systems – geared toward promoting effective creation, use and dissemination of knowledge – must be a key priority of modern society. Advancing this plea, the value added of this volume is derived from the basic understanding that Education plays the role of the key enabler in the process of navigating the challenges and problems our societies are exposed to today. The chapters included in this volume address diverse issues pertinent to the so-defined field.

Chapters 1-3 offer a strategic plan for the enhancement of higher education. The authors claim that smart education and urban society develop a long-term strategic plan to promote innovation and technology-enabled initiatives aimed at promoting inclusive growth. Technology-enhanced learning should be seen just a part of an integrated strategy geared toward that. Education should be able to respond to the challenges our societies face today by means of a variety of ubiquitous learning services and systems. Education should be able to promote well-defined pedagogical and development goals, including those related to safety, security, privacy and ethical issues. The relevant titles of the first three chapters, section A of our book is given below.

SECTION 1: SMART EDUCATION – TECHNOLOGY-ENHANCED LEARNING AND DATA-DRIVEN INNOVATION

Chapter 1: On the Way to Smart Education and Urban Society
Chapter 2: Technology-Enhanced Learning (TEL) in Higher Education – Where Are We Now?
Chapter 3: Knowledge Framework for Online Learning – Implications for Cyber-Security and Programming Education

Best practices and lessons learnt related to knowledge-intensive economies and technology driven innovation should provide international benchmarks for consultation. This is another key challenge towards sustainable growth in the global context. Inequities in development and problems in dissemination of scientific findings and bold innovations should be addresses. A critical role for science, education and research of our times is to disseminate the benefits of Innovation for maintaining a sustainable quality of life for the entire planet. In our summation volume we provide several case studies from the Gulf Cooperation Council countries, BRICS and Mexico. These can be seen as interesting case studies and can provide critical thinking and engagement. In our summation volume Chapters 4 to 7 address this issue.

SECTION 2: INITIATIVES ON KNOWLEDGE-INTENSIVE ECONOMIES AND TECHNOLOGY-DRIVEN INNOVATION

Chapter 4: The Era After Oil – Knowledge-Intensive Cities on the Arabian Gulf
Chapter 5: The National Plan of E-Government – How the UAE Government Modernized Citizen Services
Chapter 6: Exploring Knowledge Activities for Bringing Saudi Arabia and the BRICS Together
Chapter 7: Classification of Traffic Events in Mexico City Using Machine Learning and Volunteered Geographic Information

Information and Communication Technologies, and their evolution provide a new fascinating context for exploitation. In most of cases the Innovation and the Economic Growth of our times is Technology Driven. Numerous emerging technologies can be uses toward this direction including Artificial Intelligence, Cloud Computing, Analytics, 5G, Recommendation Systems, Data Science, Content Annotation and Retrieval, Big Data and Internet of Things, BlockChain, Medical Informatics, Smart Cities, etc. Section 3, and Chapters 8 to 13 summarize a very interesting debate on this.

SECTION 3: STREAMLINE AND EMERGING TECHNOLOGIES FOR TECHNOLOGY-DRIVEN INNOVATION – CASE STUDIES

Chapter 8: Exploring Key E-Services Issues Toward Future Development
Chapter 9: Improved Indoor Geo-Localization System – Bluetooth Low Energy Technology of iBeacons and Wi-Fi Network for Better IOT Systems Accuracy
Chapter 10: Managing Multi-Cloud Data Dependability Faults
Chapter 11: Retrieval of Relevant Web Pages by a New Filtering Method
Chapter 12: Review on Keyword Search and Ranking Techniques for Semi-Structured Data
Chapter 13: Extraction and Annotation of News Topics From TV Streams for Web Video Sharing – A Contribution to Produce Reliable Online Video News Content

Nothing can be done without a new well-defined mix of skills, competencies and mental models required for the Transformative Education era: Higher Education should promote feasible visions for the improvement of our societies and the cultivation of sustainable social inclusive development, growth and justice. In the Section 4 of our book our contributors elaborate on this necessity. Chapter 14 to 16 provide a good overview of strategies and models.

SECTION 4: NEW SKILLS AND NEW MENTAL MODELS FOR THE NEW ERA OF EDUCATION

Chapter 14: A Conceptual Model for Cloud-Based E-Training in Nursing Education
Chapter 15: Academicians Quality of Work Life and Its Influence on Value of Education
Chapter 16: The Imperatives of Critical Thinking, Social Norms, and Values in Africa – Pathways to Sustainable Development

Overall, the message this volume makes is twofold. First, education is piece and parcel of modern economic development model. Therefore, regulatory frameworks in the context of which education process unfolds should embrace a holistic view of education. In other words, a virtuous circle is at work when it comes to education, the use of ICT in teaching and learning process, and the outcomes well-functioning education systems can produce. Second, education is all about educating, i.e. teaching and learning excellence are only a function of the greater imperative of educating the members of a given society. Educating requires dialogue based on trust, respect, and mutual understanding. This means that the learner's and educator's agency have to be re-introduced in the debate on the shape and performance of educational systems today. Only in this way, i.e. by promoting the role of teachers, instructors, professors as, primarily, educators, will we be able to use the education systems worldwide to contribute to peace and prosperity worldwide.

Miltiadis D. Lytras
Effat University, Saudi Arabia

Linda Daniela
University of Latvia, Latvia

Anna Visvizi
Effat University, Saudi Arabia

Introduction

It is our distinct pleasure to present to the communities of academics and policy-makers the new summation volume of IJSEUS. The title of this volume, *Knowledge-Intensive Economies and Opportunities for Social, Organizational, and Technological Growth* and so its content, constitute a response to the realities of the post-industrial, information- and innovation- driven era. In this context of technological innovation, the need to adopt technology fast and develop ethically and socially sensitive policies, the imperative of inclusive and balanced socio-economic and technological growth and development is paramount for our societies. In a world were political and economic uncertainties, uneven pace of development and inequality continue to mar peoples' wellbeing, questions of knowledge production and learning acquire particular importance. Similarly, as economic development and, so personal development opportunities across countries and regions vary, it is important that that educators and policy-makers engage in a dialogue on how to collaborate to boost access to education, to improve the quality of instruction, to motivate the teachers and professors, to increase the society's and universities' administration recognition of their input, and – as a result – to create conditions conducive to growth and development in rural and urban areas, in the developed and developing world. The cases of the Millennium Development Goals MDGs) and the Sustainable Development Goals (SGDs) suggest that a lot can be achieved in relatively short period of time. From a different angle, the work of the Organization of Economic Co-Operation and Development (OECD) fills with conviction that evidenced-based policy-making, best practice sharing, and peer review can make the positive difference.

This volume inserts itself in these debates and efforts as it makes the case for the crucial role of knowledge and learning in the process of promoting sustainable development and smart education. It outlines how to structure global collaboration among international organizations, academic institutions and the business sector if balanced development is to be advanced and the opportunities the use of sophisticated technologies are to be seized.

This volume serves as an invitation to the global academic community, the business sector and policy-makers to form a global alliance that would:

- Recognizes knowledge and learning management as the key pillar of knowledge society in the postindustrial era
- Promote knowledge-intensive economy as a new model of economic development geared toward the development of both urban and rural areas;
- Set inclusive socio-economic and technological development as a priority;
- Promote the role of higher education institutions as drivers of socio-economic growth and development

- Embrace and support the vision of education geared toward the equipping students with knowledge and skills conducive to the development and exercise of worldviews and ethical stance grounded on respect for fundamental freedoms, civic responsibility, culture and tradition;
- Links and integrates academia, the business sector and the policy-making community in their attempt to define scenarios of economic growth, development and collaboration, e.g. OBOR, smart villages initiative, etc.
- Promote peace and collaboration all over the world;
- Support the work, mission and vision of international organizations, incl. the OECD, the UN, the EU;

In our capacity of academics and consultants, editors, analysts, educators, parents, we are convinced that by adopting a forward-looking, firm and active stance toward problems and challenges our societies face today, incl. poverty, exclusion, racism, bullying, inequality, harassment and discrimination of whatever kind, we can contribute to positive change in the world. By engaging in teaching we have the privilege of sharing this normative stance with our students. By engaging in research, we have the opportunity to explore the intricacies specific to these problems and challenges. As editors we have the chance to reach our broader audience and encourage meaningful debate. We truly hope that this is the case now.

Miltiadis D. Lytras
Effat University, Saudi Arabia

Linda Daniela
University of Latvia, Latvia

Anna Visvizi
Effat University, Saudi Arabia

Section 1

Smart Education:

Technology-Enhanced Learning and Data-Driven Innovation

Chapter 1
On the Way to Smart Education and Urban Society

Linda Daniela
University of Latvia, Latvia

Kwok Tai Chui
City University of Hong Kong, Hong Kong

Anna Visvizi
The American College of Greece, Greece & Effat University, Saudi Arabia

Miltiades D. Lytras
The American College of Greece, Greece

ABSTRACT

This chapter summarizes the development and transformation of the International Journal of Knowledge Society Research (IJKSR) into a new form called the International Journal of Smart Education and Urban Society (IJSEUS). In the chapter, there is used literature review methodology to analyze the 199 articles published in 32 issues from 2010 to 2017. The obtained results allow us to ascertain what were the emerging topics published in the journal, which types of research article are published more often, and which countries are represented. The aggregated data also allow us to identify future directions in developing a journal and promoting its visibility and citation. The descriptive analysis method was adopted for data analysis, the quantitative data were compiled using Excel software, and the content analysis was done descriptively by defining the research categories.

INTRODUCTION

Scientific journals are the place where researchers can inform the society about their research results, where research ideas are debated and new research directions are discussed. Publication in a peer-reviewed journal means that anyone interested will be able to see the results of the research; furthermore, the authors have to convey their ideas in a concise way so that the reader has a clear conception of the article, because there is no possibility of obscure questions being clarified in a dialogue, as is possible when

DOI: 10.4018/978-1-5225-7347-0.ch001

presenting the results of research at conferences. There are myriad magazines that provide researchers with this opportunity.

The *International Journal of Knowledge Society Research* (IJKSR) published its first issue in 2010, stating that the journal established an efficient channel of communication between academic and research institutions, policy makers, government agencies and people involved with the analysis of various aspects of the knowledge society. The journal anchors government policies and examines the emerging knowledge-intensive economy and social environment. The IJKSR investigates the key issues of the knowledge society in terms of critical aspects of human life (e.g. health, education, culture, science, business, etc.) (Lytras & Daniela, 2010). In general, the journal undertook to publish articles in which authors discuss both challenges in education and challenges in the business environment, including information technology innovations, data analysis techniques, etc. The journal's mission was to bring together academics, industry professionals, policy makers, politicians and government officials to look at the impact of information technology, and the knowledge-based era it is creating, on the key facets of today's world: the state, business, society and culture. The IJKSR investigates the 'soft' and 'hard' aspects of the knowledge society, providing organizational and cultural frameworks, as well as infrastructures enabled by the evolution of information technologies. This journal provides a global vision based on collaboration, knowledge and learning.

From 2018, the title of the journal was changed to the *International Journal of Smart Education and Urban Society* (IJSEUS) to show that the magazine is still in development and to update the journal's focus on emerging topics in today's society. These changes were logical in the focus of the journal, as evidenced by the results obtained in the literature review on the research published in the IJKSR in the period 2010–2017.

The following research questions are raised.

1. What fields of scientific research are represented in the papers published and to which of them have researchers have paid particular attention?
2. What is the range of countries represented by authors of published articles?
3. What kind of research designs are described in published articles?

To find answers to the research questions raised, the authors set out the objective of the study: to perform a literature review of articles published by IJKSR in the period 2010–2017, developing the research methodology described in the next section. This is followed by a description and analysis of the results obtained. In conclusion, the authors discuss the topical issues that have emerged during the study and draw conclusions about the research carried out, pointing to future research directions, both by analysing the data already collected and by forecasting new directions for the development of the journal.

METHODOLOGY

The literature review included 199 articles from 32 issues published in IJKSR for the period 2010–2017. This did not include articles published as an editorial preface to special issues of the journals. Several dimensions were set for the analysis of the articles:

1. The country represented by the authors of the article. The countries were determined by the information indicated in the article. If there were several authors from different countries, then it was noted that the authors of the article were from different countries; the number of countries is therefore not identical to the number of articles analysed.
2. The scientific field of research analysed in the paper. The research fields were defined on the basis of which field the research outcomes will be useful for.
3. The type of the research or article. The researches that indicated their type were grouped by the category indicated, while for those articles that did not indicate the type of research, categorization was made during the literature review of the article based on the authors' knowledge of the possible categories of research type. This does not mean that the categorization could not have been carried out according to other principles. The authors chose an intuitive approach this time.

In order to ensure that the analysis of articles followed maximally unchanged principles, all the articles included in the literature review were analysed by the first author, who created the code structure. The results of the analysis were then discussed with the other authors in order to agree on the proposed categories and defined codes. After mutual discussion, some of the codes were merged: for example, the code 'education and training' was created as a single group, as was 'engineering and physics'.

RESULTS

Altogether, 199 articles have been published in the journal by authors from 46 countries. Table 1 summarizes the results of country representation and shows that the research papers are most often published by authors from Spain (N = 32), Saudi Arabia (N = 28) and Greece (N = 21). In the analysis of national representations, the country indicated by the author of each article was taken into account. In turn, articles written by authors from more than one country are also indicated.

Subsequently, the articles were categorized according to research field, the fields being listed in Table 2. The definition of fields was a result of analysis of the articles and does not exclude the analysis of subtopics in the article. For example, if the defined field is education, this could be an article about higher education, an article about compulsory education or an article about specific learning objectives, such as learning English, so the categorization cannot be generalized. It is also important to note that during the analysis of the articles it was concluded that many articles refer to several fields of research, such as education and the use of technology in education. The breakdown by field was based on the purpose of the proposed solution. For example, if the use of different technologies (smartphones, smartboards, etc.) or technological solutions (such as learning managment platforms, social media, etc.) was intended to improve educational opportunities, it has been coded as an article on education. In contrast, if the article was about learning how to operate technology in search of new technological solutions, then the article was coded as ICT/computer sciences.

In general, it can be concluded that most articles in this period were published on education and various issues related to education (N = 69), followed by articles focusing more on computer science (N = 23) and business (N = 21).

The following analyses the dimensions that are most often addressed in published articles (see Table 2) to showcase the most recent research directions published in the journal.

Table 1. Country representation

Country	The Number of Papers Published
Spain	32
Saudi Arabia	28
Greece	21
Italy	12
United Kingdom	10
Portugal	8
United States of America, Mexico, Chile	7
Australia, Germany, Finland, Brazil, Tunisia	6
India	5
Canada, Malaysia	4
France, Japan, Check Republic	3
Pakistan, Norway, Ireland, Lithuania, Netherlands, United Arab Emirates, Nigeria, Latvia	2
Serbia, Singapore, Romania, South Africa, Poland, Hungary, Algeria, Turkey, Jordan, Taiwan, Qatar, South Korea, Hong Kong, Marocco, Uganda, Mauritus, Thailand, Indonesia	1

Table 2. Coding of research fields and number of articles

Code	Number of Articles	Principles of Coding
Education	69	Articles in which the emphasis is on educational and/or training aspects
Computer science	23	Articles that analyse the various ICT technological solutions
Business	21	Articles that analyse business, organization and management aspects
Information systems/data analysis	18	Articles that analyse aspects of information storage, compilation and security, as well as articles analysing methods for data mining, processing and analysis, offering different algorithms
Knowledge society	16	Articles that analyse the conditions for free flow of information, overall knowledge growth or community development
Urban society	10	Articles that analyse the public, public transport or environmental conditions in the context of urbanization
e-Governance	7	Articles that analyse e-governance aspects, both to ensure availability of services and to ensure a quick flow of information
Social environment	5	Articles that analyse different aspects of the social environment: conflict resolution, social media aspects
Engineering	5	Articles that analyse discoveries in engineering and physics
Culture	4	Articles that analyse aspects of culture and its preservation
Climate/agriculture	3	Articles that analyse aspects of climate change and agriculture
Medicine	3	Articles on medicine, where technological solutions for problem solving or data analysis are sought
Music	2	Articles that analyse the technological aspects of music recording
Law	1	Article on the legal aspects of a judgment

Educational Dimension

In the majority of cases, the papers published in the journal have been devoted to educational problems (N = 69). The following describes the educational research directions discussed in the journal.

Educational processes can be analysed according to the level of education under consideration—compulsory education and higher education—and from the perspective of the educational problem topic, such as the use of educational technologies and technological solutions in the teaching process (Aranguiz & Quintana, 2016; Daniela, Kalniņa, & Strods, 2017; Jdidou & Khaldi, 2016; Samra, Li, Soh, & AlZain, 2017), social problems in the education system (Cascio, 2017), analysing student satisfaction with the educational process (Adinolfi, D'Avanzo, Lytras, Novo-Corti, & Picatoste, 2016) and the virtual environment as a learning site (Vargas-Vera, 2016c).

A review of research directions in the educational dimension published in the journal suggests that researchers highlight several aspects. These include the fact that technology-enhanced learning is not an exception, but that it is a day-to-day reality. Therefore, it is necessary to focus researches on the various dimensions of education: how to teach specific subjects (Srebnaja, 2017) and how to try new teaching techniques for learning support, such as gamification ideas (Zhuhadar, Coleman, & Marklin, 2016), or what to do to attract learners' interest using already known techniques such as Power Point, quizzes, etc. (Stefanou & Kotsovoulou, 2016; Vargas-Vera, 2106b). There are also ways in which education can also be a social institution to achieve sustainable development goals (Dhiba & Prastiti, 2017).

However, Cleveland and Block (2017) point out the asynchronicity of the educational environment, emphasizing that the technology-enhanced learning process is synchronous, but the knowledge—the way knowledge is acquired, the way knowledge is constructed, the knowledge itself—can be asynchronous, and this creates a challenge in the technology-enhanced learning environment to balance pedagogical innovations and teaching strategies. This points to the need to pay more attention to the education of teachers, to provide certain quality standards (Brahimi, Sarirete, & Ibrahim, 2016) so that education is both a way for a teacher to use technology to facilitate students' learning and a way to ensure that teachers are in a position to change the life flow to ensure that the level of knowledge increases (Goburdhun & Ramsaha, 2017).

It can be concluded that smart education (Rothman, 2007; Visvizi, Lytras, & Daniela, 2018) is one of the dimensions that needs to be developed, because smart education is defined as a system in which students acquire career relevance and development; learn the significance of society and its potential impact; and operate within an appropriate pedagogy, their classes equipped with appropriate technologies and devices that provide learning opportunities and facilitate their use (Sykes, 2014).

ICT and Data Algorithmic Dimension

The next most urgent research direction was found to be ICT (N = 23) and information systems/data analysis (N = 18). The articles published in the journal relating to these scientific fields include analyses of problems, updates and proposal of solutions.

Technology, technological solutions, data mining, processing and analysis algorithmic solutions are one focus of researchers to ascertain how to improve the quality of life and solve problems that have so far not found a solution, as well as supporting the different scientific areas and providing a basis for new innovations. However, before technologies are used to strengthen one or several of these directions, it is necessary to develop, test and approbate them. Looking at the articles published in the IJKSR, it can be concluded that technology is being analysed as a support mechanism to strengthen the field of education, but there are also studies which offer algorithms for TV programme analysis (Mahdi & Zlitni, 2016); consider the development of mobile sensing applications (Torres-Ruiz & Lytras, 2016); focus on formation improvement and efficient routing algorithms for Bluetooth networks (Bakhsh, Tahir, Altalhi, Subhan, & Sheikh, 2016); and examine the development of potential technology that enables the construction and development of collaborative intelligent software systems (Pasha, 2017), as well as developing a variety of data acquisition, analysis and processing scenarios, such as the event detection approach combining linguistic features and Twitter features (Ben Kharrat, Elkhleifi, & Faiz, 2016; Cherichi & Faiz, 2017; Chianese, Marulli, & Piccialli, 2016; Dammak, Jedidi, & Bouaziz, 2017; Dayananda & Sowmyarani, 2016; D'Avanzo & Pilato, 2016; Floos, 2016; Matsumura, Miura, Komori, & Hiraishi, 2016), decisions taken during the implementation of the DSSim (similarity based on Dempster-Shafer) multi-agent ontology mapping system (Vargas-Vera, 2017), development of the Mapper module called (Vargas-Vera, 2016a) or another model (Pipitone & Pirrone 2016), and healthy and efficient storage for securely managing and preserving data (AlZain, Li, Soh, & Masud, 2016).

Business Dimension

The next most analysed field was business. Analysing articles on business aspects, it can be concluded that researchers have focused on various organizational, managerial and business aspects, emphasizing the importance of Project Management Office (PMO) in business development (Stroe, Spruit, Beltman, & Koelemeijer, 2016); recalling the need to adopt new staff profiles and employees with new business skills (Lucas de Azevedo & Borges, 2015); pointing to the need to explore internal branding as a corporate philosophy (Matiatou, 2015); and proposing an approach combining behavioural and targeting techniques for a better customer reaction to a star product by using personalized ubiquitous advertising (Noura, Ismahène, & Sami, 2017). The role of corporate social responsibility (CSR) was analysed, which is a Western concept that penetrates the Middle East (Al-Jenaibi, 2017).

Overall, it can be concluded that the focus of the articles has been more on technology and technological solutions for improving both the education environment and the business environment. There are relatively few studies in which technologies are not analysed at all. For example, in some studies, aspects of the social environment like conflicts and strategies for addressing them are analysed without emphasizing the role of technology (Čiuladienė & Katkonienė, 2017), or agricultural aspects are analysed in a particular environment (Lwin & Poungchompu, 2017).

The next step of the literature review was the analysis of articles by type of research. Where there was no such information provided, the authors categorized the article according to their understanding of the possible types of research. Articles in which the results of specific research methods were not summarized but which included some possible development models for further discussion in the academic environment were defined as concept development. Categorization was not subdivided into scientific fields. This means that in the category of concept development, for example, there can be discussion of both changes needed in the educational environment and steps that countries need to take to achieve sustainable development goals. Researches using multiple research methods were categorized according to the main or most important method used: for example, if a questionnaire was used during a case study, then the article was categorized as a case study. The results of the analysis of this research issue are summarized in Table 3.

Table 3. Type of research, number of papers and category justification

Category	Number of Papers	Category Justification
Concept development	46	Articles raising conceptual ideas for discussion in the academic environment
Experimentation/testing	45	Researches detailing different types of experiment or the development of ideas, testing ideas, and description of experiments rather than analysing the outcomes that they give to the society
Needs analysis/evaluation/ field analysis	33	Researches that have explore a particular problem, or a study of the needs of a group, or an assessment of a previously developed solution
Data mining	26	Researches using either data mining methods for identifying the research field or introducing new algorithms for data mining method improvement
Case study	15	Researches that indicate a case study was done, or research logic indicates this
Educational outcomes evaluation	12	Researches where the main focus is on the educational outcomes of learners from activities not undertaken by the researchers themselves
Questionnaire-based survey	7	Researches that use different types of survey as the main method of data acquisition
Systematic review	6	Researches where a systematic literature review or systematic documentation review are explained
Financial or other outcomes evaluation	5	Researches that analyse financial or other outcomes from activities that are not performed by the researchers themselves

In general, it can be concluded that concept development (N = 46) has been a popular way for authors to express their ideas, followed by different types of experiment (N = 45) and needs analysis (N = 33). Few published articles have used research methods such as financial and other outcomes evaluation, or systematic literature/documentation review (N = 6), and there are not many studies based on various types of survey (N = 7): although surveys have been analysed in research papers, it has not generally been the main method.

DISCUSSION AND CONCLUSION

The literature review leads to the conclusion that education is an emerging topic in most articles. Given that different experiments comprise one of the most commonly discussed types of research, it can be concluded that there are many new and up-to-date studies specifically about aspects of education; more precisely, they mainly focus on a technology-enhanced learning process that points to the need for new solutions to change the educational environment, and the meaningful use of different technologies and technological solutions. At the same time, there is clearly a growing need in the academic environment to develop new solutions, because the research logic usually flows in a way such that research is carried out in areas where there are some uncertainties and where there is a need to look for new solutions, since those previously used have become less effective. When analysing articles devoted to educational research, it can be concluded that the majority are devoted to various aspects of higher education, which may indicate that this is an area where there are more emerging problems. However, this would be a rather simplistic conclusion: it may be more a case of researchers' desire to explore the areas in which they themselves act as academic staff and therefore see the problems that are encountered and, to some extent, researchers seeking to work within their comfort zone. On the one hand, this is reasonable; on the other hand, it is possible to focus on the need to expand the research clusters so that educational studies are carried out in the broader educational environment, including compulsory education and pre-school education. Carrying out pedagogical innovations at those educational levels in which the learners have already acquired certain knowledge acquisition and construction strategies could give educators the proactive means to solve the problems that are topical in higher education.

Analysis of the articles also allows the definition of an objective for the editorial board of the journal, to address the researchers and to create wider networks of researchers that extend the range of scientific areas represented by the published article.

The literature review confirms that changing the title of the journal to the *International Journal of Smart Education and Urban Society* has been a crucial step in allowing researchers to accurately identify the focus of the journal and enable discussion on emerging issues.

FUTURE DIRECTIONS

An analysis of the articles can still be conducted in relation to other discourses, such as the interconnected areas of scientific fields, using agent theory. Similarly, a comparison of the trends that appear in the articles published in this journal with articles published in other research journals could be conducted. These articles can also be analysed from the point of view of the institutions represented by the authors of the articles by analysing the scientific capacity of these institutions. Another research could involve analysing the articles according to the context of the research, where data mining methods are used to ascertain what were the events occurring at the time of carrying out the research, and thus predict further research directions by developing data analysis and problem forecasting algorithms.

REFERENCES

Adinolfi, P., D'Avanzo, E., Lytras, M. D., Novo-Corti, I., & Picatoste, J. (2016). Sentiment analysis to evaluate teaching performance. *International Journal of Knowledge Society Research*, *7*(4), 86–107. doi:10.4018/IJKSR.2016100108

Al-Jenaibi, B. (2017). Corporate social responsibility: Case study in UAE organizations. *International Journal of Knowledge Society Research*, *8*(4), 85–100. doi:10.4018/IJKSR.2017100105

AlZain, M. A., Li, A. S., Soh, B., & Masud, M. (2016). Byzantine fault-tolerant architecture in cloud data management. *International Journal of Knowledge Society Research*, *7*(3), 86–98. doi:10.4018/IJKSR.2016070106

Aranguiz, M. S., & Quintana, M. G. (2016). ICT resources to improve learning in higher education. *International Journal of Knowledge Society Research*, *7*(4), 1–11. doi:10.4018/IJKSR.2016100101

Bakhsh, S. T., Tahir, S., Altalhi, A. H., Subhan, F., & Sheikh, M. A. (2016). Resource optimization routing and scatternet formation protocols for Bluetooth networks. *International Journal of Knowledge Society Research, 7*(2), 84–123. doi:10.4018/IJKSR.2016040106

Ben Kharrat, F., Elkhleifi, A., & Faiz, R. (2016). Improving collaborative filtering algorithms: Sentiment-based approach in social network. *International Journal of Knowledge Society Research, 7*(3), 99–118. doi:10.4018/IJKSR.2016070107

Brahimi, T., Sarirete, A., & Ibrahim, R. M. (2016). The impact of accreditation on student learning outcomes. *International Journal of Knowledge Society Research, 7*(4), 51–62. doi:10.4018/IJKSR.2016100105

Cascio, P. D. (2017). Unsettling trends in academia: A review of the predicament of minorities and members of the LGBTQ community in higher educational institutions in America. *International Journal of Knowledge Society Research, 8*(2), 36–43. doi:10.4018/IJKSR.2017040103

Cherichi, S., & Faiz, R. (2017). Leveraging temporal markers to detect event from microblogs. *International Journal of Knowledge Society Research, 8*(3), 54–67. doi:10.4018/IJKSR.2017070104

Chianese, A., Marulli, F., & Piccialli, F. (2016). Sensitivity mining in social pulses to address cultural heritage competitive intelligence. *International Journal of Knowledge Society Research, 7*(2), 40–56. doi:10.4018/IJKSR.2016040103

Čiuladienė, G., & Katkonienė, A. (2017). Framing the conflict: How students see it. *International Journal of Knowledge Society Research, 8*(4), 75–84. doi:10.4018/IJKSR.2017100104

Cleveland, S., & Block, G. (2017). Toward knowledge technology synchronicity framework for asynchronous environment. *International Journal of Knowledge Society Research, 8*(4), 23–33. doi:10.4018/IJKSR.2017100102

D'Avanzo, E., & Pilato, G. (2016). The good, the ugly and the bad situation awareness in the big data: A cognitive architecture for social forecasting. *International Journal of Knowledge Society Research, 7*(2), 25–39. doi:10.4018/IJKSR.2016040102

Dammak, S. M., Jedidi, A., & Bouaziz, R. (2017). Filtering method for the annotated and non-annotated web pages. *International Journal of Knowledge Society Research, 8*(1), 1–22. doi:10.4018/IJKSR.2017010101

Daniela, L., Kalniņa, D., & Strods, R. (2017). An overview on effectiveness of technology enhanced learning (TEL). *International Journal of Knowledge Society Research, 8*(1), 79–91. doi:10.4018/IJKSR.2017010105

Dayananda, P., & Sowmyarani, C. N. (2016). Comprehensive study on keyword search on semi structured data. *International Journal of Knowledge Society Research, 7*(3), 62–85. doi:10.4018/IJKSR.2016070105

Dhiba, S. F., & Prastiti, D. C. (2017). Social education institution a way to achieve sustainable education in remote areas: A case study of 'Harapan' Bajulmati, Indonesia. *International Journal of Knowledge Society Research, 8*(3), 21–33. doi:10.4018/IJKSR.2017070102

Floos, A. Y. (2016). Arabic rumours identification by measuring the credibility of Arabic tweet content. *International Journal of Knowledge Society Research, 7*(2), 72–83. doi:10.4018/IJKSR.2016040105

Goburdhun, S., & Ramsaha, J. (2017). The relevance of social sciences teacher education programme in preparing effective secondary school educators. *International Journal of Knowledge Society Research, 8*(2), 1–12. doi:10.4018/IJKSR.2017040101

Jdidou, Y., & Khaldi, M. (2016). Increasing the profitability of students in MOOCs using recommendation systems. *International Journal of Knowledge Society Research, 7*(4), 75–85. doi:10.4018/IJKSR.2016100107

Lucas de Azevedo, V. L., & Borges, M. (2015). More collaboration, more collective intelligence. *International Journal of Knowledge Society Research, 6*(3), 1–18. doi:10.4018/IJKSR.2015070101

Lwin, S., & Poungchompu, S. (2017). Determinant of food security on upland agriculture households in Paletwa Township, Chin State of Myanmar. *International Journal of Knowledge Society Research, 8*(2), 44–53. doi:10.4018/IJKSR.2017040104

Mahdi, W., & Zlitni, T. (2016). Navigating news programs in large TV-streams: A knowledge-based approach. *International Journal of Knowledge Society Research, 7*(1), 80–98. doi:10.4018/IJKSR.2016010106

Matiatou, M. (2015). From internal branding to cultural transformation: A virtuous circle. *International Journal of Knowledge Society Research, 6*(2), 1–19. doi:10.4018/IJKSR.2015040101

Matsumura, N., Miura, A., Komori, M., & Hiraishi, K. (2016). Media mediate sentiments: Exploratory analysis of tweets posted before, during, and after the Great East Japan earthquake. *International Journal of Knowledge Society Research, 7*(2), 57–71. doi:10.4018/IJKSR.2016040104

Noura, A., Ismahène, H. K., & Sami, F. (2017). Process of personalizing the ubiquitous advertisements. *International Journal of Knowledge Society Research, 8*(2), 13–35. doi:10.4018/IJKSR.2017040102

Pasha, M. (2017). Semantically enabled communication framework for autonomous systems. *International Journal of Knowledge Society Research, 8*(2), 54–68. doi:10.4018/IJKSR.2017040105

Pipitone, A., & Pirrone, R. (2016). Aligning relational schema and OWL ontologies with hidden Markov model. *International Journal of Knowledge Society Research, 7*(2), 1–24. doi:10.4018/IJKSR.2016040101

Rothman, R. (2007). Building 'smart education systems'. *Education Week, 26*(44), 25–27.

Samra, H. E., Li, A. S., Soh, B., & AlZain, M. A. (2017). A cloud-based architecture for interactive e-training. *International Journal of Knowledge Society Research, 8*(1), 67–78. doi:10.4018/IJKSR.2017010104

Srebnaja, J. (2017). The role and implementation of new technologies in the ELT (English language teaching). *International Journal of Knowledge Society Research, 8*(3), 68–83. doi:10.4018/IJKSR.2017070105

Stefanou, V., & Kotsovoulou, M. (2016). Use of PowerPoint in the classroom: A participatory research project. *International Journal of Knowledge Society Research, 7*(4), 38–50. doi:10.4018/IJKSR.2016100104

Stroe, A., Spruit, M. R., Beltman, B., & Koelemeijer, S. J. (2016). PMOMM: The Project Management Office maturity model. *International Journal of Knowledge Society Research, 7*(3), 47–61. doi:10.4018/IJKSR.2016070104

Sykes, E. (2014). New methods of mobile computing: From smartphones to smart education. *TechTrends, 58*(3), 26–37. doi:10.100711528-014-0749-2

Torres-Ruiz, M. J., & Lytras, M. D. (2016). Urban computing and smart cities applications for the knowledge society. *International Journal of Knowledge Society Research*, *7*(1), 113–119. doi:10.4018/IJKSR.2016010108

Vargas-Vera, M. (2016a). Data integration framework: A children and parents cohort case study. *International Journal of Knowledge Society Research*, *7*(1), 99–112. doi:10.4018/IJKSR.2016010107

Vargas-Vera, M. (2016b). Methodology for the elaboration of quizzes using propositional logic calculus in an e-learning environment. *International Journal of Knowledge Society Research*, *7*(4), 63–74. doi:10.4018/IJKSR.2016100106

Vargas-Vera, M. (2016c). Reflections on the Second Life platform used in the development of a virtual university campus. *International Journal of Knowledge Society Research*, *7*(4), 12–23. doi:10.4018/IJKSR.2016100102

Vargas-Vera, M. (2017). The implementation of DSSim: A multi-agent ontology mapping system. *International Journal of Knowledge Society Research*, *8*(3), 34–53. doi:10.4018/IJKSR.2017070103

Visvizi, A., Lytras, M. D., & Daniela, L. (2018). (Re)Defining smart education: Towards dynamic education and information systems for innovation networks. In M. D. Lytras, L. Daniela, & A. Visvizi (Eds.), *Enhancing knowledge discovery and innovation in the digital era* (pp. 1–12). Hershey, PA: IGI Global. doi:10.4018/978-1-5225-4191-2.ch001

Zhuhadar, L., Coleman, P., & Marklin, S. (2016). Applicative personalized learning: How gamification is driving learning. *International Journal of Knowledge Society Research*, *7*(4), 24–37. doi:10.4018/IJKSR.2016100103

Chapter 2
Technology-Enhanced Learning (TEL) in Higher Education:
Where Are We Now?

Linda Daniela
University of Latvia, Latvia

Raimonds Strods
University of Latvia, Latvia

Daiga Kalniņa
University of Latvia, Latvia

ABSTRACT

As technologies become more exciting, interactive, and reachable, various technological solutions are used in higher education. On the one hand, there is the conviction that technologies are indispensable, both for improving learning and for making learning process more effective, both in terms of learning outcomes and in terms of costs. Additionally, technology and technological solutions can provide sustainability of knowledge because students develop the competences that they will need in their future professional work. This chapter summarises the systematic literature review (SLR) carried out by the authors in analysing research that has been done on the impact of technology-enhanced learning on learning outcomes in order to understand what emerging research has been done, as the authors published a similar study one year ago. The SLR was conducted for the period of 2010 to 2018 by selecting peer-reviewed articles on specified terms. The selected articles were then analysed following sub-purposes. The descriptive analysis method was adopted for the data analysis.

DOI: 10.4018/978-1-5225-7347-0.ch002

INTRODUCTION

With ever-increasing technological advancement, the educational environment is transforming. Educators must introduce various technological innovations to prepare students for their professional career, where they will have to deal with different technology to make the learning process more interesting, keep students more engaged in the learning process, and, in order to improve motivation, facilitate self-directed learning. On the other hand, solutions are always being sought on using technology to make education more sustainable and to achieve Sustainable Development Goals that are put forward for education. Another important aspect is the desire to use technology to reduce the financial burden that lies on the education system by organising day-to-day activities, as well as paying for lecturers' hours of work and managing the premises in which studying takes place. Educational technologies are perceived as a means to improve student Learning Outcomes (LO) and, at the same time, optimise financial resources. In the context of this article, the authors focus more on LO without analysing aspects that may signal positive Financial Outcomes.

The study process in Higher Education (HE) is a transitional line between general compulsory education and the work environment, in which employers seek highly-qualified and innovative young specialists. This is one of the arguments in support of emerging technologies and looking for new ways to ensure that HE keeps up with the innovations and transfers accumulated knowledge. This leads to the following research questions:

1. What technologies and technological solutions have been the focus of research in the context of higher education from 2010 to 2018?
2. What are the main conclusions about the benefits of Technology-Enhanced Learning (TEL) in promoting LO in the context of HE?
3. What challenges in using TEL have been revealed through research?

To determine TEL's effectiveness in the HE discourse, this study aims to carry out a Systematic Literature Review (SLR) to analyse the articles available on the EBSCOhost web database, with the results of the research on the efficacy of TEL in the HE discourse.

Limitations

This article focuses only on full-text articles available on the EBSCOhost web search platform that were published in English in peer-reviewed journals. Selected studies were published in the period from 2010 to 2018.

BACKGROUND

One year ago, the authors published a study analysing the articles published on TEL over the period of 2010 to 2016, where it was concluded that few articles had addressed LO in the long term. In most cases, studies have been conducted with little research basis or with little research time. The principles that need to be taken into account in order make a more student-centred learning process and help achieve higher learning outcomes have not been analysed (Daniela, Kalniņa, & Strods, 2017). However, despite the fact that such results have not been convincingly discovered so far, with the rapid change in technology and technology solutions, it is necessary to accept that it is almost impossible to carry out long-term research in order to to be sure of the positive impact of the use of specific technologies, before their implementation, since the long-term research process is too slow in relation to technological development. It requires continually monitoring ongoing technology in the context of an enriched learning process, and synthesysing different information units to seek answers to the use of technological advancement in education. This leads to neccessity to acept that one must learn what is now unknown, acquire competences that are not yet necessary and it is unknown, what will be needed. This will make researchers more open to new information, develop new research competencies, and develop new research methods that enable immediate information being put together as a puzzle that forms the overall picture and shows future directions for educational development.

At present, studies have focused on the learning experience, while other studies focus on LO (Goodyear & Ellis, 2008). Most research on the use of TEL is of a small sample, specific in content, and it is difficult to see convincing evidence of TEL's impact (Kirkwood & Price, 2014). It is also necessary to analyse the aspects that can lead to various social exclusion risks in the process of using technology (Di Serio, Ibanez, & Kloos, 2013), to promote self-handicapping motivation (Migdley & Urdan, 2001), and so forth. According to Kirkwood and Price (2014), sometimes the TEL term is related only to technology-like devices, with less thought of how these technologies can be used in the educational process to promote learning. Every year, numerous conferences are held on using TEL in education, educational materials, and scientific publications. Some of these conferences highlight not only the invaluable impact of technology on the learning process, but also emphasise that using technology must be purposeful and structured in order to develop creativity, promote learning, and reduce both physical and social risks. Technology in education is often introduced because of the belief that students like it and it encourages learning. However, these statements are often not based on real evidence of their effectiveness. In addition to the risks mentioned above, there are related costs, such as the introduction and use of technology, copyright, uniqueness of theory and practice, assumptions and expectations about the expected results, lack of academic staff knowledge about the use of TEL in the study process, and work overload of academic staff. (Tynan et al., 2012; Walker et al., 2012; Walker et al., 2014). This suggests that it is necessary to analyse and evaluate the research that has been done to date on the benefits and risks of using technology in order to understand the current research areas, discover the latest findings, and conceptualise knowledge of the TEL process.

METHODOLOGY

A systematic literature review of 110 published research papers was undertaken between February and May, 2018 in the EBSCOhost web search platform, including Academic Search Complete and ERIC. The review process followed the three phases as set out in the Joanna Briggs Institute Reviewers Manual (2014). In the first phase, initial key words were identified from the researchers' knowledge of the field. These included: 'learning outcomes', 'technology', and 'Higher education'.

In the second phase, database-specific searches were conducted. These included the terms: 'Learning Outcomes' and 'Technology', combined with the term 'Higher Education', using Boolean phrases. All searches were limited to full-text, peer-reviewed articles that were published in English between 2010 and 2018. This subsequent search identified a further 110 papers; during the reading of the selected articles, however, it turned out that the authors did not have access to the full text versions. Articles were selected by matching keywords during the selection, but later it was found out at the time of reading that they some papers were not about HE, so the statistics summarised for 81 articles are below.

Countries represented: Australia (15), USA (16), USA/Equador (1), Malaysia (1), Pakistan (1), Turkey (3), Lebanon (1), India (1), Serbia (1), United Kingdom (5), Ireland/France/Belgium/Finland (1), Greece (1), Sweden (1), South Africa (1), Kuwait (1), Portugal (1), Finland (3), Nepal (1), Spain (2), Canada (2), Germany (1), Israel (2), Taiwan (1), Saudi Arabia (1), Hong Kong (2), United Arab Emirates (1), and Algeria (1). Eleven articles did not indicate a represented country.

These papers were evenly divided among all three authors. Initially, the authors read each part of their respective articles, then jointly discussed the read articles and agreed on the articles to be included in the analysis. After the first step of data analysis, the authors divided the data into two groups:

1. Articles focusing on specific technologies (e.g., mobile phones, ipads, clickers), technological solutions (e.g., virtual reality, e-portfolio, wikis), and the use of online facilities (e.g., Learning Management Systems, Social Media), and their impact – 47; and
2. Articles focusing on LO, where technologies are used only as a means (learning achievements, improved motivation, interest) – 34.

Many articles did not precisely define the research group, or did not explicitly state which innovations were introduced and which technologies had been used. Some articles also did not define the research context, and it was therefore not possible to identify whether improvements could be found after the application of particular technologies.

Thematic analysis principles were used to analyse the content of the selected articles (Braun & Clarke 2006). Each article was read several times to become acquainted with the evidence revealed in the article and the methods used in obtaining this evidence. Each author read the articles individually to mark the topical information, then the obtained results were cross-compared and discussed. The discussions resulted in reaching the agreement on how to analyse the selected articles and whether to exclude others from further analysis (see Table 1). After excluding irrelevant articles, 67 were selected for further analysis.

RESULTS

Results were analysed in accordance with the defined criteria in a previous publication by the authors (Daniela, Kalniņa, & Strods, 2017), where the first step was analysing the articles according to their research design.

Table 1. Exclusion & inclusion criteria

Exclusion	Inclusion
• Special needs education • Not a scientific paper about the research • Analysis of PISA data • Using technology for assessment and/or feedback purposes • Financial outcomes of technology used	• Learning outcomes of technology use measured and discussed • Clearly described process how technologies were used in the learning process • Primary study published in peer review journal • English language

Research Design

The authors, in analysing the selected articles, wished to find out what kinds of studies were more frequently used when analysing the use of technologies in the study process; however, the majority of the published studies had not indicated what kind of research had been performed. Therefore, the authors of the present article classified the studies according to a particular kind of research design if it was mentioned in the paper (see Table 2). The obtained results show that 25 papers did not indicate the study design, but of the indicated design studies the most popular *Sase study* was conducted with 15 studies, followed by *Literature review* with 14 studies, and *Experimental research* with 11 studies. The studies also included one *Action research* and two *Ethnographic research* studies. Subsequently, the authors excluded articles with no clear study design and articles where SLR was carried out.

Research Methods

Subsequently, by following the above criteria, the authors analysed 28 articles. The next criterion used was identifying research methods (see Table 3). Table 3 shows more research methods than papers analysed because some articles used and specified several methods. It appears that research using pre-approved research tests (2 papers) have established positive benefits from TEL. This demonstrates that outcomes have already been measured over a longer period of time. Some studies have analysed LO by comparing student knowledge in different groups at different time periods, with variables in using technology or technological solutions. This points to the positive trend that while in the context of TEL it is difficult to carry out long-term research because the technology is constantly changing and evolving. There are, however, studies analysing long-term outcomes that differ from the the results of a previous study when there was little evidence found on outcomes of TEL (Daniela, Kalniņa, Strods, 2017). That said, it should be mentioned that some of these studies had weaknesses. Some studies contained a qualitative analysis section, which did not analyse long-term TEL benefits, and the obtained results could not be generalised. Research surveys using different kinds of questionnaires mostly have been used to find out the respondents' opinions, but did not measure academic achievement or improved learning outcomes. Studies have been performed in a relatively short time period, where the longest research period was one study year. Some studies had a pre-test and post-test, but they also were performed over a short duration. Some research was carried out with methods that had not been approved. Some papers mentioned a lot of research methods, but contained no data analysis. Therefore, the authors excluded from further analysis those articles in which research methods had been applied, but the never analysed, or papers in

Table 2. A categorisation of the reviewed articles

Nature of Intervention or Study	Number of Article(s)
Case study	15
Experimental research	11
Ethnographic study	2
Action research	1
Not defined research desig	25
Systematic literature review	14

which the conclusions expressed were unrelated to the results obtained, thus calling into question the legitimacy of the conclusions drawn. Also, articles that did not analyse the study process were excluded, because these articles could not be attributed to the TEL field.

Nature of Conclusions and Period of Research

After excluding articles from further analysis as set forth in the preceding section, 19 were analysed in-depth. The authors read these articles again to summarise their conclusions regarding the positive benefits from applying TEL principles, as well as evidence stressing other pedagogical aspects that have to be considered in applying TEL (see Table 4). However, there is also a relatively high degree of challenge among researchers in making positive statements, because studies have been carried out in a relatively short period of time, with the longest study spanning only one year. In some studies, the current situation was detected without checking TEL's influence over a longer period of time; these relied on students' opinions or statements defined in documents. This points to many needs, including: developing the pedagogical competence of academic staff so that different technologies are used meaningfully; focusing on LO; focusing on scaffolding students on the knowledge construction process; and spending less time on the effect of fascination by technologies. The fact that a fairly large number of studies show students' increasing willingness to use TEL principles in the study process confirms this necessity. It is also important to take into account that quality of learning is enhanced if the learning environment is controlled and structured according to learning goals (Xia & Rekola, 2013).

Risks Found Out in the Research

The criterion – risks found during the research – has also been included in the analysis of the summarised data (see Table 5), which concluded that in only six studies the authors mentioned risks that could emerge when applying different TEL strategies.

Table 3. A categorisation of the research methods

Research Methods Used	Number of Articles Involving This Research Method
Questionnaire/survey	18
Activity monitoring/observation	7
Pre test/post tests	9
interviews	7
End of course evaluation	3
Student feedbacks	3
Online activity monitoring/LMS	2
Focus group	3
Content analyses	1

Table 4. Positive conclusions

Period of the Research	Main Conclusions	Articles
One semester	The results showed that using digital games enhanced students' Learning Satisfaction and Learning Performance. This indicated that students who played games to support their education were more satisfied with their learning outcomes and with this instruction method. They believed that this tool and technology would bring significant change in their learning performance.	Ellahi, A., Zaka, B., & Sultan, F. (2017)
Not indicated	This study showed that the vast majority of student participants perceived the integration of Information Communication Technology (ICT) into learning activities as useful, convenient, flexible, encouraging reflection, and good for knowledge sharing and asynchronous interactions.	Arenas, E. (2015)
One year	A high percentage (88%) of students thought that the online environment offered increased opportunities to interact with experts in the field. In general, students found that the hybrid nature of the course increased the quality of interaction, with both the instructor and their peers Many students (87%) agreed that the instructional technology (IT) was largely beneficial in their learning process.	Palmer, M. M., Shaker, G., & Hoffmann-Longtin, K. (2014)
Not indicated	The article concluded that a holistic approach was required to enable a successful institutional transition into enhanced blended learning. There should be more explicit guidelines for staff on blended learning design, a networking event for stakeholders to share experiences and expertise, and staff–student partnerships working around blended course development.	Adekola, J., Dale, V. M., & Gardiner, K. (2017)
One semester	Use of the Virtual Learning Environment (VLE) together with additional Web 2.0 tools enhanced the student learning experience. Students were also decisive in their view that the technology did enhance their learning experience. Students ranked the keynote lectures as the most effective tool for learning. In a very large class the VLE provided an efficient mechanism for ensuring that all students gained access to all course material. The learning outcomes and assessment criteria must be well thought out in advance to encourage usage of a particular learning tool. This is particularly important in a very large class, where minor mismatch between tool and task can lead to frustration and lack of student engagement. Students did not radically change their learning methods as a result of the provision of new technology tools, although there was some evidence in the use of podcasts that technology provided individual learners with more personalised tools for learning.	Saunders, F. C., & Gale, A. W. (2012).
One semester	To examine the learning outcomes and mental effort, instruction materials should include multiple levels of complexity; meaningful learning is a result of efficient instruction that produces good transfer and retention on future assessments. In this case, the efficient instruction was in the form of flipped classroom instruction. Efficient learning not only reached the instructional goals as explored in this study, but also encouraged faster learning.	Lee, D., Morrone, A. S., & Siering, G. (2018)
Not indicated	Active student-centred teaching and learning methods, such as group work, online tutorials, games, and interactive websites should be used to accommodate both digitally-advantaged and digitally-disadvantaged students. These active student-centred methods could alleviate boredom on the part of digitally-advantaged students and at the same time help digitally-disadvantaged students to learn from their advantaged peers or to revise and reinforce lessons covered in the classroom.	Naidoo, S., & Raju, J. (2012)
One study year	Today's adaptive learning systems have negligible impact on learning outcomes, which is one aspect of quality. There was also evidence that adaptive systems positively impacted other aspects of quality, such as student persistence and engagement.	Coffin Murray, M., & Pérez, J. (2015)
9 weeks	When students were given the possibility of using Google + Discussion Circles (GDCs) in learning processes, the majority took advantage of the academic, social, and structural dynamics created by these technologies in many ways, which supported their learning, assessment activities, and overall academic outcomes. The use of social media technologies created opportunities for students to learn experientially.	Kivunja, C. (2015)

continued on following page

Table 4. Continued

Period of the Research	Main Conclusions	Articles
One study year	Using web-based technologies in teaching requires clear instructions and user-friendly systems that motivate students to interact and actively participate, even in a virtual setting. Web-based technologies in teaching provided self-oriented learning and offered flexibility for the participants. Using web-based technologies in teaching could possibly lessen the mismatch between teaching goals and learning goals of the students. Web-based technologies in teaching supported critical thinking by offering benefits of information sharing and interactive educational experiences where students were encouraged to learn from one another. The main challenge of using web-based technologies in teaching was establishing a virtual environment where knowledge sharing and social interaction would satisfy a purposeful educational experience	Xia, B. S., & Rekola, M. (2013)
Not indicated	Findings from the study indicated no significant difference, in terms of gender, among the average scores for advantages, limitations, and independence of m-learning. Nonetheless, female students had a more positive attitude toward the practicality factor of m-learning compared with male students. Students having a mobile device exhibited a more positive attitude toward practicality of m-learning.	Gezgin, D. M., Adnan, M., & Acar Guvendir, M. (2018)
Not indicated	There was a need to: - provide support training, orientation, and research for ICT integration and mobile learning practices on a wider scale; - devise appropriate teaching methods that demand the use of mobile learning in addition of face-to-face learning; - develop flexible learning and assessment methods; and - record class lectures and share with students, so that students focused on information processing.	Parajuli, K. P. (2016)
4 weeks	Students were motivated by the use of a hybrid game-based app for learning purposes and perceived a high degree of usefulness and added value. Use of a hybrid game-based app had a positive impact on learning outcomes. Hybrid game-based apps significantly improved proficiency.	Berns, A., Isla-Montes, J., Palomo-Duarte, M., & Dodero, J. (2016)
Not indicated	Online instruction methods are likely to continue proliferating in the future, and educators and students are less fearful about online courses than they once were. It is of the utmost importance to continue to look at how technological and human factors influence student experiences and learning outcomes. Online courses are important for fostering a positive and productive learning environment.	Limperos, A. M., Buckner, M. M., Kaufmann, R., & Frisby, B. N. (2015)
Not indicated	The information packages of undergraduate programs generally reflected intentions for integrating technology into educational programs. However, these intentions were not yet very manifested in the programs' curricula, particularly in the learning outcomes.	Coskun, Y. D. (2015)
One semester	The model of flipped learning could be used in a wide range of higher education courses, regardless of whether it is delivered face-to-face, as hybrid learning, or fully online.	Blau, I., & Shamir-Inbal, T. (2017)
One week	Online students felt that they learned more and were more satisfied with their instructors when they used audio and text feedback, rather than written feedback.	Rockinson-Szapkiw, A. J. (2012)
Not indicated	The ubiquitous learning environments should be used as a supportive tool.	Virtanen, M. A., Kääriäinen, M., Liikanen, E., & Haavisto, E. (2017)
Not indicated	Most students felt they became more confident in expressing their opinions and ideas using communication technologies. Students also responded that using online learning had improved their technical skills. Students still needed the guidance of instructors.	Lansari, A., Tubaishat, A., & Al-Rawi, A. (2010)

Table 5. A categorisation of risks

Risks Found Out in the Research	Articles Where Risks Were Found
Some students demonstrated anxiety as they started to create and engage with GDCs for the first time. But there was also evidence that they tolerated ambiguity as they sought clarification from their peers.	Kivunja, C. (2015)
Having students with diverse digital expertise in the same classroom posed a problem in teaching Information Literacy (IL), especially when the lesson included online lessons. For example, through boredom those students who were proficient in the use of computers tended to 'surf the net' and 'play' on social networking tools during the IL lesson, while the educator was assisting disadvantaged students with basic computer skills. Digitally-disadvantaged students, in the meantime, were likely to experience a degree of anxiety or feelings of inadequacy, especially with being seated together with their digitally-advantaged peers in the same IL classroom.	Naidoo, S., & Raju, J. (2012)
Teachers felt anxiety during the classes where they had to collaborate becaue they did not have enough proficiency in using digital media.	Lee, D., Morrone, A. S., & Siering, G. (2018)
Online learning cannot replace face-to-face learning because students needed the possibility of speaking with an experienced tutor.	Saunders, F. C., & Gale, A. W. (2012)
Some papers also stress out the necessity to be aware about copyright issues	Adekola, J., Dale, V. M., & Gardiner, K. (2017)
Some students reported perceptions that the ICT-mediated learning environment did not give them the options to choose from when learning tasks, and that somehow they were misaligned with their preferences and needs. Some teachers and students were dissatisfied with the use of these technologies, specifically in relation to supporting reflection, discussion, and collaboration. For some students, their willingness to learn collaboratively and to participate in the communicative environments was inhibited by a number of socio-cultural factors, such as cultural differences and attitudes toward cultural diversity.	Arenas, E. (2015)

DISCUSSION AND CONCLUSION

It is essential to remember that technologies themselves are not necessary for the study process. The main condition for successful student learning and knowledge development is still the students' brains and neural connections, which help to construct knowledge. Technology and technological solutions, however, can be a great tool in helping them master the competences that they will need in the future to create new, innovative solutions that help others, solve problems, and provide access to services for those who need to learn remotely. Technology can help to personalise the study process and can ensure that students themselves take responsibility for their learning.

However, it is important to understand that ensuring the availability of technologies in the study process also requires understanding that the level of competence in the use of these technologies may differ. Therefore practitioners in higher eductation must foresee that the technological and pedagogical competence of accademic staff should be improved to scaffold students in using technology, and learning outcomes should not be dependent on students' gaps in digital competencies. Further research is also needed to provide academic staff with research tools, such as rubrics, that can be used to evaluate specific technologies from the point of view of learning outcomes.

Out of the 110 selected articles, only 19 articles qualified for inclusion in the analysis. This indicates that there is still a rather divergent understanding among academic staff about research structure and research logic.

When selecting the articles for analysis, it was possible to select 110 full-text articles in the EBSCOhost database available from the University portal; however not all articles could be fully accessed, indicating that the unavailable studies were being published. That being said, the fact that not all of the ideas summarised here are available poses a number of risks in the academic field, such as the availability of knowledge only to university professors and students, who have purchased access to specific databases. This in turn leads to the risk that the latest findings are not known to other researchers, thus continuing the cultivation of outdated knowledge. At a time when technological development adversely affects the academic environment on the ability to quickly adapt to change, research should also become more accessible to help create new knowledge.

Answering the research questions:

1. When evaluating which technologies and technological solutions have been the focus of research in the context of higher education from 2010 to 2018, the research has concluded that there are no particular trends. The situation is diverse and suggests that researchers have analysed what is available at the moment, such as clickers, mobile phones, and LMS, but there is limited research on technological solutions challenging both the academic competence of professors and the academic comfort of students; these limited articles focused on Google Discussion Circles and Virtual Reality.
2. In addressing the benefits of TEL in promoting LO in the context of HE, it can be concluded that the benefits are more qualitative than quantitative, since there is no convincing evidence that the TEL process contributes to knowledge growth. There is evidence that if TEL is used in accordance with students' needs, understanding, and competence levels, it enhances the motivation of their studies and provides them with an opportunity to study at a convenient time and appropriate pace. This in turn can positively influence knowledge development, as a positive and motivating learning environment is a condition for students to use their own cognitive potential.
3. In looking at the challenges of using TEL as represented in research, it can be concluded that the main challenges are – the digital gap, when the level of digital competence differs; if the student's individual learning needs are not taken into account, it can lead to anxiety, which may result in lowered motivation.

In conclusion, academic professions should focus on a number of ideas:

- Improving research competence both in carrying out research at a time when technological advancement is ahead of research logic, and providing meaningful and data-based information in the preparation of research papers;
- Improving the digital competence of academic staff to scaffold students whose individual digital competences can be very different. If they are not supported, it can lead to students with high cognitive potential, but poorly-developed digital competences. As a result, these students can choose to perform in the field where they feel comfortable but not suceed in the field where they could achieve high levels of innovation; and
- Starting a discussion on the opening of research space so that the latest scientific research is available to a wider range that would support the creation of new innovations.

REFERENCES

Adekola, J., Dale, V. M., & Gardiner, K. (2017). Development of an institutional framework to guide transitions into enhanced blended learning in higher education. *Research in Learning Technology*, 25.

Arenas, E. (2015). Affordances of learning technologies in higher education multicultural environments. *Electronic Journal of E-Learning*, *13*(4), 217–227.

Berns, A., Isla-Montes, J., Palomo-Duarte, M., & Dodero, J. (2016). Motivation, students' needs and learning outcomes: A hybrid game-based app for enhanced language learning. *SpringerPlus*, *5*(1), 1–23. doi:10.118640064-016-2971-1 PMID:27547679

Blau, I., & Shamir-Inbal, T. (2017). Re-designed flipped learning model in an academic course: The role of co-creation and co-regulation. *Computers & Education*, *115*, 69–81. doi:10.1016/j.compedu.2017.07.014

Braun, V., & Clarke, V. (2006). Using thematic analysis in psychology. *Qualitative Research in Psychology*, *3*(2), 77–101. doi:10.1191/1478088706qp063oa

Coffin Murray, M., & Pérez, J. (2015). Informing and performing: A study comparing adaptive learning to traditional learning. *Informing Science*, *18*, 111–125. doi:10.28945/2165

Coskun, Y. D. (2015). Promoting digital change in higher education: Evaluating the curriculum digitalisation. *Journal of International Education Research*, *11*(3), 197–204.

Daniela, L., Kalniņa, D., & Strods, R. (2017). An overview on effectiveness of technology-enhanced learning (TEL). *International Journal of Knowledge Society Research*, *8*(1), 79–91. doi:10.4018/IJKSR.2017010105

Di Serio, A. D., Ibanez, M. B., & Kloos, C. D. (2013). Impact of an augmented reality system on students' motivation for a visual art course. *Computers & Education*, *68*, 586–596. doi:10.1016/j.compedu.2012.03.002

Ellahi, A., Zaka, B., & Sultan, F. (2017). A Study of supplementing conventional business education with digital games. *Journal of Educational Technology & Society*, *20*(3), 195–206.

Gezgin, D. M., Adnan, M., & Acar Guvendir, M. (2018). Mobile learning according to students of computer engineering and computer education: A comparison of attitudes. *Turkish Online Journal of Distance Education*, *19*(1), 4–17. doi:10.17718/tojde.382653

Goodyear, P., & Ellis, R. (2008). University students' approaches to learning: Rethinking the place of technology. *Distance Education*, *29*(2), 141–152. doi:10.1080/01587910802154947

Joanna Briggs Institute Reviewers' Manual. (2014). Joanna Briggs Institute.

Kirkwood, A., & Price, L. (2014). Technology-enhanced learning and teaching in higher education: What is 'enhanced' and how do we know? A critical literature review. *Learning, Media and Technology*, *39*(1), 6–36. doi:10.1080/17439884.2013.770404

Kivunja, C. (2015). Innovative methodologies for 21st century learning, teaching and assessment: A convenience sampling investigation into the use of social media technologies in higher education. *International Journal of Higher Education*, *4*(2), 1–26. doi:10.5430/ijhe.v4n2p1

Lansari, A., Tubaishat, A., & Al-Rawi, A. (2010). Using a learning management system to foster independent learning in an outcome-based university: A gulf perspective. *Issues in Informing Science & Information Technology*, *7*, 73–87. doi:10.28945/1193

Lee, D., Morrone, A. S., & Siering, G. (2018). From swimming pool to collaborative learning studio: Pedagogy, space, and technology in a large active learning classroom. *Educational Technology Research and Development*, *66*(1), 95–127. doi:10.100711423-017-9550-1

Limperos, A. M., Buckner, M. M., Kaufmann, R., & Frisby, B. N. (2015). Online teaching and technological affordances: An experimental investigation into the impact of modality and clarity on perceived and actual learning. *Computers & Education*, *83*, 1–9. doi:10.1016/j.compedu.2014.12.015

McIntyre, S., Watson, K., & Larsen, S. (2009). Strategies for large scale blended learning initiatives: Training, teaching and management. In E. Terry, A. Jefferies, & A. Bracq (Eds.), *Proceedings of the 4th International Blended Learning Conference* (pp. 295–308). Hatfield: University of Hertfordshire. Retrieved from unsworks.unsw.edu.au

Migdley, C., & Urdan, T. (2001). Academic self-handicapping and performance goals: A further examination. *Contemporary Educational Psychology*, *26*(1), 61–75. doi:10.1006/ceps.2000.1041 PMID:11161640

Naidoo, S., & Raju, J. (2012). Impact of the digital divide on information literacy training in a higher education context. *South African Journal of Library and Information Science*, *78*(1), 34–44. doi:10.7553/78-1-46

Palmer, M. M., Shaker, G., & Hoffmann-Longtin, K. (2014). Despite faculty skepticism: Lessons from a graduate-level seminar in a hybrid course environment. *College Teaching*, *62*(3), 100–106. doi:10.1080/87567555.2014.912608

Parajuli, K. P. (2016). Mobile learning practice in higher education in Nepal. *Open Praxis*, *8*(1), 41–54. doi:10.5944/openpraxis.8.1.245

Rockinson-Szapkiw, A. J. (2012). Should online doctoral instructors adopt audio feedback as an instructional strategy? Preliminary evidence. *International Journal of Doctoral Studies*, *7*, 7245–7258. doi:10.28945/1595

Saunders, F. C., & Gale, A. W. (2012). Digital or didactic: Using learning technology to confront the challenge of large cohort teaching. *British Journal of Educational Technology*, *43*(6), 847–858. doi:10.1111/j.1467-8535.2011.01250.x

Tynan, B., Ryan, Y., Hinton, L., & Lamont Mills, A. (2012). *Out of hours: Final report of the project e-teaching leadership: Planning and implementing a benefits-oriented costs model for technology enhanced learning*. Retrieved from http://www.olt.gov.au

Virtanen, M. A., Kääriäinen, M., Liikanen, E., & Haavisto, E. (2017). The Comparison of students' satisfaction between ubiquitous and web-based learning environments. *Education and Information Technologies*, *22*(5), 2565–2581. doi:10.100710639-016-9561-2

Walker, R., Voce, J., & Ahmed, J. (2012). *2012 survey of technology enhanced learning for higher education in the UK*. Retrieved from Oxford Universities and Colleges Information Systems Association: http://www.ucisa.ac.uk/groups/ssg/surveys.aspx

Walker, R., Voce, J., Ahmed, J., Nicholls, J., Swift, E., Horrigan, S., & Vincent, P. (2014). *2014 survey of technology enhanced learning: Case studies*. Retrieved from Oxford: Universities and Colleges Information Systems Association: http://www.ucisa.ac.uk/groups/dsdg/asg/~/media/7BCB3F2FF0E14 1A79A66BC87DDB34A14.ashx

Xia, B. S., & Rekola, M. (2013). Learning outcomes and knowledge sharing using web-based technologies in Finnish forest education from an educational experience point of view. *E-Learning and Digital Media*, *10*(1), 95–106. doi:10.2304/elea.2013.10.1.95

Chapter 3
Knowledge Framework for Online Learning: Implications for Cybersecurity and Programming Education

Simon Cleveland
City University of Seattle, USA

Greg Block
City University of Seattle, USA

ABSTRACT

While online education continues to grow, virtual instructors face certain asynchronous uncertainties when it comes to knowledge exchange with students. These challenges are especially prevalent in the cybersecurity and programming domains. To counteract such uncertainties and minimize teaching deficiencies expected to occur in asynchronous learning environments, this chapter assesses knowledge factors that impact virtual knowledge transfer and absorption processes. Synchronicity framework is proposed to integrate knowledge-seeking behavior, knowledge properties, knowledge domains, knowledge types, knowledge tools, and technology synchronicity. A real-life case is provided to integrate the framework in practice.

INTRODUCTION

A key component of distance learning platforms is an asynchronous learning environment, which promises to allow both students and their instructors to engage in learning at their own pace and schedule, without having to adhere to a fixed meeting time or location (Moller, 1998; Swan, 2001). Asynchronous learning can be effective for tasks where steps and outcomes are well-defined, such as an assignment to review academic literature to participate in a discussion forum (Loncar, Barrett & Liu, 2014; Nandi, Hamilton & Harland, 2012); however, in tasks where the steps are not well-defined or outcomes are not known in advance, a student may struggle when questions or uncertainties arise and there is no guidance on how to proceed (e.g. software engineering tasks)(Arkorful & Abaidoo, 2015; Dowling, Godfrey & Gyles, 2003).

DOI: 10.4018/978-1-5225-7347-0.ch003

Two domains that are particularly challenging for online students are cybersecurity and programming. For example, Hoffman, Burley and Toregas (2012) argued that universities are slow to respond to the cybersecurity needs of the industry due to their academic silos and challenges in meeting the institution's mission through online programs. Furthermore, these challenges are compounded since universities' online learning systems are often compromised by cyber-attacks and hindered by inefficient IT policies and procedures (Bandara, Ioras & Maher, 2014).

Research has identified numerous techniques for addressing these deficiencies in asynchronous learning (Chen & Wang, 2004; Johnson & Altowairiki, 2016; Underdown & Martin, 2016; Cleveland & Block, 2017); however, many of the approaches require some form of synchronous intervention (cognitive apprenticeships) (Jonaseen et al., 1995), or are limited to addressing finite problems that can be anticipated in advance (scaffolding)(Sims, Dobbs & Hand, 2002). These approaches do not improve a student's own problem solving skills for indeterminate problems without resorting to synchronous interactions, thus representing a challenge for online programs that offer 'self-paced' competency-based degrees with minimum intervention by instructors. As a result, the research question for this study is: how can knowledge be organized to minimize teaching deficiencies expected to occur in asynchronous learning environments?

In this paper, the authors present an asynchronous knowledge technology synchronicity framework that integrates knowledge form, types, dimensions, and technology synchronicity required for completing knowledge exchange in a virtual classroom. An example is provided to illustrate a deficiency in knowledge transfer and to integrate the framework.

The rest of the paper is structured as follows. First, an examination of knowledge exchange in the virtual classroom is performed and a set of building blocks is analyzed in detail. Next, a framework to organize the knowledge exchange properties are proposed. The paper concludes with a summary and a call for further research.

KNOWLEDGE EXCHANGE IN ONLINE ENVIRONMENT

In this study, the authors define asynchronous learning as the process of acquiring and converting knowledge within an online environment that is independent of space and time. The process, which was made possible by advancements in the technology for computer-mediated communication (CMC), has revolutionized the education industry. It has bridged the interaction between students and educators through virtual classrooms hosted within online learning systems (OLS).

The modern virtual classroom is a space where students interact with one another and with the instructor through forums, simulations, and collaborative assignments. Such interaction has been found effective in increasing the motivation to learn, mastery of course material, and greater quality of educational experience (Hiltz and Wellman, 1997). The interaction between instructor and students consists of complex processes that involve the seeking, acquisition, conversion, and integration of knowledge between parties. In the following section, we examine specific behaviors, knowledge properties, knowledge domains, and knowledge tools that will be used to formulate the proposed framework.

Knowledge Seeking Process

First, an examination of the process of knowledge seeking is performed. A key factor in the process of knowledge exchange is knowledge seeking behaviors. Such behaviors are defined as scanning for data and information in order to satisfy certain information needs (Xu, Tan, & Yang, 2006). These activities are best explained by the information foraging theory that postulates that individuals would forage for information when they encounter vague problems requiring solutions (Pirolli & Card, 1999).

The virtual classroom poses specific disadvantages to learning when it comes to knowledge seeking behaviors (Posey et al., 2010). For example, the asynchronous discussion forums have been found to limit problem solving learning in courses on physics and statistics (Hong et al., 2003; Kortemeyer, 2006). Furthermore, asynchronous discussions have been perceived as unreal and less meaningful (Merryfield, 2001), as well as contributing to the feelings of isolation and not being part of learning communities (Hrastinski, 2008).

Behavioral Categories

Studies show that such behaviors fall into various categories. For example, Vandenbosch and Huff (1997) proposed four such categories:

- **Undirected:** Purposeless exposure to information;
- **Conditioned:** Exposure that is not based on specific search for information;
- **Informal:** Attempt to discover information that has no structure;
- **Formal:** Directed attempt to discover precise information.

Furthermore, studies found that knowledge seeking behavior encompasses focused search which "occurs when organizational members or units actively search in a narrow segment of the organization's internal or external environment, often in response to actual or suspected problems or opportunities,"(Huber, 1991, p. 97).

Finally, according to Belkin (1980), knowledge seeking behaviors consist of:

- Seeker's awareness of knowledge disparity;
- Quest for gathering relevant information;
- Awareness of reduced knowledge disparity.

Behavioral Triggers

Savolainen (2006) argued that specific triggers evoke knowledge seeking behaviors. Once the trigger occurs, a seeker considers potential sources of information, retrieves the knowledge, and evaluates its relevance. The outcome of this behavior consists of assessment of the acquired knowledge, whether it has fulfilled the information need, or further knowledge seeking is necessary.

Student Motivation and Learning Style

Taylor et al. (2018) proposed two additional factors that should be considered. These include student motivation and learning style. The researchers argued that engagement in a course is driven by the student's motivation, especially when it comes to cybersecurity students. For example, the increasing labor demand for skilled cybersecurity professionals can lead students to enroll in easily accessible cybersecurity courses that can be completed quickly and used as "stepping stone to gaining almost immediate employment in the security industry." (Taylor, et al., 2018, p. 105).

The authors argued that the learning style for cybersecurity students is different from those of other disciplines. This is most likely due to the fact that cybersecurity courses include assessments that lead to conclusive answers. Furthermore, learning style may be impacted by the students' background. For example, students with background in cybersecurity will welcome "epistemological and ontological stance which posits that understanding of phenomena is reached though objective study and experimental methods, and there is a finite set of solutions to any problem." (p. 105). Additionally, cybersecurity students will tend to shy away from lessons that contain greater subjectivity and uncertainty in the training methods, while focusing on specific theories that can provided a single and precise description of events and behaviors.

In the software engineering discipline, a student may be directed to complete a task involving a programming language. Slight, undetectable missteps in a programming assignment may involve syntactical, lexical, or logical errors that cannot be easily anticipated at the time a course is designed. Successful engineering students rely on problem solving skills, which generally involve a series of heuristics that conclude when a solution has been identified; however, in the virtual classroom, students cannot easily distinguish between intrinsic errors (that is, unanticipated conditions that are inherent to the assignment) and hygienic errors (that is, unexpected errors that are inherent to the platform). Students are expected to reason through intrinsic errors, but can be expected to rely on their instructor to help them resolve hygienic errors. As a result, in the virtual asynchronous learning classroom, the instructor is not available to provide guidance to the student, so the student may consider all errors to be hygienic and request the instructor solve the problem the student should be equipped to solve.

Students enrolled in programming courses may not be used to thinking abstractly, and as a result cannot reorganize a problem or question into a programming language (Truong, Roe, & Bancroft, 2005, January). Writing a program consists of two critical components: crafting source code following the syntactical rules of the chosen programming language, and the implications of executing the program (Hung, 2012). Novice programmers too often focus on details rather than view a program holistically; fix programs incrementally rather than refactor or rewrite the program; focus on localized changes in an effort to minimize the impact or side-effects of a change; and are unable to capitalize on or expand their understanding of programming concepts. An early study of novice programmers characterized these learners into 3 groups: stoppers, movers, and extreme movers. Stoppers would respond to obstacles and unexpected conditions by giving up, or expecting outside help; extreme movers would make random, out-of-sequence changes to their programs, resulting in little forward progress; while movers would make progressive changes that allowed them to improve their comprehension of the problem, and guide them to an eventual solution (Perkins, Hancock, Hobbs, Martin, & Simmons, 1986). Hwang et al (2012) identified 6 types of learning problem-solving behavior in the context of programming: completely independent, performing poorly without assistance, self-improving using assistance, confident after enlightenment, imitating and plagiarizing. The success of each behavior is gradient, with the greatest degree of success achieved by those who are completely independent, and the least degree of success borne by the imitators and plagiarizers. Students who practiced the central behaviors (self-improving and confident) tended to shift to either extreme (complete independence or imitating and plagiarizing).

Programming errors are presented by a language compiler, which displays information about the line number of the source text where the error as identified, the character position in the line, and an error message that may include contextual information regarding the error. However, the actual error may occur in a previous line than the line of text indicated in the error message. Further, a single syntactical error condition can produce multiple error messages with such volume that the programmer cannot easily distinguish which error message represents the original error. Novice programmers may be challenged in comprehending the error message, traversing from the error message to the line or lines of source code that originated the error, and postulating the correct fix for the error condition (Patel & Kanani, 2013).

Identifying and remediating problems in a programming context is a form of problem-solving that encourages the development of cognitive processes. Novice programmers by definition lack sufficient experiential acumen, and do not know how to apply the knowledge they have accrued. In most teaching modalities, novice programmers suffer from lack of immediate, individualized feedback to guide them in problem-solving (Shadiev, et al., 2014). When a novice programmer does not receive feedback, they may become frustrated, begin to doubt their own skills and abilities, and disengage from the goal of developing a program.

In the context of programming, the problem-solving exercise can be characterized by the following steps (Su, Yang, Hwang, Huang, & Tern, 2014):

1. Comprehending the problem;
2. Gathering information in order to define and plan a solution;
3. Assimilating the information to design and deploy the solution; and
4. Verifying the solution and presenting the solution to others

In addition to understanding the problem representation, the programmer must adopt a goal-oriented strategy to solve the problem. Not all problems are the same: some problems are well-structured, while others are ill-structured or unstructured. Well-structured problems have finite outcomes based on a limited set of parameters and fixed constraints. An example of a well-structured programming problem is a program that grades mathematical exams where there is only a single correct answer to each exercise. Ill-structured problems, on the other hand, may have multiple outcomes, a wide range of parameters, and few or now constraints. An example of an ill-structured problem is a program that corrects grammar in student essays. Novice students learn more productively when they are introduced to well-structured programs, and gradually work their way to ill-structured programs through a series of steps or exercises (Uysal, 2014).

As important as problem-solving skills are to successful programming, the programmer must be able to express the solution in the context of a programming outcome. This emphasis blends the generalized skill of problem-solving with competency in programming. The successful problem-solver must be able to visualize a program design that resolves the problem, express the solution in the chosen language and architecture, and be able to debug, test and deploy the developed solution (Hwang, Wang, Hwang, Huang, & Huang, 2008). Computational thinking describes another set of strategies a successful programmer needs as they transform problems into solutions. Computational thinking strategies accrete problem-solving knowledge through experience to simplify the problem space. These strategies include decomposition as a technique for breaking larger problems into smaller ones; abstraction and pattern detection to recognize previously-observed problems; and automation to develop a self-contained solution (Quinn, 2016).

As studies have indicated, novice programmers adopt a variety of roles in a learning environment, including stoppers, movers, and extreme movers (Perkins, Hancock, Hobbs, Martin, & Simmons, 1986), as well as a gradient of behaviors, from completely independent to imitators and plagiarizers (Hwang, Shadiev, Wang, & Huang, 2012). Hung (2012) applied two distinct learning types, sensors and intuitors, to programmatic learning. For example, sensors preferred learning facts and enjoyed well-structured problems, while intuitors preferred to discover possibilities and favored innovation over repetition. Some of the deficiencies in learning programming may be due to cognitive issues if a novice programmer lacks the acumen to structure programmatic solutions, while other deficiencies in learning programming may be motivational or behavioral, such as the "stoppers" who quit rather than challenge themselves to solve a problem, or the "imitators" who try to free-ride on the work of others. As noted by Hwang et al (2012) learners may transition from one behavior to another. This may have been caused by some impetus in the learning environment that initiated the behavioral change.

Learning Techniques

Many techniques have been proposed to improve learning outcomes in the context of programming. Hung (2012) proposed adapting learning program design to match student learning styles; for example, a diagram method was used to instruct students who were input-oriented, while an analogy method was selected for students who were perception-oriented. Students who participated in the method that matched their learning style significantly outperformed a control group. Some of the techniques could be adapted to a distance learning platform but would not be successful in an asynchronous environment. Asynchronous learning networks (ALNs) are computer-mediated platforms that encourage students to collaborate on problem-solving. Student perceptions of the discussion quality of ALNs are mixed; students do not participate equally, frustrating the students who actively participate (Benbunan-Fich & Hiltz, 1999). Further, students may contribute one-sided, closed-ended discussions because the overhead of participating in a collaborative effort causes them to lose motivation, resulting in a "least collaborative effort" contribution (Hall, 2014).

Scaffolding is a technique where an instructor offsets a student's deficiencies by completing portions of a task the student is not able to complete. The instructor coaches the student by providing help, instructions, or encouragement, or by completing steps on the student's behalf. As the student gains competency, the instructor gradually removes the scaffolding. Scaffolding derives from a concept known as the zone of proximal development (ZPD), which describes the distance between "the actual development level determined by the independent problem solving and the level of potential development as determined through problem solving under adult guidance or in collaboration with more capable peers" (Vygotsky, 1980, p. 86). However, scaffolding is effective only when the help is immediate (Black, 2006).

Properties of Knowledge

In order to understand the disadvantages of asynchronous learning, examination is performed of the properties of knowledge. Knowledge exchange involves the process of capturing, organizing, transferring, and reusing experiential knowledge (Lin, 2006). In this study, the authors adopt the view that knowledge in the virtual classroom resides in two forms: tacit (non-codifiable, rooted in the minds of experts, capable of transfer via shadowing and mentoring) and explicit (codifiable and capable of transfer using techniques, such as stories, metaphors, and models) (Polanyi, 1966, Nonaka, 1994). In the virtual classroom, instructors should be capable of using asynchronous technology to transfer the explicit form of knowledge by articulating key theoretical concepts and their application in practice. In contrast, the tacit knowledge form would be impossible to transfer via asynchronous mode due to the limitations of interactions and technology.

Extant literature suggests several categories of knowledge properties. For example, Miller and Jablin (1991) introduced three categories:

- **Referent:** Associated with specific function;
- **Appraisal:** Associated with selected performance;
- **Relational:** Associated with acceptance of social behavior.

Other researchers have added technical type, which was related to information that is used for "defining a problem/task; learning techniques applicable to dealing with the problem/task; finding solutions; or identifying a piece of missing data," (Madzar, 2001, p. 222). Cross and Sproull (2004) proposed five additional categories to include:

- Solutions;
- Meta-knowledge;
- Problem reformulation;
- Validation of plans or solutions; and
- Legitimation from contact with a respected person.

Finally, researchers categorized the wanted knowledge as task information (related to technical skills, or feedback associated with performance) and social information (related to social feedback) (Xu, Kim & Kankanhalli, 2010).

Domains of Knowledge

In addition to knowledge behaviors and knowledge properties, it is prudent to discuss the impact of knowledge domains in the exchange of knowledge in asynchronous mode. First, the authors examine the knowledge exchange that occurs between parties within a shared knowledge domain. Cleveland (2016) argued that experienced parties can use ICTs to easily share knowledge with knowledge seekers who have prior mental model and experiential understanding of the exchanged knowledge. This occurs as a result of the previously formulated mental models in each participant with regards to the transferred knowledge. As a result, in scenarios where knowledge is exchanged in the same domain of understanding, ICTs in a virtual classroom will facilitate the transfer of knowledge.

In contrast, if the parties lack understanding of the domain of the exchanged knowledge, the sharing process will be impacted due to ambiguity. In such a situation, knowledge seekers will not be able to leverage "mental models through which to filter the new information and to make sense of it," (Cleveland, 2016, p. 4).

As a result, the greater the disparities between the domains of knowledge between the instructor and students in the virtual classroom, the less knowledge exchange will take place.

Types of Knowledge

To combat such disparities, the main focus is on the types of knowledge. In addition to properties and domains of knowledge, it is argued that knowledge in the virtual classroom should be categorized for easier transmission. For this purpose, in this study the authors adopt the three knowledge types proposed by Shulman (1986):

- Subject matter content knowledge (SMCK);
- Pedagogical content knowledge (PCK);
- Curricular knowledge (CK).

Instructors who possess SMCK knowledge will be able to demonstrate why and how key concepts relate to the topic of the course. For example, when it comes to software engineering, an instructor should be able to explain why specific software requires detailed testing and how the testing should be conducted, via a case-based example.

PCK involves the use of a variety of methods (e.g. stories, examples, analogies) by the instructor to demonstrate specific subject matter knowledge. An instructor with substantial PCK will successfully help students integrate the subject matter knowledge.

Finally, the CK includes the creative use of textbooks and materials to help students fully master a specific discipline. For example, an instructor can incorporate a specific software package, along with a best practices software handbook, to help students reflect, brainstorm, and combine the subject matter lessons.

Dimensions of Knowledge

In addition to form and type, it is argued that knowledge in the virtual classroom consists of four dimensions (as proposed by De Jong and Ferguson-Hessler, 1996): 1) situational, 2) procedural, 3) strategic, and 4) conceptual. Each of these knowledge dimensions involves different properties and as a result, will require different mediums of exchange. This leads to the conclusion that both the situational and procedural knowledge dimensions are related. For example, the situational knowledge dimension is related to a specific scenario that may require sequential steps to be executed in order for knowledge to be transferred. Furthermore, the procedural knowledge dimension is associated with a set of requirements and rules related to a situation.

In the software engineering scenario described earlier, an instructor has provided a set of steps for students to install a specific software program on their computers. As a result of the instructions, a student is experiencing a software-related malfunction and reaches out for specific information on how to resolve the errors.

To help with the software error, the instructor uses a conceptual knowledge dimension to relay principles associated with the software engineering domain. The conceptual knowledge dimension consists of a set of symbols and models that allow the instructor to tap into the student's mental model (in our case, understanding of programming modules and code) and explain the cause of the errors.

Lastly, the instructor will convey specific actions to tie theory together with the practical example, via the strategic knowledge dimension.

Knowledge Tools

In addition to the types of knowledge, an examination of the tools that instructors can leverage to facilitate the knowledge exchange is performed. In this study, the focus is on three such tools: stories, reflections, and lessons learned.

Stories

A story represent an account of an event that allows the storyteller to share an identity, help the audience make decisions, and share an identity (Cleveland, 2013). Extant literature demonstrates that stories have been proven to transfer knowledge used to solve specific problems through the application of the experiences acquired through the narratives (Jonassen & Hernandez-Serrano, 2002; Newell, 2004). Some researchers report that stories are some of the most frequent tools to exchange tacit knowledge (Goffin & Koners, 2011). For example, one study found that storytelling and sharing of metaphors occurred every 20 minutes among personnel during post-implementation review meetings. Furthermore, Whyte and Classen (2012) found that storytelling was widely used to distribute information related to best practices and ideas, while Cleveland (2013) argued that "stories are powerful mechanisms for joining, sorting, and recombining personal and shared knowledge. They can also be used as input into the collective and self-reflection processes of the learning cycle," (p. 3).

Reflection

When it comes to reflection, studies report that this technique assists in the exchange of valuable lessons learned, support learning, and channel innovation (Knipfer, Kump, Wessel & Cress, 2012). Instructors can facilitate knowledge sharing through reflection by requiring students to use personal journals. Such journals, also known as reflective guides, leverage precise questions that allow students to document practices and identify relevant knowledge sources.

For example, extant literature reports that the use of reflective journals in organizations leads to the exchange of knowledge, conflict management, strengthening of interpersonal communication, and fortification of leadership skills (Loo, 2002). Moreover, studies on the use of reflective guides among students showed that the use of such guides significantly improved the processes of documenting lessons, and the capturing of valuable explicit knowledge (Matturro & Silva, 2010).

Lessons Learned

Finally, the authors introduce lessons learned documentation, a technique used in the project management field during post-project review meetings for knowledge creation. According to Williams (2008), 86% of the 522 participants in his study reported an increase in competency levels as a result of the application of this technique. The participants reported that lessons learned documentation helped them build their soft skills. Among the surveyed, 75% stated that learning diaries can be the tool to document lessons, while over 80% believed that such lessons can be captured and shared via stories. This study was supported by Tukel, Rom and Kremic (2008), who found that lessons learned documents prevented the loss of knowledge. Their key recommendation was for the wide application of such "methodical documentation" to facilitate learning practices.

Virtual Tools

In the realm of cybersecurity, new approaches to teaching cybersecurity education have been introduced. For example, North (2016) conducted a study among graduate students in a cybersecurity course. He compared the use of simulations versus virtual machines in order to determine effectiveness of each technique. The outcomes of the study showed that the use of virtual environments among students led to the development of higher level mastery of cybersecurity topics than simulations. Furthermore, the researcher found that students experienced higher degree of enjoyment and trust in the courses that leverage virtual machines.

In another study, Black, Chapman and Clark (2018) leveraged the constructivist learning theory, which posits that leaners are responsible for the seeking and development of their own knowledge based on persona experience. The researchers studied the use of Enhanced Virtual Laboratory (EVL), a tool developed by the Department of Defense to enhance education for cyber security. The tool uses web-based laboratories to allow students to gain knowledge virtually, while giving instructors sufficient platform to share knowledge on the topics of cyber security. The results of the study showed that majority of students completed large portion of the assignments successfully using the tool. The authors argued that the virtual environment tool was sufficient to deliver "a new and improved method of remotely delivering cyber security content to different users with variable depths of security knowledge." (p. 4).

Another tool, which can be used by instructors of cybersecurity is SCAMPER (Gould & Cleveland, 2018). SCAMPER was originally introduced by Eberle (1996) and is an abbreviation for Substitute, Combine, Adapt, Modify, Put, Eliminate, and Reverse. Instructors for virtual cybersecurity courses can introduce the tool to students for the purposes of cyber threat assessment. Since the tool can assist students in following a structured learning processes, it lends itself for use in virtual setting. As Gould and Cleveland (2018) proposed, instructors can teach students to ask informative questions during the Substitute in order to assess the security level of a system. In this stage, instructors can teach students to make a proper decision among alternatives solutions.

Next, students can explore applicable steps to elect the best security solution for a system during the Combine state. Instructors can split the students into groups and allow them work together to leverage various techniques and technologies to neutralize a possible cyber-attack.

Continuing to work together through potential solutions, the students will leverage innovative tools and techniques to challenge the status quo in the Adapt stage. Here, instructors will urge students to find alternative solutions that are more flexible than previously envisioned.

Once a solution has been selected, students will learn how to secure a system using the new innovative approach during the Modify stage. Next, instructors can expand on the lessons by asking students to apply the solution to other areas in the Put to another use stage. Gould and Cleveland (2018) cited examples such as security policies and argued that eliminating current measures of security during the Elimination stage can help increase productivity with half of the resources. In the Rearrange stage, instructors can challenge students to achieve a better result by "reversing the existing security measures." (p.4).

Technology Synchronicity

Finally, it is argued that knowledge exchange in the virtual classroom demands either low or high synchronicity as proposed by the media synchronicity theory (MST) (Dennis et al., 2008).

The theory suggests that individuals leverage a set of communication patterns to exchange information in order to either minimize knowledge ambiguity within a specific domain, or validate pre-existing knowledge (Cleveland et. al, 2015). Each of the patterns is characterized by a specific process. First, the conveyance process occurs when individuals share brand new knowledge, which necessitates the creation or revision of a mental model, in order for the knowledge seeker to absorb such knowledge. The second process is convergence. During this process, the knowledge that is exchanged, has been encountered in the past, and is used to validate a pre-existing mental model in the knowledge seeker.

The theory also stipulates that the facilitation of the conveyance demands one-way broadcasting of knowledge in the direction of the knowledge seeker for the purposes of mental model formulation. On the other hand, the convergence process involves rapid and reoccurring transfer of small knowledge amounts in order to validate the knowledge seeker's pre-existing mental model (Cleveland et al., 2015).

Due to the differences in the amount and frequency of information transfer required to fulfill each processes' demands, the theory explains that "the conveyance process will require technology channels to facilitate rich information flow, thus leading to low synchronicity between the participants (e.g. instant messaging, screen sharing, and file sharing), while convergence requires technology mediums that facilitate high synchronicity (video and audio conferencing)." (Cleveland et al., 2015, p. 7).

As a result, it is argued that in asynchronous environments, students who lack conceptual mental models in order to understand specific lessons, will require a highly synchronous communication medium of interaction with the instructor to formulate new mental models and absorb new knowledge (e.g. real-time video and audio sessions) (Cleveland 2016). Therefore, asynchronous virtual classrooms without rich medium capabilities (e.g. live conferencing, synchronous instant messaging) would be ill-equipped to allow the exchange of knowledge with conceptual and strategic knowledge dimensions.

PROPOSED KNOWLEDGE TECHNOLOGY SYNCHRONICITY FRAMEWORK FOR ASYNCHRONOUS ENVIRONMENT

Based on the analysis in this study, in Table 1 the authors propose a knowledge technology synchronicity framework (KTSF) for asynchronous environment that integrates the various building blocks of knowledge exchange, mainly: knowledge seeking behaviors, knowledge properties, knowledge domains, knowledge types, knowledge dimensions, knowledge tools, and technology synchronicity.

DISCUSSION

To illustrate the use of KTSF in transferring knowledge in the virtual classroom, the authors provide a specific case experienced in a graduate software engineering course taught in an asynchronous learning environment at a university in the northeastern area of the United States. The class took full advantage of discussion forums for students to prepare and discuss concepts related to predictive analytics. In addition, students were asked to complete exercises related to data preparation and analysis tasks in the

Table 1. Asynchronous Knowledge Technology Synchronicity Framework

Building Blocks	Concepts	Sources
Knowledge Seeking Behaviors	Categories include undirected, conditioned, informal, and formal; Information foraging theory; Awareness of disparity, quest for information gathering; Motivation; Learning style, Virtual	Xu, Tan, & Yang, 2006; Pirolli & Card, 1999; Vandenbosch & Huff, 1997; Huber, 1991; Belkin (1980), Savolainen, 2006; Taylor et al. 2018
Knowledge Properties	Experiential knowledge; Tacit Knowledge; Categories such as referent, appraisal, and relational; solutions, meta-knowledge, problem reformulation, validation of plans or solutions, legitimation from contact with a respected person	Lin, 2006; Polanyi, 1966; Nonaka, 1994; Miller & Jablin, 1991; Madzar, 2001; Cross & Sproull, 2004; Xu, Kim & Kankanhalli, 2010
Knowledge Domains	Within and Outside of Shared Knowledge Domains	Cleveland, 2016
Knowledge Types	Subject matter content knowledge, Pedagogical content knowledge, Curricular knowledge	Shulman, 1986;
Knowledge Dimensions	Situational, procedural, strategic, and conceptual;	De Jong and Ferguson-Hessler, 1996
Knowledge Tools	Stories, Reflective Journals, Lessons Learned Documents, Virtual Tools, SCAMPER	Cleveland, 2013; Gould & Cleveland, 2018; Jonassen & Hernandez-Serrano, 2002; Newell, 2004; Goffin & Koners, 2011; Whyte & Classen, 2012; Knipfer, Kump, Wessel & Cress, 2012; Loo, 2002; Matturro & Silva, 2010; Williams, 2008; Tukel, Rom & Kremic, 2008.; Black, Chapman & Clark, 2018
Technology Synchronicity	High and Low	Dennis et al., 2008; Cleveland 2016, Cleveland et al., 2015

programming language R. Finally, assessments were given to measure the students' grasp of terms and concepts presented throughout the course. Class sizes usually ranged from 5-10 students. The majority of students were adult learners with professional careers, with an even mix of male and female students.

Discussion questions related to the material taught in each weekly lesson (explicit knowledge form with subject matter, pedagogical and curricular knowledge types), and students could refer to background information using the course textbook. Students had little difficulty submitting posts on time, with the requisite word and reference count. Discussions were generally one or two levels deep.

Five exercises were assigned in the course. The first exercise involved downloading and installing R Studio, a programming environment for writing and executing R scripts (procedural knowledge dimension). The instructions specified a web address where the student could choose the download package appropriate to their desktop device and instructions for installing the software. In subsequent exercises the students would create text files that contained R statements to manipulate data sources, transform data frames, and plot the results. Students who were familiar with the R Studio environment, including the process of downloading and installing software, but encountered installation errors, needed assistance (situational knowledge dimension) via asynchronous discussions (low synchronicity medium).

Students with very little or no knowledge of the R language, who had considerable challenge understanding the complexity of constructing R statements (conceptual knowledge dimension), encountered error messages that resulted from malformed commands. A common reaction to these unforeseen circumstances was for the student to reach out to the instructor to "intervene"; that is, to identify the source of the problem and instruct the student with what steps were required to remediate the problem. However, due to the asynchronous nature of the exchange, the communications were frequently cryptic ("You need to enclose the string variable in matching quotes"; "What is a string variable?") and increasingly more frustrating to the participants. In such instances, the limitation of the asynchronous environment imposed considerable challenges to the ability to exchange knowledge via high synchronicity.

CONCLUSION

In this study, the authors provided arguments for the deficiencies of the asynchronous virtual classroom. Knowledge technology synchronicity framework for asynchronous environment was proposed to characterize and categorize the building block form knowledge exchange in asynchronous classrooms. The research question for this study was: How can knowledge be organized to minimize teaching deficiencies expected to occur in asynchronous learning environments? To answer this question, seven distinct building blocks for knowledge exchange in asynchronous environments were examined. The authors assessed what drives knowledge exchange (knowledge seeking behaviors) and the properties, domains, types and dimensions of exchanged knowledge. Moreover, a set of knowledge tools and technology mediums were analyzed that can help facilitate the knowledge exchange in asynchronous classrooms. The authors leveraged two domains for virtual learning: cybersecurity and programming.

Finally, the authors provided a specific case to illustrate the application of the framework in a virtual classroom. It is prudent to note that a limitation of this study is the single case used to support the framework. As a result, future research will focus on the application of the framework in hybrid learning environments and the examination of specific strategies that can help build student problem solving skills for open-ended problems in an asynchronous environment. Future studies should also focus on developing specific asynchronous approaches to teaching students soft skills, such as cyber leadership for preventing threats (Cleveland & Cleveland 2018a) as well as handling of post-incident activities in the aftermath of a cybercrime (Cleveland & Cleveland 2018b).

REFERENCES

Arkorful, V., & Abaidoo, N. (2015). The role of e-learning, advantages and disadvantages of its adoption in higher education. *International Journal of Instructional Technology and Distance Learning*, *12*(1), 29–42.

Bandara, I., Ioras, F., & Maher, K. (2014). Cyber security concerns in e-learning education. *Proceedings of ICERI2014 Conference*.

Belkin, N. J. (1980). Anomalous states of knowledge as a basis for information retrieval. *Canadian Journal of Information Science*, *5*(1), 133–143.

Benbunan-Fich, R., & Hiltz, S. R. (1999). Impacts of Asynchronous Learning Networks on Individual and Group Problem Solving: A Field Experiment. *Group Decision and Negotiation*, *8*(5), 409–426. doi:10.1023/A:1008669710763

Black, M., Chapman, D., & Clark, A. (2018). The Enhanced Virtual Laboratory: Extending Cyber Security Awareness through a Web-based Laboratory. *Information Systems Education Journal*, *16*(6), 4.

Black, T. (2006). Helping Novice Programmers Succeed. *Journal of Computing Sciences in Colleges*, *22*(2), 109–114.

Chen, X., & Wang, Y. (2004). Use synchronous chat to improve online learning experience. *AMCIS 2004 Proceedings*, 365.

Cleveland, M., & Cleveland, S. (2018b). Cybercrime Post-incident Leadership Model. *Proceedings of the Thirteenth Midwest Association for Information Systems Conference*.

Cleveland, S. (2013). On-the-job learning practices for IS students. *Proceedings of the 16th Southern Association for Information Systems Conference (SAIS)*, 16-21.

Cleveland, S. (2016). Social Media Systems in the Workplace: Toward Understanding Employee Knowledge Creation via Microblogging within Shared Knowledge Domains. *Informatics*, *3*(3), 1–19. doi:10.3390/informatics3030011

Cleveland, S., & Block, G. (2017). Toward Knowledge Technology Synchronicity Framework for Asynchronous Environment. *International Journal of Knowledge Society Research*, *8*(4), 23–33. doi:10.4018/IJKSR.2017100102

Cleveland, S., & Cleveland, M. (2018a). Toward Cybersecurity Leadership Framework. *Proceedings of the Thirteenth Midwest Association for Information Systems Conference*.

Cleveland, S., Ellis, T., & Hinojosa, C. (2015). Knowledge exchange via unified communication services: A grounded theory approach. *Proceedings of the 21st Americas Conference on Information Systems*.

Cross, R., & Sproull, L. (2004). More than an answer: Information relationships for actionable knowledge. *Organization Science*, *15*(4), 446–462. doi:10.1287/orsc.1040.0075

De Jong, T., & Ferguson-Hessler, M. G. (1996). Types and qualities of knowledge. *Educational Psychologist*, *31*(2), 105–113. doi:10.120715326985ep3102_2

Dennis, A. R., Fuller, R. M., & Valacich, J. S. (2008). Media, tasks, and communication processes: A theory of media synchronicity. *Management Information Systems Quarterly*, *32*(3), 575–600. doi:10.2307/25148857

Dowling, C., Godfrey, J. M., & Gyles, N. (2003). Do hybrid flexible delivery teaching methods improve accounting students' learning outcomes? *Accounting Education*, *12*(4), 373–391. doi:10.1080/0963928032000154512

Goffin, K., & Koners, U. (2011). Tacit knowledge, lessons learnt, and new product development. *Journal of Product Innovation Management*, *28*(2), 300–318. doi:10.1111/j.1540-5885.2010.00798.x

Gould, D., & Cleveland, S. (2018). Evolutionary Systems: Applications to Cybersecurity. *Proceedings of the Thirteenth Midwest Association for Information Systems Conference*.

Hall, B. (2014). Designing Collaborative Activities to Promote Understanding and Problem-Solving. *International Journal of e-Collaboration*, *10*(2), 55–71. doi:10.4018/ijec.2014040104

Hiltz, S. R., & Wellman, B. (1997). Asynchronous learning networks as a virtual classroom. *Communications of the ACM*, *40*(9), 44–49. doi:10.1145/260750.260764

Hoffman, L., Burley, D., & Toregas, C. (2012). Holistically building the cybersecurity workforce. *IEEE Security and Privacy*, *10*(2), 33–39. doi:10.1109/MSP.2011.181

Hong, K.-S., Lai, K.-W., & Holton, D. (2003). Students' satisfaction and perceived learning with a web-based course. *Journal of Educational Technology & Society*, *6*(1), 116–124.

Hrastinski, S. (2008). Asynchronous and synchronous e-learning. *EDUCAUSE Quarterly*, *31*(4), 51–55.

Huber, G. P. (1991). Organizational learning: The contributing processes and the literatures. *Organization Science*, *2*(1), 88–115. doi:10.1287/orsc.2.1.88

Hung, Y. (2012). The Effect of Teaching Methods and Learning Style on Learning Program Design in Web-Based Education Systems. *Educational Computing Research*, *47*(4), 409–427. doi:10.2190/EC.47.4.d

Hwang, W., Shadiev, R., Wang, C., & Huang, Z. (2012). A Pilot Study of Cooperative Programming Learning Behavior and its Relationship with Students' Learning Performance. *Computers & Education*, *58*(4), 1267–1281. doi:10.1016/j.compedu.2011.12.009

Hwang, W., Wang, C., Hwang, G., Huang, Y., & Huang, S. (2008). A Web-Based Programming Learning Environment to Support Cognitive Development. *Interacting with Computers*, *20*(6), 524–534. doi:10.1016/j.intcom.2008.07.002

Johnson, C., & Altowairiki, N. (2016). Developing Teaching Presence in Online Learning Through Shared Stakeholder Responsibility. Handbook of Research on Innovative Pedagogies and Technologies for Online Learning in Higher Education, 151.

Jonassen, D., Davidson, M., Collins, M., Campbell, J., & Haag, B. B. (1995). Constructivism and computer-mediated communication in distance education. *American Journal of Distance Education*, *9*(2), 7–26. doi:10.1080/08923649509526885

Jonassen, D., & Hernandez-Serrano, J. (2002). Case-based reasoning and instructional design: Using stories to support problem solving. *Educational Technology Research and Development*, *50*(2), 65–77. doi:10.1007/BF02504994

Knipfer, K., Kump, B., Wessel, D., & Cress, U. (2012). Reflection as a catalyst for organizational learning. *Studies in Continuing Education*. doi:10.1080/0158037X.2012.683780

Kortemeyer, G. (2006). An analysis of asynchronous online homework discussion in introductory physics courses. *American Journal of Physics, 74*(6), 526–536. doi:10.1119/1.2186684

Lam, A. (2000). Tacit knowledge, organizational learning and societal institutions: An integrated framework. *Organization Studies, 21*(3), 487–513. doi:10.1177/0170840600213001

Lin. (2006). Impact of organizational support on organizational intention to facilitate knowledge sharing. *Knowledge Management Research & Practice, 4*(1), 26-35.

Loncar, M., Barrett, N. E., & Liu, G. Z. (2014). Towards the refinement of forum and asynchronous online discussion in educational contexts worldwide: Trends and investigative approaches within a dominant research paradigm. *Computers & Education, 73*, 93–110. doi:10.1016/j.compedu.2013.12.007

Loo, R. (2002). Journaling: A learning tool for project management training and team-building. *Project Management Journal, 33*(4), 61–66.

Madzar, S. (2001). Subordinates' information inquiry: Exploring the effect of perceived leadership style and individual differences. *Journal of Occupational and Organizational Psychology, 74*(2), 221–232. doi:10.1348/096317901167325

Matturro, G., & Silva, A. (2010). Using reflective guides to capture software projects experience. *Proceedings of the International Conference on Information and Knowledge Engineering*, 202-207.

Merryfield, M. (2001). The paradoxes of teaching a multicultural education course online. *Journal of Teacher Education, 52*(4), 283–299. doi:10.1177/0022487101052004003

Miller, V. D., & Jablin, F. M. (1991). Information seeking during organizational entry: Influences, tactics, and a model of the process. *Academy of Management Review, 16*(1), 92–120. doi:10.5465/amr.1991.4278997

Moller, L. (1998). Designing communities of learners for asynchronous distance education. *Educational Technology Research and Development, 46*(4), 115–122. doi:10.1007/BF02299678

Nandi, D., Hamilton, M., & Harland, J. (2012). Evaluating the quality of interaction in asynchronous discussion forums in fully online courses. *Distance Education, 33*(1), 5–30. doi:10.1080/01587919.2012.667957

Newell, S. (2004). Enhancing cross-project knowledge. *Engineering Management Journal, 16*(1), 12–20. doi:10.1080/10429247.2004.11415234

Nonaka, I. (1994). A dynamic theory of organizational knowledge creation. *Organization Science, 5*(1), 14–37. doi:10.1287/orsc.5.1.14

Patel, K., & Kanani, H. (2013). Dynamic Error Handling Tool. *International Journal of Engineering and Innovative Technology, 2*(11), 34–36.

Perkins, D., Hancock, C., Hobbs, R., Martin, F., & Simmons, R. (1986). Conditions of learning in novice programmers. *Journal of Educational Computing Research, 2*(1), 37–55. doi:10.2190/GUJT-JCBJ-Q6QU-Q9PL

Pirolli, P., & Card, S. (1999). Information foraging. *Psychological Review, 106*(4), 643–675. doi:10.1037/0033-295X.106.4.643

Polanyi, M. (1966). *The tacit dimension*. Gloucester, MA: University of Chicago Press.

Posey, G., Burgess, T., Eason, M., & Jones, Y. (2010). The Advantages and Disadvantages of the Virtual Classroom and the Role of the Teacher. *Southwest Decision Sciences Institute Conference*, 2-6.

Quinn, B. (2016). Computational Thinking Guiding Change in Online Education. *Journal of Medical Biomedical and Applied Sciences*, *4*(12), 8–17.

Savolainen, R. (2006). Time as a context of information seeking. *Library & Information Science Research*, *28*(1), 110–127. doi:10.1016/j.lisr.2005.11.001

Shadiev, R., Hwang, W., Yeh, S., Yang, S., Han, L., & Hsu, G. (2014). Effectts of Unidirectional vs. Reciprocal Teaching on Web-Based Computer Programming Learning. *Journal of Educational Computing Research*, *50*(1), 67–95. doi:10.2190/EC.50.1.d

Shulman, L. S. (1986). Those who understand: Knowledge growth in teaching. *Educational Researcher*, *15*(2), 4–14. doi:10.3102/0013189X015002004

Sims, R., Dobbs, G., & Hand, T. (2002). Enhancing quality in online learning: Scaffolding planning and design through proactive evaluation. *Distance Education*, *23*(2), 135–148. doi:10.1080/0158791022000009169

Su, A., Yang, S., Hwang, W., Huang, C., & Tern, M. (2014). Investigating the Role of Computer-Supported Annotation in Problem-Solving-Based Teaching: An Empirical Study of a Scratch Programming Pedagogy. *British Journal of Educational Technology*, *45*(4), 647–664. doi:10.1111/bjet.12058

Swan, K. (2001). Virtual interaction: Design factors affecting student satisfaction and perceived learning in asynchronous online courses. *Distance Education*, *22*(2), 306–331. doi:10.1080/0158791010220208

Taylor, J., Thackray, H., Hodge, S. E., & McAlaney, J. (2018). Introducing Psychological Concepts and Methods to Cybersecurity Students. In *Psychological and Behavioral Examinations in Cyber Security* (pp. 98–108). IGI Global. doi:10.4018/978-1-5225-4053-3.ch006

Truong, N., Roe, P., & Bancroft, P. (2005, January). Automated Feedback for “Fill in the Gap” Programming Exercises. In *Proceedings of the 7th Australasian conference on Computing education* (pp. 117-126). Australian Computer Society, Inc.

Tukel, O., Rom, W., & Kremic, T. (2008). Knowledge transfer among projects using a learn-forget model. *The Learning Organization*, *15*(2), 179–194. doi:10.1108/09696470810852339

Underdown, K., & Martin, J. (2016). Engaging the Online Student: Instructor-Created Video Content for the Online Classroom. *Journal of Institutional Research*, *5*(1), 8–12. doi:10.9743/JIR.2016.2

Uysal, M. (2014). Improving First Computer Programming Experiences: The Case of Adapting a Web-Supported and Well-Structured ProblemSolving Method to a Traditional Course. *Contemporary Educational Technology*, *5*(3), 198–217.

Vandenbosch, B., & Huff, S. L. (1997). Searching and scanning: How executives obtain information from executive information systems. *Management Information Systems Quarterly*, *1997*(1), 81–107. doi:10.2307/249743

Vygotsky, L. (1980). *Mind in Society: The Development of Higher Psychological Processes*. Cambridge, MA: Harvard University Press.

Whyte, G., & Classen, S. (2012). Using storytelling to elicit tacit knowledge from SMEs. *Journal of Knowledge Management*, *16*(6), 950–962. doi:10.1108/13673271211276218

Williams, T. (2008). How do organizations learn lessons from projects-and do they? *Engineering Management. IEEE Transactions on*, *55*(2), 248–266.

Xu, K., Kim, H.-W., & Kankanhalli, A. (2010). Task and social information seeking: Whom do we prefer and whom do we approach? *Journal of Management Information Systems*, *27*(3), 211–240. doi:10.2753/MIS0742-1222270308

Xu, Y., Tan, C., & Yang, L. (2006). Who will you ask? An empirical study of interpersonal task information seeking. *Journal of the American Society for Information Science and Technology*, *57*(12), 1666–1677. doi:10.1002/asi.20339

Section 2

Initiatives on Knowledge-Intensive Economies and Technology-Driven Innovation

Chapter 4
The Imperatives of Critical Thinking, Social Norms, and Values in Africa:
Pathways to Sustainable Development

Essien D. Essien
University of Uyo, Nigeria

ABSTRACT

This chapter examines critical thinking as a veritable tool in the usage of values, norms, and beliefs for sustainable development in contemporary societies. The problem statement raises the need for knowledge of what critical thinking, social norms, and values are, as well as the competence in how to employ and apply its tools in social policy designed for development. Illustrating from an extensive contemporary literature on social norms and values, this chapter examines how social norms, values, and beliefs can enhance the development and transformation of the society. Findings hypothesize that critical thinking, social norms, and values, based on thoughts and experience, are capable of guiding human behavior towards truth, development, and functional transformation toward solution to many societal problems. The study suggests collaborative measures to encourage inculcating social norms and values in the society.

INTRODUCTION

In every society, the strength of social values lies in the conformity to the societal norms and values of the social system where individuals and groups participate in social activities (Essien, 2017). This requires critical thinking because changes in the evaluative norms and values in the society are imperative with any social transformation, standards of behaviour, stability, personality development and consistency to the prevalent cultural values (Turkkahramana, 2013). In many African societies, just as in any society across the globe, norms and values provides the substratum of social order and behaviour. In this regard, some social practices and unapproved behavioural pattern are proscribed by social norms. It is therefore, not surprising that some individuals who emphasize the validity of fundamental moral

DOI: 10.4018/978-1-5225-7347-0.ch004

principles, norms, or laws (nonrelativists) have relatively negative attitudes about proscribed forms of behaviour (Ullah & Shah, 2016).

In contrast, individuals who challenge the value of universal moral norms endorse more permissive attitudes. Social values in every society have sociological, psychological, religious, and ethical bases, even though their sociological foundations were becoming much stronger. This explains why social values under transition tends to display disparity in its observance by various social layers of our society, creating confusion and increasing social problems in society, at home and schools (Aydun, 2011). The increasing gap between ways of thinking and behavior of people especially the youths in our contemporary and ideal society is aptly considered an indicator to moral gulf, social violence and intolerance problems. Norms and values are responsible for providing basis to the prevalence of a healthy cultural tradition in any society as well as devising the rules and laws which governs the relative society (Bacanli, 2011).

Although changes are inevitable in every society as has been witnessed in the history of both human beings and societies, nonetheless always in a subtle manner which guarantees consistency to the social equilibrium of the society, any deviation or disregard to the existing norms and value system always lands such society into a social disintegration and disorganization of social fabrics. Thus, the perverse manifestations experienced in our society today are the result of permissive anti-social practices and normlessness (Cagel, 2011). This has given birth to deviant behaviours such as violence conflicts, extramarital and premarital sex, terrorism, militancy, insurgency, drug addiction and many more all associated with decrease in positive values and normlessneass. This is symptomatic to a strong violation of code of ethics, social values and norms in our contemporary society. This study therefore, set out to examine the imperatives of critical thinking and social norms as a tool and pathways to positive values and sustainable development in Africa (Essien, 2017).

THEORETICAL FRAMEWORK

This scholarly discourse is founded on Martin Fishbein and Icek Ajzen theory of Reasoned Action (TRA) and Planned Behaviour (TPB) which posits that "a person's behavior is determined by his/her intention to perform the behavior and that this intention is, in turn, a function of his/her attitude toward the behavior and his/her subjective norm" (Ajzen, 1991, p. 179). According to this theory, the best predictor of behavior is intention. Intention is the cognitive representation of a person's readiness to perform a given behavior, and it is considered to be the immediate antecedent of behavior. This intention is determined by three things: their attitude toward the specific behavior, their subjective norms and their perceived behavioral control. The theory of planned behavior holds that only specific attitudes toward the behavior in question can be expected to predict that behavior (Ajzen, 1991, p. 180). In addition to measuring attitudes toward the behavior, we also need to measure people's subjective norms – their beliefs about how people they care about will view the behavior in question.

This study also finds premise on Vygotskian's socio-cultural model which cogitates that superior order functions and develops out of every social interaction. According to this theory, "every function in an individual's cultural development becomes visible in life in twofold" (Adamson & Chance, 1998). Firstly, it appears on the social level between people, and secondly, it emerges later on in an individual level from inside the person involved. This implies that the individual in the society must first relates with people which includes the parents, siblings, elders and peers in order to develop culturally before the norms and values are molded and deposited in such an individual. The most important goal worthy of note here is the act of learning through socialization. It follows that, cultural growth of an individual depends to a large extent on the social interactions with the societal norms, values and the wisdom of the elders to serve as agents of articulating societal values, norms and standard of behaviour.

Conceptual Clarification

Critical Thinking

In trying to understand the concept of critical thinking, there are diverse explanations, although resulting from proximal assumptions and maintaining some similarity amongst them (Allen, Rubenfeld, & Scheffer, 2004; Halpern, 1999, 2006; Yanchar, Slife, & Warne, 2008). The conceptual diversity comes from the fact that critical thinking is studied in different scientific subjects and applied in multiple contexts (Philley, 2005). In this sense, this area has benefited from the interest of researchers in the fields of Education, Psychology or Philosophy, ethics, sociology (Phan, 2010; Yanchar et al., 2008). Seeking some level of convergence from the different definitions available in the literature, critical thinking can be defined as a more complex and significantly demanding logic form of higher-order reasoning (Brady, 2008; Philley, 2005). In terms of its operationalization, critical thinking presumes a repertoire of faculties: articulation of ideas; meaning elicitation; consideration of divergent arguments and search of evidence to evaluate the legitimacy of each one; formulation of hypothesis; justification of personal arguments and beliefs; decision making; problem solving; monitoring and evaluation of personal cognitions and actions (Facione, 2010; Halpern, 1998, 1999, 2006). Critical thinking is a multifaceted cognitive construct, with an inductive, deductive and creative nature, comprising a heterogeneous set of skills and necessarily implying the motivation to use them (Bailin et al., 1999; Facione, 2010; Halpern, 2006; Philley, 2005).

Thinking critically is a defense against a world of too much information and too many people trying to convince us (Epstein, 2006, p. 1). It is also widely accepted that the development of critical thinking skills is a top goal for development (Browne & Keeley-Vasudeva, 1992). Despite efforts to transform African society developmentally into a dynamic environment capable of promoting positive developmental strides and growth, the current policy is expected to focus more on the enhancement of critical thinking skills for a more proactive and impactful outcomes. Nonetheless, there is no uniform, clear-cut and concise definition of critical thinking, and an overwhelming number of definitions can be found in the literature. For instance, Dewey's (1933, p. 12) definition of "reflective thinking" is one of the forerunners of what has come to be known as critical thinking: "reflective thinking, a distinction from other operations to which we apply the name of the thought. He asserts that critical thinking involves (1) a state of doubt, hesitation, perplexity, mental difficulty, in which thinking originates, and (2) an act of searching, hunting, inquiring to find material that will resolve the doubt, settle and dispose of perplexity." Also Richard Paul, one of the leaders in the field, views critical thinking as "learning how to ask and answer questions of analysis, synthesis and evaluation (Paul, 1985, p. 37), and ''the ability to reach sound conclusions based on observations and information" (Paul, 1988, p. 50). In the same vein, Brookfield (1987, p. 229) asserts that critical thinking comprises two interrelated processes: "identifying and challenging assumptions, and imagining and exploring others."

Accordingly, Beyer (1988, p. 61) states that critical thinking "involves precise, persistent and objective analysis of any claim, source, and belief'' and "to judge its accuracy, validity or worth." In the opinion of Chaffee (1992, p. 3), critical thinking involves a variety of cognitive activities which include solving problems and making informed decisions, developing evidence and arguments to support views, critically evaluating the logic and validity of information, applying knowledge to various contexts and new circumstances and exploring issues from multiple perspectives. Freely (1993, p. 1) and Paul (1993, p. 22) share the definition that critical thinking involves the ability to ''analyze, criticize, advocate ideas, reason inductively and deductively, and to reach factual or judgmental conclusions based on sound inferences." Pithers and Soden (2000, p. 239) concur that critical thinking includes the following abilities: identifying a problem and its associated assumptions, clarifying and focusing the problem, analysing, understanding and making use of inferences, inductive and deductive logic and judging the validity and reliability of assumptions, sources of data or available information. As the researchers applied the Watson–Glaser Critical Thinking Appraisal (2002a) for determining the critical thinking abilities of the prospective development or developing societies, the definition formulated by these authors obviously forms the focus of this research. Epstein (2006, p. 1) supports this view by stating that critical thinking involves evaluation, namely to be convinced that some claim is true or some argument is good, as well as being able to formulate good arguments. According to Drewett (1995, p. 72), it is clear that critical thinking is a holistic activity which incorporates both theory of arguments and the context in which those arguments occur. Against the background of an African scenario, this is the type of thinking expected of prospective society anxious for development. They should be capable of identifying arguments, draw conclusions within arguments, draw conclusions about arguments and construct their own arguments about what should make life sustainable.

Social Norms

Norms have been conceptualized in a variety of ways. Social norms are customary rules of behavior that coordinate our interactions with others (Bicchieri, 2006). Once a particular way of doing things becomes established as a rule, it continues in force because we prefer to conform to rule given the expectation that others are going to conform too (Lewis, 1969). Social norms like other ethical, sociological and psychological phenomena are a societal construct that has widespread usage because it helps in describing and explaining human behaviour. By characterization, social norms are rules and standards that are understood by members of a group as constituting a benchmark for appropriate and inappropriate values, beliefs, attitudes and behaviours which guides and/or constrain social behaviour without the force of laws. Scholars such as Sumner and Sherif (2010) wrote of norms as folkways or habitual customs exhibited by a group jointly and negotiated for the sake of social behaviour. It involves the customs, traditions, standards, rules, values, fashions, and all other criteria of conduct which are standardized as a consequence of the contact of individuals.

Normatively speaking, norms emerge within a social system; hence, social behaviour in every society is the characteristics of a socio-cultural collective unit (Champagne, 2007). This explains why norms in every society perform the role of invoking commonly accepted rules for desirable behaviours. In every society, expected attitudes and behaviours are called "norms." That explains why being a member of a particular society requires one to act according to the norms of respect, responsibility, trustworthiness, fairness and citizenship. When someone breaks a norm, the people around him or her are expected to point out, or "confront," such a fellow with the error in a helpful manner (Hechter & Karl-Dieter, 2001). The person being confronted is expected to accept helpful feedback in a respectful manner. Social norms therefore, include societal expectations for our behaviour, the expectation of others about our behaviour, our own expectations for our behaviour as well as standards that develop out of our observations of others' behaviour (Fehr, Urs & Simon, 2002). This includes rules forbidding unacceptable social behaviours such as societal taboos and government established laws. Social norms represent rules of behaviour that are valued or reinforced by the culture of a particular people. They are however, culturally specific and capricious in nature, and its power is derived from its value to the culture within which it operates (Rohr, 1978). This is because the substance of any norm is neither inherently good nor inherently valuable; its power is granted by its acceptance within the culture it functions.

Ethics

Despite daily exposure to ethical issues and beliefs, defining "ethics" as a formal activity can be difficult. This difficulty is only exacerbated with the recognition that so-called "ethical-decisions" can vary widely among people and between societies. This problem, to an important extent, applies to professional ethics as well. All of this implies that while there is a high level of exposure to ethics, it does not follow that people are effective decision makers regarding important ethical issues (Carroll, 2004). 'Ethics' might initially be defined in two ways; first, ethics refers to well-based standards of right and wrong that prescribe what humans ought to do, usually in terms of rights, obligations, benefits to society fairness, or specific virtues. Secondly, it refers to the study of development of one's ethical standards (Velasquez et al., 1987). It is important to take into account the distinction between morals and ethics, ethics being practical reflection of some moral values. Morals refer to generally accepted standard of right and wrong in society, often learned during childhood, but ethics are learned at the time of confrontation of the problems. 'Ethics', by contrast are learned at the time of confrontation of the problems.

Ethics as field of study is universal but perception of ethical correctness differs. Ethics are fundamental building blocks in determining either individual actions or the success or failure of any institution and/or society (Schudt, 2000). This however, is why Essien asserts that, "no sustainable improvement in human well being is possible without sound ethical consideration" (2012). They affect institution's reputation and help to define an organization's model that will thrive even in adversity. In every society, individuals, leaders and policy makers make decisions that often have a critical impact on their environment and society. Ethics in the context of this discourse means behaviours or attitudes that are considered moral. Before an action or attitude is considered ethical or moral, it must have to conform to the code of conduct in the society and satisfy the expectations of the generality of the people. Contemporary African society is today lamenting a moral world fallen apart. African society seems to be in a state of near chaos in the realm of morality hence conflicts which is the result of absence in social values that helped society to remain integrated, like peace, harmony, respect for authority, respect for and fear of the supernatural realities (Kinoti, 1992, p. 80).

Values

Another conceptual issue that requires consideration in relation to ethics and norms in the society is the concept of values. The human being is the only living creature that can distinguish between good and bad, and right and wrong (Schudt, 2000). The word "values" therefore, refers to the human being's idea of what is acceptable or unacceptable, virtuous or without virtue. Values thus indicate the importance allocated by the individual to behaviours, activities, experiences or phenomena, and provide the individual with a guideline for his personal conduct. (Dwivedi, 1978). The values held by any society, a group of people, and individuals could be one of the causes of ethical dilemmas, norms defiance and social disorder especially because of the fallibility of man, his greed for power, personal aggrandizement, wealth and status, rather than his endeavour to remain unbiased to others and a good citizen for the society by adhering to societal norms.

The fallibility of man can undermine the effective personal relationship, performance of his civic duties and could have a spill-over effect on the functionality of the society as well as the absence of moral leadership and communal accountability (Waldo, 1994). This development could also serve as an example to other communities, groups and/or individuals to indulge in unethical practices (Waldo, 1994). However, owing to the presence of different cultures in some societies conflicting value judgements can be made which could affect the developmental activities of the community, and also directly influence the quality relationship, leadership, and impact of the society on the citizens. Like anything else, human beings are affected by value judgements and generally empathize with the values held by groups and/or other individuals in society. The obligation therefore faced by the individuals is to bear in mind that the values of the society are always "normative" hence the "starting point of their ethical reflections" (Rohr, 1978, pp. 59-60). Because of different personal judgements, individual are most likely to have different interpretations of societal value. This however, should not interpreted as a discomfiture or an indication of unethical behaviour, but rather, it demands that those affected endeavour to understand societal values through an interpretation of and reflection on their personal values. (Rohr, 1978, p. 75)

Culture

There is no question that culture is now a ubiquitous concept in social discourse and scholarship in many societies. It is a normative concept. And as a normative concept, culture seems to have acquired the status of an ontological model that goes right into the heart of human beings (Kenrick, 2000). Normative Culture is a way of behaving rather than a system of rules. It uses peer pressure in a positive way among both the young and old to create shared expectations regarding attitudes and behavior. Nevertheless, culture is an extremely difficult concept to define. Scholars have more than 100 definitions for the term, and nowhere close to reaching a consensus on what should be the generally acceptable definition. While there is no commonly accepted definition of culture, most have agreed that culture has something to do with a shared set of beliefs, practices, and values that often result from shared experiences or heritage; and that it is passed on from parent to their children (Gomez, 2007). Culture is described by some social scientists as "the patterned behaviors or way of life shared by a group of people or which a social group develops to understand, use, and survive its environment".

Culture is said to have the capacity to influence almost all aspects of our lives. It encompasses all that human beings have and do to produce, relate to one another and adapt to their physical environment. It also includes agreed-upon principles of human existence such as values, norms and sanctions which constitutes the techniques of survival (Naomi, 1993). Culture is also that aspect of our existence which makes us similar to some people, yet different from the majority of the people in many societies. Culture involves a collection of beliefs and attitudes, as well as shared understandings which allow people to live together in relative harmony, shaped by human and natural forces. It encompasses both ideas and artifacts and includes such things as technology, language, beliefs, and values. Culture is transmitted both consciously and unintentionally. It perpetuates itself as learned behavior, molding the ways societies behave across generations (Rossi, 1980). Though individuals make use of inherited cultural knowledge to guide their actions and interpret their experiences, cultures are not permanently fixed. They undergo change as members of a society learn new things and encounter and respond to new experiences, ideas, and peoples. In this way cultures reconstruct themselves, enhance our lives, brings wonderful traditions, informs our beliefs about the world, and structures the ways in which we act and behave (Naomi, 1993).

Taboos

Taboos constitute an important part of the characterization of every society. It makes for the different social norms and values that govern the members' behavior. Taboo is generally considered as a socio-cultural phenomenon affiliated with beliefs, values, custom, and hierarchical power (Chigidi & Shona, 2009). Every culture has its own taboos. As a cultural construct, every human society prohibits or restricts certain kinds of behavior, although those prohibited in one society are not necessarily the same as in another. Taboo is a foremost element of African culture (Esema, 2002). In Africa, taboo is a cultural production which emphasizes prohibitions and dictates behavioral and/or conversational re-orientation with implications on the sacred (religious) or magical (profane) component of African culture. In African societies, like many other societies of the world, several activities are characterized by externally sanctioned prohibitions, and human behavior is not governed solitary through rational decision making alone. But societies often have shared values and standards of acceptable behaviour that members of the society are encouraged to follow.

Specifically, cultural norms or a societal value guides and regulates behavioral patterns and thoughts of their members by agreed upon expectations and rules. These rules constitute the list of behavioural guidelines which is typically referred to as societal norms, customs and taboos (Essien, 2013). These include not only illegal or criminal behaviors sanctioned by law, but also norms of conduct enforced through social sanctions. However, the customs and taboos governing interactions between people and their environment in African society are complex and dynamic. From anthropological perspective, taboo-related prohibitions are largely behavioral in nature (Van Gennep, 1904). The taboo's role in any society it exists is to prevent certain behaviors from happening, such as the killing of totemic animals, desecrating sacred places and engaging in certain activities against societal norms (Frazer, 1994). However, prohibitions in every taboo can also be conversational in nature. It is in this light that conversational nature of taboos encompasses the restriction of peoples' freedom to talk about certain topics due to social, moral or religious conventions (Frazer, 194). A good example of this kind of taboo is exemplified in deathrelated taboos. African people have an obsession with the desire to inculcate right ethos in an individual. Taboos are among a number of methods through which the character of an individual is shaped in Ibibio cosmology. Taboos in African is teleological in character in that they involve sanctions that are meant to inculcate the most appropriate traits in a person that would make him a worthy member of the community (Essien, 2013).

Sustainable Development

The concept of sustainable development can be interpreted in many different ways, but at its core, it is an approach to development that looks to balance different, and often competing, needs against an awareness of the environmental, social and economic limitations we face as a society. Sustainable development is development that meets the needs of the present, without compromising the ability of future generations to meet their own needs (Bell, 2009). All too often, development is driven by one particular need, without fully considering the wider or future impacts. We are already seeing the damage this kind of approach can cause, from large-scale financial crises caused by irresponsible banking, to changes in global climate resulting from our dependence on fossil fuel-based energy sources (HRW, 2009). The longer we pursue unsustainable development, the more frequent and severe its consequences are likely to become, which is why we need to take action now. The notion of sustainable development is both normative and teleological, containing the road ahead, the line forces and the noble purpose as well. Thus, human beings, so complex and also so different, must act, individually and collectively as a family, organizations, social groups, states, unions of states, and society according to a set of moral values, norms, beliefs and principles in order to live a good life, to grow, develop and to preserve the Planet.

Social Norms and Human Behaviour

Human behavior everywhere is not governed only by rational decision making. A culture or a society guides the behavior and the thoughts of their members by agreed upon expectations and rules. Norm is a proscription of behaviour that affects everyday life. Communal norms for instance exist in invariably all cultures throughout the world, and represent a class of informal institutions, where traditional, social and religiously governed norms define the human behaviour. These norms remain the prime factor guiding their conduct and behaviour towards fellow man and society (Allan & Kate, 2006: 40). Scholars are of the opinion that norms symbolizes the prohibition of an action based on the belief that such behaviour is either too sacred and consecrated or too dangerous and accursed for ordinary individuals to undertake hence caution is required. Under this interpretation, a norm is a form of "thought police" that governs not just human behavior, but also the human thoughts. For example, merely thinking or considering incest, necrophilia, or cannibalism is a violation of the relevant norms.

Norms can be repugnant and appalling actions or behavior which includes the display of some bodily functions (Masaka, & Chemhuru, 2011, p. 132). From anthropological perspective, norm-related prohibitions are largely behavioral in nature (Van Gennep, 1904, p. 48). The norms' role in any society it exists is to prevent certain behaviors from happening, such as the killing of totemic animals, desecrating sacred places and engaging in certain activities against societal norms (Frazer, 1994, p. 99).

However, prohibitions in every norm can also be conversational in nature. It is in this light that conversational nature of norms encompasses the restriction of people's freedom to talk about certain topics due to social, moral or religious conventions (Frazer, 1994, p. 97). A good example of this kind of norm is exemplified in death-related norms. This is why many researchers see death as the paragon of a conversational norm (Masaka & Chemhuru, 2011, p. 58), because it is a painful subject that people find difficult discussing. Yet, death-related norms are founded on a general prohibition of killing. For instance, murdering someone or oneself (through suicide or assisted suicide) is a behavioral norm in many modern societies (Martindale, 2005, p. 22).

Configuring Social Norms in African Worldview

Norms exist in every society and governs how members of such society behave, think, make judgments and perceive their world. The hidden dimension of African social norms institution revolves around the moral sanctions that help in shaping an African person's (eti-uwem) virtue in the human community. African people have an obsession with the desire to inculcate right ethos in an individual. Social norms are among a number of methods through which the character of an individual is shaped in Africa cosmology. Norms in Africa is teleological in character in that they involve sanctions that are meant to inculcate the most appropriate traits in a person that would make him a worthy member of the community (Esema,

2002, p. 98). In African worldview, as exemplified in the culture of the Akwa Ibom people, the concept of norms is predicated on the premise of the "trinity". The trinity principle in regards to social norms presupposes that there are three most important norms in Africa. These most important social norms give birth to other societal norms. They are: (a) Norms connected with (ukot) inlaw otherwise known as "abasi ukot". (b) Norms connected with (ayeyen) grandchild or else "abasi ayeyen". (c) Norms connected with (esenowo inuaesit) visitor otherwise called "abasi "esenowo inuaesit" (Esema, 2002, p. 102).

Similarly, according to Chigidi (2009), "The avoidance rules in these norms are restrictive and not directive in the sense that they only tell the individual what not to do and not what to do" and by implication one is made to pick up desirable behavioral traits otherwise acting contrary to the dictates of this norms invites undesirable consequences. A good character is a solid weapon against various anti-social behaviors. The outcome of good character is good reputation whereby a person becomes the envy of many because of his commendable dispositions (Masaka, & Chemhuru, 2011, p. 133). Though the inculcation of commendable character traits in individuals is a lifelong process, it is believed, among the Africans, that such moral education makes an indelible impression in one's formative years. In this light, children are taught the difference between good and bad behavior and they also learn to avoid a number of norms. Strong and severe warnings for those who violate Africa moral code are quite visible in the taboos. The veracity of punishment and a host of other unwanted consequences if one dares to violate the moral code help to instill commendable moral behavior in African society (Udo, 1984, p. 22). It is in light of this fact that Gelfand (1979, p. 33) contends that "the purpose of these norms is to instil a sense of discipline into the children as well as one of fear."

The aspect of fear that is normally associated with African norms is a way of dissuading people from performing immoral acts. Hence, it has an instrumental value in that it discourages people from engaging in certain behaviors that run contrary to the ethos of African society.

Although African taboos are fear inducing, this fear has no intrinsic worth, but is a means to an end, that is, promotion of good behavior. Though this means of achieving a virtuous life is morally questionable, the philosophy in African norms is that the end justifies the means. Fear may not be the best and ethically appropriate tool to achieve the end of a virtuous life in Ibibio society, but it is believed that the goodness of the end trivializes the badness of the means (Esema, 1984, p. 25). It is important to note that the individual within African society, just as in other African societies, does not live in a moral island. A human being can only be fully comprehended as an inseparable part of the whole (Menkiti, 2006, p. 98). This communitarian view of the individual emanates from the realization that the moral life in every human being is shaped not only by the community of physical beings alone but also by spiritual forces. In this regard, Gelfand (1979, p. 35) makes an important observation that though "the origin of Ibibio norms is unknown, they bring home to the young the realization that other forces exist besides the physical ones" in the moral education of members of African society. Since enforcement of African moral code has a spiritual dimension, the learners of African morality come to know of "the existence of spiritual powers, so important in the African religious belief" (Gelfand, 1979, p. 35; see also Bourdillon, 1987, p. 88) such as the ancestral spirits as well as African deity who provide the living with, among others, direction and appropriate moral guidance (Asante, 2000, p. 44).

African people believe that spiritual forces are custodians of their moral code. Ancestral spirits play a crucial role in making sure that one picks up desirable character traits and avoid vices. The African believe that ancestral spirits help in ensuring that one's character is good provided that that person does not offend them through, among others, failure to perform periodic rituals in their honor as well as a host of other social misdemeanors such as incest (Udo, 1984, p. 26). Therefore, the violation of norms can be seen as a direct provocation of ancestral spirits who are the custodian of the moral code. Tatira (2000, p. 61) concurs with Gelfand (1979, p. 37) when he notes that "an act that breaches a taboo triggers a reaction supposedly at the supernatural level. Without this fear of the unknown, young people are generally adventurous, full of doubts and questions, and like experimenting with things." It is pertinent to note that taboos are effective moral tools because their violation invites the ire of ancestral spirits who are one of the key pillars of traditional religion. It is worthy of note that African norms do not reveal the correct consequences for performing certain actions but give a consequence that a human being naturally loathes and fears. Therefore, for one to comprehend the complexity of African norms, one has to look at their common and hidden meaning.

The Cultural Context of African Social Norms

Social norms are generally considered as a socio-cultural phenomenon affiliated with beliefs, values, custom, and hierarchical power. Every culture has its own norms. As a cultural construct, every human society prohibits or restricts certain kinds of behavior, although those prohibited in one society are not necessarily the same as in another. In African society, the shared norms are what defined the culture or subculture of African people. In this regard, prohibitions concerning issues like stealing, adultery, killing, illegal possession of farmland and other properties, and violating the norms regulating relationships with in-laws, visitors, and grandchildren attracts societal disapproval and punishment which serves as a deterrent to others and thus engender positive values through abstention. Norm is a foremost element of African culture (Esema, 2002, p. 101). It is one way in which African society expresses its disapproval of certain kinds of behaviour believed to be harmful to its members, either for supernatural reasons or because such behaviour violates a moral code. A culture or a society guides the behavior and the thoughts of their members by agreed upon expectations and rules.

In line with this understanding, Osei (2006, p. 20) posits that "norms represent the main source of guiding principles regulating and directing the behaviour of individuals and the community towards the Supreme Being and especially the gods and the ancestors in African traditional societies". This is why Freud (1912, p. 73) showed that norms are productions of culture insofar as they are established by a group's recognized authorities to regulate the group. It is the prohibition against touching, saying, or doing something for fear of immediate harm from a supernatural force. Akindele and Adegbite (1999, p. 592) further explain that normative words and expressions reflect social customs and views of the people's culture. It can be characterized as being concerned with behaviour which is believed to be supernaturally forbidden or regarded as immoral and improper. Every society has a culture and norms hold the society together. Although some norms can be traced to apparent danger in health and safety, no common explanation has been given for most others; most authorities agree that they tend to relate to objects and actions that are significant for the maintenance of social order.

In African society, a norm is a powerful social interdiction system relating to any area of human activity or social custom that is venerated and or forbidden based on moral judgment, religious beliefs and or scientific consensus. They are recognized to avoid disrespect to any given authority, be it legal, moral and/or religious (Udo, 1984, p. 37).

The Responsibilities of Norms in African Society

Norms in every society are fundamental to understanding social order as well as variation in human behavior in such society (Campbell, 1964, p. 12; Durkheim, 1951, p. 84). In African society, norms provide expectations shared by the members of a group about appropriate ways to behave in given situations. This provision of expected ideas of how to behave is what makes up the culture of African people. It is a common feature to find that in African, norms constitute the potential "pressure" in situations that: help to define the nature of social reality; form the foundation upon which people base their interaction; and provide a common medium for members' self-evaluation. By means of these mechanisms, norms increase feelings of personal and group identity in the entire Africa (Ekong, 1983, p. 132). It is worthy of note that in Africa like its counterpart in other part of the world, norms shape behavior of people by providing limits within which people receive social approval for their behavior or actions. These guidelines establish an informal basis for estimating how far one may go before experiencing the normative power of ridicule, rejection, and loss of status among group members, friends, acquaintances, and co-workers. That is why group norms reflected in the dominant or most typical attitudes, expectations and behaviors not only characterize these groups but also regulate group members' actions to perpetuate the collective norm.

Indeed, norms can be powerful agents of control as "choices" of behavior are framed by these norms and as the course of behavior most commonly taken is typically in accordance with normative directives of "reference groups" that are most important to the individual (Esema, 2002, p. 108). Although group norms that are backed by powerful punishments for violations in African society can restrict behavioral freedom such as debarring the community members from entering the violator's premises, buying any goods from him/her, and selling any product to him/her, they tend to promote excessive uniformity, even though it serves the critical responsibility of understanding and predicting actions and behaviours of people in the society. African norms contain the collective power to create and regulate social reality (Ekong, 1983, p. 131). This is done through the process where norms serve as a mechanism to create order and predictability in social relationship and the understanding of each other's action. It is apparent that the awareness of the norms operating in a particular community or group situation enables each member to anticipate how others will enter or attend to a particular situation, for example, what kind of cloth will they wear? And what are they likely to say and do, as well as what behavior on one's own part will be expected and approved (Esema, 2002, p. 109). Norms in African society also help to interlock the roles that the people or members of the society perform in social situations. Sometimes in a group setting, a new norm usually will emerge that was not operating before or that challenges the prior standard operating procedure. In this case, the norms that have been already internalized and integral to the society are called into question. These roles and duties of African norms in this thesis constitute the capacity to which critical responsibilities of African norms are executed (Udo, 1983, p. 97).

The Significance of Norms in Africa

Social norms are rules of behavior that society uses to assess the population. How people respond to a violation of social norms depends on a number of different factors. Norms are an expression of interconnectedness of two inseparable dimensions in the African worldview: the visible world and the invisible one. That interconnectedness can also be seen as interdependence whereby the quality of life of the ancestors and of people depended on each other's actions (Menkiti, 2006, p. 48). Norms are an expression and a means of perpetuating what was considered as the most important features in African culture: preservation of life and well being of people. Life and its quality were seen as crucial and the society applied a variety of methods to preserve it and transmit it, including norms. Norms in Africa showed the communal dimensions of one's actions. In the culture where one was defined by belonging to a community and where the community was far more important than its individual members, one's actions affected that community. Norms therefore were helping the people to recognize their own importance. In this setting, if one could affect his or her community negatively, then such a fellow was not unimportant. To a certain extent, norms could be viewed as 'self-esteem-enhancing beliefs' (Mbiti, 1969, p. 83).

Norms in Africa helped the people to realize that an improper behaviour would always have consequences for them, the community and the nature. Norms were an expression of a quite sophisticated moral system ruling the lives of the people in the community and the life of an individual. Even though formulated as 'negative' principles stressing (Ku) 'do not...' and teaching people about what was not acceptable in the community, by implication, they were also pointing out to the actions that were supposed to be done. By preventing people from doing wrong things, they were helping them to focus on what was encouraged in the community. Norms in Africa are an expression of the 'natural law', a general set of rules commonly understood by all human beings (Esema, 2002, p. 102). Similarities among various tribes, norms concerning prohibitions against such acts as incest, murder, stealing etc. show the common ground of various norms. Often formulated in a way that implied more than their original wording, they were helping people to exercise their common sense and moral responsibility in interpreting them as well as applying them to various situations. In a society where there was no police or access to police is remote, norms served as a guardian of moral values. To a certain extent, they were better than modern law enforcement agencies, because, in most cases, breaking of a norm was associated with an automatic punishment – one did not have to be caught to be punished. Through norms one was made aware that an improper action as defined by the society one was a member of, and how such action would result in negatively implications affecting the harmony of the person concerned, the family and the society at large. Transgressing it would also result in punishment (Parrinder, 1969, p. 20). It was a way of teaching people that each action entails consequences. Such consequences/punishment was usually automatic, personal (creating feelings of fear and guilt), affecting, in one way or another family and the community. That punishment could also be administered by the community (Miller, 1983, p. 21).

The Influence of Norms and Values on Sustainable Development

Social norms and values are highly influential dynamics shaping individual behaviour in the society. In Africa, social norms and values constitute rules or expectations of behaviour within a specific cultural or social group (Essien, 2013). Even though it is oftentimes unspoken, these norms offer social standards of appropriate and inappropriate behaviour, governing what is (and is not) acceptable and co-ordinating our interactions with others (Durlauf & Blume, 2008). Social norms and values persist within African societies because of individuals' preference to conform, given the expectation that others will also endeavour to conform (Illika, 2005). In atypical African society, a variety of external and internal pressures are thought to maintain values, cultural and social norms (Durlauf & Blume, 2008). Thus, individuals are discouraged from violating norms by the threat of social disapproval or punishment and feelings of guilt and shame that result from the internalization of norms. In a typical traditional African setting, societal norms, values and beliefs does not necessarily have to correspond with an individual's attitudes be it positive or negative, even though they may influence these attitudes and beliefs whenever norms becomes internalized (Illika, 2005). This presupposes that values and social norms may vary such that behaviour acceptable to one social group, ethnic group, family, gang and/or culture may not be tolerated by another. This however, explains why dissimilar cultural and social norms support different types of behaviour such violence, aggressive tendencies like the use of matchet or knife to let out blood in the circumstances of slight provocation. Another good example is the traditional frame of mind that men have the right to control or discipline women through physical means irrespective of status and size. This standpoint or traditional belief makes women vulnerable to violence and assault by intimate partners and other men in the society, and also girls are intentionally placed at risk of sexual abuse and harassment (Jewkes, Penn-kekana & Rose-Junius, 2005). Norms can also protect against violence and encourage development. People with positive societal norms and values exhibit hardwork spirits and innovativeness which can promote social tolerance and supportive character to foster growth and development.

CONCLUSION

The study epitomizes the relevance of critical thinking as a pathway in the enterprise of sustainable development. The research results show that social norms, values and beliefs are the evaluative standards which help people sensitize through experience the proper understanding of their surroundings. It noted that African norms are meant to instill correct dispositions in people through fear-inducing moral sanctions. Norms, understood as unwritten laws restraining people from breaking communal custom, is used in Ibibio societies to preserve harmony in the different communities as well as a good relationship with spiritual beings. The "ku" or "kunam" principle in African culture serves as vital regulator in inculcating commendable moral traits in individuals (Esema, 2002, 103). The paper also noted that even though African norms have limitations in that they do not disclose the true consequences of certain human character defects, the most important aspect of these norms is to inculcate commendable character traits in their apprentices that would make them worthy members of society that would not only behave in a desirable way towards fellow human beings, but also relate to the environment in a manner that embodies respect for biodiversity as well as sustainable exploitation of nature's resources (Masaka & Chemhuru, 2011, p. 133).

African norms provide prohibitions that forbid people from behaving in such manners that are a threat to the welfare and wellbeing of fellow human beings as well as the environment. Even though norms foster commendable character traits among people through threat of severe reprisals for the ones who violate them, they have desirable utility because they help to keep the wicked in check. In this regard, though the means of enforcing desirable behavior among people are morally questionable, norms constitute one among a number of sanctions employed to ensure proper behavior in the society, the end has justified the means (Essien, 2017). Nonetheless, norms are formulated in forms of (ku's) or (kunam's) 'don'ts' and sometimes being ambiguous, they on the other hand enabled people to maintain the moral order and hierarchy in the society (Udo, 1984, p. 40). In this regard, though the contemporary society is quite different from the traditional one, there is a need to enforce norms or to come up with an alternative way that will promote traditional values.

RECOMMENDATIONS

The research indicates that alternative ways of nurturing and encouraging critical thinking skills among Africans should be explored. In this regard, it is worth mentioning that critical areas of concern are development and progress. Another department that merits attention is that efforts should be made to enforce and inculcate traditional values and social norms on the people to enhance functional development.

There is the need for the involvement of elders in the inculcation and construction of knowledge regarding traditional values and norms that would foster proper growth and development of the society. There is a need for encouragement of communication and verbalizing of thoughts concerning values and norms by educators and the media in the society.

From this research it is clear that the cultural environments tend to impact positively and directly on the realization of the functional values and norms of African social values for effective transformation in Africa. This provides the rational for the importance of accepting the challenge to develop and implement strategies for improving social norms and values that are constructive for the development of the society.

There is a great necessity for the overhaul of the curriculum to incorporate civic education and ethical or moral education which will lay the foundation for the teaching of social, cultural, and traditional African norms, values, and beliefs of the African that has the capability of transforming the society. Realizing the ideal of quality education will make it possible for the young people to become competent thinkers who can identify and solve problems and make decisions useful for the society by using creative and critical thinking faculty.

In this regard therefore, it is essential not only to convince the people that social values and norms are important for the smooth running of the society, but that they also need to be appropriately articulated and integrated to provide a healthy and functional society.

REFERENCES

Abasiattai, M. B. (1987). *Akwa Ibom and Cross River State: The land, the people and the culture, Calabar*. Wusen Press.

Adamson, L. B., & Chance, S. E. (1989). Coordinating Attention to People, Objects, and Language. In A. M. Wetherby, S. F. Warren, & J. Reichle (Eds.), *Transitions in Prelinguistic Communication* (pp. 15–38). Baltimore, MD: Paul H. Brookes.

Adamu, S. O. (1978). *The Nigerian Statistical System: Retrospect and Prospect*. Ibadan: University of Ibadan Press.

Agboola, T., & Mabawonku, A. O. (1996). Indigenous knowledge environmental education and sanitation: application to an African city. In D. M. Warren, L. Egunjobi, & B. Wahab (Eds.), Indigenous knowledge in education. Ibadan: Indigenous Knowledge Study Group.

Ahn, J. B.-K. (2003). Sexual norms and morality among the Agikuyu people of Kenya. In M. Katola (Ed.), MIASMU Research Integration Papers to Moral Teaching and Practices of African Religion. Academic Press.

Ajzen, I. (1991). The theory of planned behavior. *Organizational Behavior and Human Decision Processes, 50*(2), 179–211. doi:10.1016/0749-5978(91)90020-T

Allan, K., & Kate, B. (2006). *Forbidden Words: Taboo and the Censoring of Language*. Cambridge, UK: Cambridge University Press; doi:10.1017/CBO9780511617881

Andemariam, M. (2001). Place of norms in Gikuyu morality. In L. Magesa (Ed.), MIASMU Research Integration Papers to Moral Teaching and Practices of African Religion. Academic Press.

Asante, K. W. (2000). *Zimbabwe Dance: Rhythmic Forces, Ancestral Voices-An Aesthetic Analysis*. Trenton: African World Press, Inc.

Aydın, M. (2011). Values, functions and morals. *Eğitime Bakış, 19*, 39–45.

Bacanlı, H. (2011). Value is valuable. *Eğitime Bakış, 19*, 18–21.

Beyer, B. K. (1988). *Developing a thinking skills programme*. Boston: Allyn & Bacon.

Bicchieri, C. (2006). *The Nature and Dynamics of Social Norms*. New York: Cambridge University Press.

Boss, J. A. (1999). *Analyzing Moral Issues*. Belmont: Mayfield Publishing Company.

Bourdillon, M. F. C. (1987). *The Shona Peoples: An Ethnography of the Contemporary Shona, with special Reference to their Religion*. Gweru: Mambo Press.

Brookfield, S. (1987). *Developing critical thinkers*. Open University Press.

Browne, M. N., & Keeley-Vasudeva, M. L. (1992). Classroom controversy as an antidote for the spage model of learning. *College Student Journal, 26*, 368–373.

Carroll, A. B. (2004). Managing ethically with global stakeholders: A present and future challenge. *Academy of Management Review, 18*(2), 114–120.

Çengel, Y. (2011). Human values for a safety future. *Eğitime Bakış, 19*, 13–17.

Chaffee, J. (1992). Critical thinking skills: The cornerstone of development education. *Journal of Development Education, 15*(3), 2–8.

Champagne, D. (2007). *Social Change and Cultural Continuity among Native Nations*. New York: Altamira Press.

Chigidi, W. L. (2009). Shona Taboos and Norms: The Language of Manufacturing Fears for Sustainable Development. *The Journal of Pan African Studies, 1*(3), 174–188.

Dewey, J. (1933). *How we think*. Boston: Heath.

Drewett, M. (1995). There must be some way out there: A case study in the teaching of critical thinking. *Suid Afrikaanse Tydskrif vir Wysbegeerte, 14*(2), 72–76.

Durkheim, E. (1963). *Incest: The nature and the origin of the taboo*. New York: Lyle Stuart.

Durlauf, S. N., & Blume, L. E. (2008). *New Palgrave Dictionary of Economics* (2nd ed.). London: Macmillan.

Dwivedi, O. P. (1987). Ethics, the Public Service and Public Policy. *International Journal of Public Administration, 10*(1), 21–50. doi:10.1080/01900698708524556

Ekong, E. (1983). *Sociology of Ibibio*. Calabar: Scholars Press Nig. Ltd.

Epstein, R. L. (2006). *Critical thinking*. Belmont, CA: Wadsworth Thomas Learning.

Esema, J. D. (2002). *Norms, Culture, Customs and Traditions of Akwa Ibom people of Nigeria, Uyo*. Mocomp Publishing Co.

Essen, A. J. A. (1982). *Ibibio Profile, Calabar*. APCON Press Ltd.

Essien, D. E. (2012). Ethics in the Nigerian Public Service: Enhancing Role, Professionalism and Ethical Value. *Journal of Religious Studies, 11*(2), 40–48.

Essien, D. E. (2013). The Hidden Dimension of Ibibio Taboos. *Uniuyo Journal of Humanities, 16-17*, 256-275.

Essien, D. E. (2017). Toward a Normative Ethical Claims of Ibibio Social Norms, Values, and Beliefs. *International Journal of Knowledge Society Research, 8*(1), 51–66. doi:10.4018/IJKSR.2017010103

Faithman, N. U. (1999). *Ibibio Jews of Nigeria*. Uyo: Menorah Publishing Coy.

Fehr, E., Urs, F., & Simon, G. (2002). Strong reciprocity, human cooperation, and the enforcement of social norms. *Human Nature (Hawthorne, N.Y.), 13*(1), 1–25. doi:10.100712110-002-1012-7 PMID:26192593

Fisher, R. B. (1997). West African Religious Traditions. New York: Maryknoll.

Fraser, J. G. (1994). *The Golden Bough: A Study in Magic and Religion*. New York: Oxford University Press.

Freely, A. J. (1993). *Argumentation and debate*. Woodsworth.

Freud, S. (1955). Totem and Taboo: Some Points of Agreement between the Mental Lives of Savages and Neurotics. In J. Strachey (Ed.), The Standard Edition of the Complete Psychological Works of Sigmund Freud (Vol. 13). London, UK: Hogarth Press. (Orig. pub. 1913)

Freud, S. (1985). *The Origins of Religion: Totem and Taboo, Moses and Monotheism and Other Works*. Harmondsworth, UK: Penguin Books Ltd.

Freud, S. (2000). *Totem and Taboo. Translated into Chinese by Yung-yi Yang*. Taipei: Chih-wen Publishing Co.

Gelfand, M. (1968). *African Crucible: An ethno-religious study with special reference to the Shona-speaking people*. Capetown: Juta & Company Limited.

Gelfand, M. (1979). *Growing up in a Shona Society: From Birth to Marriage*. Mambo Press.

Gomez, N. (2007). Healing Hidden Wounds. *Cultural Survival Quarterly, 31*(3).

Gorer, G. (1965). *Death, grief and mourning in contemporary Britain*. London: Cresset Press.

Hechter, M., & Karl-Dieter, O. (Eds.). (2001). *Social Norms*. New York: Russell Sage Foundation.

Holden, L. (2000). *Encyclopedia of Taboos*. Oxford, UK: ABC CLIO Ltd.

Human Development Report. (2004). *Globalization and Cultural Choice*. Washington, DC: UNDP.

Ifemesia, C. C. (1978). Comments on the South Province of Nigeria for the year 1938. *Southern Eastern State: Heritage Magazine*, *3*(2), 1–16.

Ilika, A. L. (2005). Women's perception of partner violence in a rural Igbo community. *African Journal of Reproductive Health*, *9*(3), 77–88. doi:10.2307/3583414 PMID:16623192

Jewkes, R., Penn-Kekana, L., & Rose-Junius, H. (2005). "If they rape me, I can't blame them": Reflections on gender in the social context of child rape in South Africa and Namibia. *Social Science & Medicine*, *61*(8), 1809–1820. doi:10.1016/j.socscimed.2005.03.022 PMID:15913860

Jin, Z. (1994). *Explore the Secret of Norms*. Taipei: Shu-xin.

Kenrick, J. (2000). The Forest Peoples of Africa in the 21st Century. *Indigenous Affairs, Hunters and Gatherers, 2000*(2), 10-24. Retrieved from http://www.iwgia.org

Kinoti, H. W. (1992). African morality: past and present. In J. N. K. Mugambi & A. Nasimiyu-Wasike (Eds.), *Moral and ethical issues in African Christianity* (pp. 73–82). Nairobi: Initiatives.

Li, X-J. (1994). *Norms and Laziness*. Beijing: International Culture Publishing Co.

Martindale, D. (2005). Are sex and death taboos in advertising? *International Journal of Advertising*, *25*, 1, 9–33.

Masaka, D., & Chemhuru, C. (2011). Moral Dimensions of Some Shona Taboos and Norms (Zviera). *Journal of Sustainable Development in Africa*, *13*(3).

Mbiti, J. S. (1969). *African religions and philosophy*. Nairobi: West African Publishers.

Menkiti, I. A. (1984). Person and Community in African Traditional Thought. In R. A. Wright (Ed.), *African Philosophy: An Introduction* (pp. 171–181). New York: University Press of America.

Menkiti, I. A. (2006). On the Normative Conception of a Person. In K. Wiredu (Ed.), *A Companion to African Philosophy* (pp. 324–331). Oxford, UK: Blackwell Publishing.

Miller, H. B. (1983). Platonists and Aristotelians. In H. B. Miller & W. H. Williams (Eds.), *Ethics and Animals* (pp. 1–14). Clifton, NJ: Humana Press. doi:10.1007/978-1-4612-5623-6_1

Naomi, K. (1993). Culture: Perspectives and Challenges of Indigenous People. *Cultural Survival Quarterly*, *31*(4), 52–81.

Noah, M. E. (1980). *Ibibio Pioneers in Modern Nigerian History*. Scholars Press.

Northrop, D. (1973). Pre-colonial references to the Annang Ibibio. *Ikenga, 2*(1).

Nssien, F. U. (1991). *The People called Ibibios, An annotated Biography*. Calabar: De-Rio Press.

Obio-Offiong, U. E. E. (1958). *A first step to the study of Ibibio History*. Aman Press.

Oku, E. E. (1989). *The Kings and Chiefs of Old Calabar (1785-1925)*. Glad Tiding Press Ltd.

Osei, J. (2006). The Value of African Taboos for Biodiversity and Sustainable Development. *Journal of Sustainable Development in Africa*, *8*(3), 42–61.

Ouidade, F. (2010). *Social Norms and Taboos*. Oxford, UK: Clarendon Press of Oxford University Press.

Parrinder, G. (1969). *Religion in Africa*. New York: Praeger Publishers.

Paul, R. W. (1988). Critical thinking in the classroom. *Teaching*, *K-8*, 49–51.

Paul, R. W. (1993). The logic of creative and critical thinking. *The American Behavioral Scientist*, *37*(1), 21–40. doi:10.1177/0002764293037001004

Pithers, R. T., & Soden, R. (2000). Critical thinking in education. *Review of Educational Research*, *42*(3), 237–249.

Posner, E. (2000). *Law and Social Norms*. Cambridge, MA: Harvard University Press.

Rohr, J. (1978). *Ethics for Bureaucrats*. New York: Marcel Dekker.

Rossi, I. (Ed.). (1980). *People in Culture: A Survey of Cultural Anthropology*. New York: Praeger.

Scanlan, R. (2003). The concept of taboos in African society with reference to the Kikuyu of Kenya. In M. Getui (Ed.), MIASMU Research Integration Papers to African Culture: An overview. Academic Press.

Schudt, K. (2000). Taming the Corporate Monster: An Aristotelian approach to corporate virtue. *Business Ethics Quarterly*, *10*(3), 711–723. doi:10.2307/3857900

Sherif, J., & Sumner, O. (2010). Beyond Rules, Laws and Social Norms. *Journal of Personality and Social Psychology*, *34*(3), 599–607.

Talbort, P. A. (1969). *The People of Southern Nigeria*. London: Frank Cass and Co. Ltd.

Tatira, L. (2000a). *Zviera ZvaVaShona*. Gweru: Mambo Press.

Tatira, L. (2000b). The Role of Zviera in Socialisation. In E. Chiwome, Z. Mguni, & M. Furusa (Eds.), *Indigenous Knowledge and Technology in African and Diasporan Communities* (pp. 146–151). Harare: University of Zimbabwe.

Türkkahramana, M. (2013). Social values and value education, *Procedia - Social and Behavioral Sciences, 116*(2014), 633 – 638.

Udo, E. A. (1984). *Who are the Ibibio?* Onitsha: Africana FEP Publishers.

Ullah, A., & Shah, M. (2016). Perception of Social Values and Its Role in Controlling Corruption Geniein Peshawar Region. *Journal of Accounting & Marketing*, *5*, 177. doi:10.4172/2168-9601.1000177

Ullmann-Margalit, E. (1977). *The Emergence of Norms*. Oxford, UK: Oxford University Press.

Uya, O. E. (1983). *A History of Oron People of the lower Cross River Basin*. Manson Publishing Company.

Van den Heever, G. (2001). On how to be or not to be: Theoretical reflection on religion and identity in Africa. *Religion and Theology*, *8*(1), 1–25. doi:10.1163/157430101X00017

Van-Gennep, A. (1904). *Tabou et totémisme à Madagascar: Étude descriptive et théorique*. Paris: Ernest Leroux.

Waldo, D. (1994). Public Administration and Ethics. In F. S. Lane (Ed.), *Current Issues in Public Administration*. New York: St. Martin's Press.

Watson, G., & Glaser, E. (2002). *Watson–Glaser critical thinking appraisal—UK edition. Test booklet*. The Psychological Corporation.

Chapter 5
The Era After Oil:
Knowledge-Intensive Cities on the Arabian Gulf

Wolfgang G. Stock
Heinrich Heine University Düsseldorf, Germany

Julia Barth
Heinrich Heine University Düsseldorf, Germany

Julia Gremm
Heinrich Heine University Düsseldorf, Germany

ABSTRACT

This chapter investigates seven Gulf cities (Kuwait City, Manama, Doha, Abu Dhabi, Dubai, Sharjah, and Muscat) that have grown rich due to large reserves of oil and natural gas. Now, with the threat of ending resources, governments focus on the development towards a knowledge society with knowledge-based industries and knowledge-intensive cities. The authors analyzed the cities in terms of their "smartness" or "informativeness" by field research on-site, a quantitative survey and in-depth qualitative interviews (N = 34). They studied prototypical building blocks of a city of the knowledge society, namely infrastructures (digital city, smart, green and sustainable city, creative city, and knowledge city), economy, politics and administration, location factors, as well as physical and digital spaces. Especially Doha in Qatar is well on its way towards becoming an informational city, but also Dubai and Sharjah (both in the United Arab Emirates) received good scores.

INTRODUCTION

In the heat of the Gulf region 50 years ago, the desert dominates the landscape, the beaches are almost empty, and the few people living there work as pearl divers, fishermen, traders or peasants. Replaced by glittering facades, high-end hotels, artificial islands, huge shopping malls, and the tallest constructions of the world, the region nowadays attracts people from all over the world. The catalyst for this development was the detection of huge amounts of oil and gas resources in the 1930`s leading to prosperity (Figure 1).

DOI: 10.4018/978-1-5225-7347-0.ch005

The states of the Gulf Cooperation Council (GCC) are nowadays aware of ending resources and declining demand of oil- and gas-burning countries due to electric mobility and enhanced ecological awareness. Additionally, new oil production methods as hydraulic fractioning, increasingly used, for instance, in the United States of America, intensify competition on the oil markets. The oil price fluctuation and its decline over years jeopardize the wealth of the GCC States.

How do the countries of the Gulf region and their cities respond to this drastic change? Are they, like so many other cities in the world, trying to reach the status of a 'smart' or an 'informational' city? Is it indeed a target of these very wealthy oil-based cities to set up a knowledge economy and become a knowledge society? We are going to look behind the glimmering facades of the Gulf cities and describe and analyze their status as informational cities, i.e. as cities of the knowledge society. It is no methodological problem to concentrate on cities (instead of the entire state or emirate) because in all states and emirates there is only one big city region dominating the whole country. Our study was done by field research on-site and by applying a customized questionnaire and interviews (N = 34) concerning questions about important aspects of cities of the knowledge society, namely infrastructures, economy, politics and administration, location factors as well as (physical and digital) spaces (Barth et al., 2017a).

BACKGROUND

What do we refer to when we speak of 'smart cities', 'informational cities' or 'cities of the knowledge society?' The field of research concerning the smart city concept is still emerging (Nam & Pardo, 2011b). A consistent definition of 'smart city' does not exist due to a variety of conceptual variants (Nam & Pardo, 2011a). Many definitions focus mainly on the technological perspective including the use of ICT. There are two concepts available in the latest literature: (1) the concept of a smart city in a broader sense and (2) the concept of a smart city in a narrower sense (Fietkiewicz & Stock, 2015). The concept in the

Figure 1. Oil as the source of wealth in the GCC States.

broader sense is defined by six characteristics including smart economy, smart people, smart governance, smart mobility, smart environment, and smart living (Giffinger et al., 2007). This concept is referring to Castells' (1989) 'informational city' where the flows of money, power and information, summarized as the 'space of flows,' dominate the geographical space of places. The smart city in the narrower sense is an ICT-driven green city which focusses on the environment (Hall et al., 2000). According to Chourabi et al. (2012), the smart city can be seen as a sustainable and at the same time livable city. The concept of a smart city in the narrower sense, which is one of the partial aspects in our research, is included in the concept of a smart city in the broader sense or—as we refer to it—the 'informational city' (Stock, 2011). In our terminology, 'informational city' and 'city of the knowledge society' are used synonymously. The scientific discipline, which studies cities of the knowledge society, is 'informational urbanism' (Stock, 2015; Barth et al., 2017a; Barth et al., 2017b; Barth et al., 2018). This term was coined by Stallmeyer (2009) to analyze "spatial transformations brought about by informational developments" (Stallmeyer, 2011, p. 2). In our interdisciplinary and holistic framework of informational urbanism we study all aspects of knowledge and information, be it digital or physical, man or machine generated, which has implications for cities, their spaces, their institutions and—most important—their people. Albeit ICT is the heart of a smart city, we may not forget face-to-face information dissemination and knowledge both in the form of tacit knowledge (bound to persons) as well as in the form of explicit knowledge (bound to documents) (Stock & Stock, 2013).

The presented results cover all larger Gulf cities of the Gulf Cooperation Council (GCC) States: Kuwait City (Kuwait), Manama (Bahrain), Doha (Qatar), Abu Dhabi, Dubai, Sharjah (all United Arab Emirates) and Muscat (Oman). Since except Damman, a center of oil production, no Saudi Arabian metropolis is located at the Gulf coast, we bracket cities in Saudi Arabia out. Likewise, we did not study UAE's smaller emirates (Ajman, Fujairah, Ras al-Khaimah, and Umm al-Quwain). The case studies of the seven analyzed cities were conducted on-site, i.e., all cities were visited by the research teams.

The states in the Gulf region have grown rapidly due to their large reserves of oil and gas. Being aware of ending resources, the era after the oil is already being planned by the governments in the Gulf region (Höselbarth, 2010; Ewers & Malecki, 2010).

In the development plan of *Kuwait*'s government we read, "Statement: [To] transform Kuwait into a financial and trade center, attractive to investors, where the private sector leads the economy, creating competition and promoting production efficiency, under the umbrella of enabling government institutions, which accentuates values, safeguards social identify, and achieve human resource development as well as balanced development, providing adequate infrastructure, advanced legislation and inspiring business environment" (Kuwait, 2015, p. 10). Strategic directions are to "encourage increasingly sophisticated production processes, using knowledge-intensive processes as opposed to oil technology or labour-intensive processes" and "move towards a knowledge economy in Kuwait" (Kuwait, 2015, p. 22).

Bahrain's vision for 2030 is strongly related to the establishment of a knowledge economy. "Bahrain's economy will attain increasing levels of sophistication and innovation, enabling the country to claim an attractive position in the global value chain. As a result, we will capture emerging opportunities, particularly by expanding to knowledge-based sectors and by increasing our output of high-value-added goods and services. This will mean greater wealth for companies, employees and society. The Government and private sector will act as partners in this by:

- Creating an environment highly conducive to entrepreneurship and innovation
- Improving access to capital for SMEs, and access to venture capital for business in general
- Establishing a platform for advancing innovations to market readiness, which will help Bahrain tap into the global knowledge economy" (Bahrain, 2008, p. 16).

In accordance with *Qatar*'s National Vision 2030 (GSDP, 2008), the main goal of this country is to build on knowledge: "As the Qatar economy diversifies more from its reliance on gas and oil, success will increasingly depend on the ability to compete in a global knowledge economy. Educating and training Qataris to their full potential will be critical to continuing progress" (GSDP, 2011, p. 13).

The government of the *United Arab Emirates* promotes the development of a competitive knowledge economy: "We want the UAE to transform its economy into a model where growth is driven by knowledge and innovation. Productivity and competitiveness will come to rival the best in the world, as a result of investment in science, technology, research and development throughout the fabric of the UAE economy. … This shift to knowledge economy can only be accomplished within the entrepreneurial environment that harnesses the talent and creativity of Emiratis" (UAE, 2014, p. 18).

One pillar of *Abu Dhabi*'s economic vision is a "sustainable knowledge-based economy" (Abu Dhabi, 2008, p. 5). A benchmark country for the emirate is Ireland, because "the success of which in developing a diversified knowledge-based economy and experience in investment attraction is relevant to Abu Dhabi's drive for diversification" (Abu Dhabi, 2008, p. 9). Universities "will also eventually lead to a more vibrant research and development community, which will ultimately drive the knowledge economy that the Emirate seeks to establish" (Abu Dhabi, 2008, p. 93).

The emirate of *Dubai* describes two knowledge-based aims: (1) "Preparing Dubai's workforce for the high-value, knowledge-driven economy, which requires attracting and retaining highly skilled employees, improving Nationals' qualifications and increasing their motivation" and (2) "Turning Dubai into a vibrant science and technology hub in targeted sectors, by supporting the development of existing sectors, and establishing the right environment for nurturing the post 2015 economy" (Dubai Strategic Plan 2015, 2007, p. 22).

Due to only small oil reserves, *Sharjah*'s economy has already "one of the highest levels of diversification in the region, a strong manufacturing presence, and an abundance of small and medium-sized enterprises" (Sharjah, 2018). In this emirate there is no need to reduce the share of the oil industry.

In *Oman*, the Industrial Innovation Center will activate a "support program for establishing and expediting the development of industrial enterprises that are based on innovative products that will eventually form the base of the 'knowledge economy'" (TANFEEDH, 2017, p. 56). Also in Oman, reduction of the influence of the oil industry on the country's economy is targeted. "Economic diversification is considered a strategic objective which will lead to economic growth in light of the global economic conditions. This focus creates an obligation to shift from a commodity-based economy—namely oil and its derivatives—to an economy focusing on growth in other sectors to propel economic development" (TANFEEDH, 2017, p. 13).

The propositions of the diverse master plans of the Gulf States are clear: They prepare for unlearning the reliance on both oil as well as natural gas and for the transitioning towards knowledge societies.

METHODS

As part of a large research project on informational cities (Barth et al., 2017a; Barth et al., 2017b; Barth et al., 2018; Fietkiewicz & Pyka, 2014; Fietkiewicz & Stock, 2015; Mainka, 2018; Stock, 2011; Stock, 2015), we analyzed the Gulf states and emirates in terms of informational cities (smart cities in the broader sense). Our research model is concerned with eight aspects of cityness and informativeness: digital city, knowledge city, creative city, smart city (in the narrower sense), economy (sectoral mix and labor market), politics and administration (political willingness to build an informational city and e-government), (hard and soft) location factors, and spaces (space of flows as well as space of places) (Figure 2).

In this chapter, we will report about our 34 interviews in the seven Gulf cities. We applied a quantitative questionnaire and conducted semi-standardized interviews with inhabitants between the 18th of February and the 6th of March 2013 in Abu Dhabi, Doha, Dubai and Manama and between the 28th of March and the 9th of April 2014 in Kuwait City, Muscat and Sharjah. The duration of the interviews varied between 30 and 120 minutes. We based our questionnaire on the method of ServQual (Parasuraman, Zeithaml, & Berry, 1988) which is the standardized instrument to measure the quality of service in retailing and service organizations, but can be used in different research areas as well. We provided our interviewees with 18 questions that were split into two parts. The left hand side contains questions concerning the expectation (E) of the interviewees about an informational city in general and the right hand side contains the same question but refers to the experience (P) of the interviewees in the city in question. For each question there is a seven-ary scale where 7 represents the strongest form of agreement and 1 represents the strongest form of disagreement. To get an idea about the gap (Q) between the expectation and experience of a given question, the difference value $Q = E - P$ is calculated. For each question and for each city and also for the whole region, we calculate the arithmetic mean to create a ranking later on. We are aware of the fact that Likert-type scales are ordinal and therefore do not allow using the arithmetic mean. However, since Likert (1932) reasons that consecutive natural numbers are equidistant, we interpret our scale as an interval scale and point this out to all our survey participants. Consequently we can calculate the arithmetic mean.

Since the original ServQual questions do not fit our needs, we designed our questionnaire with the help of literature. The question *Is the ICT infrastructure in an Informational City more important than automotive traffic infrastructure?*, which belongs to the dimension of digital infrastructure, arose from the theory of the fifth Kondratieff according to which the economic wave of automobiles and petrochemistry is replaced by the wave of information technology (Nefiodow, 1991). The dimension of knowledge infrastructure contains three questions *Does an Informational City have to have knowledge-intensive companies and knowledge institutions, e.g. universities and science parks?* This question was generated by thinking about Sassen's (2001) proposition that the city, where the headquarters of a company that operates on a global scale is located, becomes a center of information. For Mainka et al. (2013), besides the physical library space the offering of digital libraries is important, inducing the questions *Is a fully developed content infrastructure, e.g. supported by digital libraries, a characteristic feature of an Informational City?* and *Are libraries important in an Informational City as a physical place for face-to-face communication?*

Figure 2. Research framework for analyzing informational cities (in red: essential characteristics of prototypical cities of the knowledge-intensive economy).

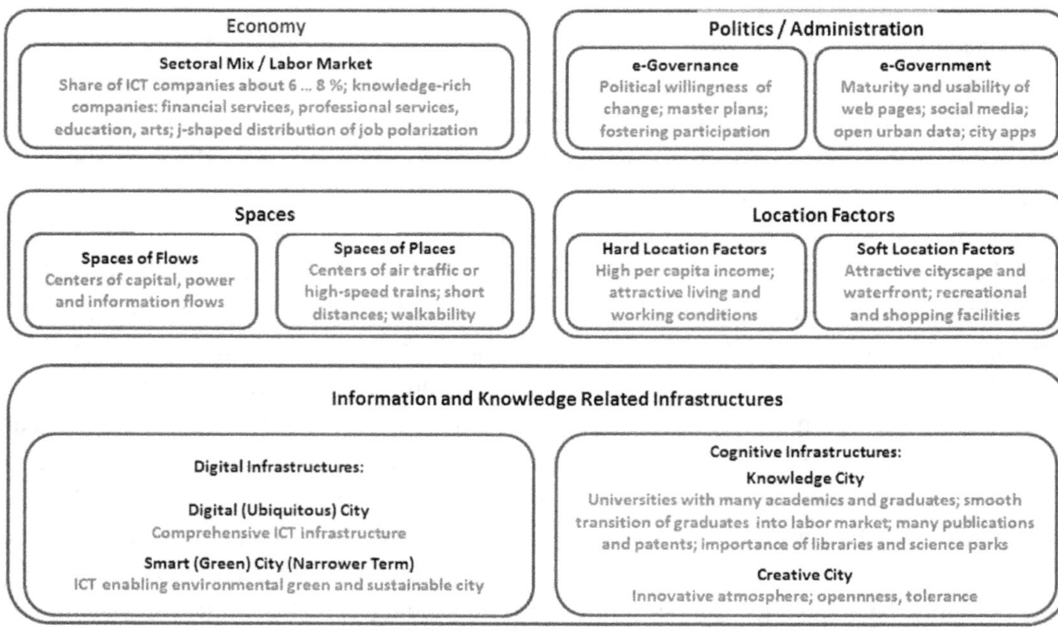

For the creative infrastructure dimension, we thought of the three T's mentioned by Florida (2002) and formulated the question *Does an Informational City have to be a creative city?* The T's refer to the terms technology, talent and tolerance, which are all required components for a city to be creative and economically successful (Murugadas, Vieten, Nikolic, Fietkiewicz, & Stock, 2015). We also included the questions *Are physical places where knowledge workers or creatives can meet each other important for an Informational City?* in this dimension, since an informational city needs creative people who "want to meet and socialize with people unlike themselves, trade views and spare over issues" (Florida, 2002, p. 21).

Does an Informational City have to have an urban structure which is characterized by short distances? is an interesting question for the dimensions of physical spaces and of smart infrastructure, since the reduction of traffic causes better quality of life which allures creative people. "The key is to concentrate residences, work areas and amenities so as to produce the shortest possible trip distances, most being possible by bicycle and public transport" (Hall, 1997b, p.15).

Two questions belong to the dimensions of sectoral mix and labor market which we include together in one indicator. The first one is: *Does an Informational City have to have companies with information market activities, e.g. telecommunications companies?* According to Stock (2011), especially companies with information market activities should have their headquarters in an informational city. Furthermore, the question *Is the labor market in an Informational City characterized by job polarization (loss of routine middle-class jobs)?* arose when we thought about the automatization of large economic processes and thus the occurrence of the two opposites (1) well-paid analytic tasks and (2) badly-paid manual labor (Goos & Manning, 2007). In contrast to general job polarization with a typical U-shaped distribution of the development of work force, some informational cities exhibit for the native population a J-shaped curve. There is no increase of low-income jobs for the natives, but a high increase of well-paid occupations (Barth et al, 2017a). However, who will perform the low-income jobs?

The dimension of political willingness contains two questions as well. Since the political willingness is an essential aspect to realize the idea of an informational city and to get financial support for the initial period (Stock, 2011), we asked, *Does the political willingness to create a knowledge economy play a significant role for establishing an Informational City?* The second question *Is an Informational City characterized by e-governance (incl. e-government, e-participation, e-democracy)?* combines Moon's (2002) five steps or pillars of e-Government (Fietkiewicz, Mainka, & Stock, 2017): (1) information dissemination (one-way communication), (2) two-way communication where the user gets a response, (3) service and financial transactions, (4) vertical and horizontal integration, and (5) political participation.

The three questions *Does an Informational City have to have many leisure time facilities?*, *Does an Informational City have to have an offer of a variety of shopping malls?* and *Does an Informational City have to have architainment, e.g. spectacular skyline, amazing waterfront, big wheels (is such a city an event in itself)?* deal with soft location factors which comprise the diversity of cultural and leisure activities. An informational city offers hence a variety of services, which cover different interests of individuals (Van den Berg, van der Meer, & Otgaar, 2007). With the transition to knowledge economies, the location factors are shifting. In the past, hard location factors seen from companies' perspective dominated; in informational cities preference is given to soft location factors from the point of view of the employees (Stryjakiewicz, 2010). It is therefore crucial to invest in soft location factors to inspire the desired knowledge workers and creatives for the city (Thite, 2011).

Further three questions are generated to cover the dimension of spaces. The question *Does an Informational City have to have capital-intensive companies, e.g. banks and insurances?* refers to Castells' (1989) space of flows. By inventing the question *Is an Informational City supposed to be a global city (world city)?* we thought of "The World City Hypothesis" by Friedmann (1986, 1995) where cities are defined economically rather than by administrative borders. Given the fact that free access to information and freedom of communication are basic rights of anyone (Balkin, 2005), we asked: *Is it important for an Informational City that there is a free flow of all kinds of information (incl. mass media information)?* The last question (Q18) *Do you think that your city is an Informational city?* shall give an impression of our interview partner to which extend the discussed city corresponds to the prototypical informational city.

RESULTS

Table 1 outlines the results of the ServQual questionnaire for the seven Gulf cities.

Digital City

Renowned as a part of the informational city, digital cities are based on ICT, which build a new kind of infrastructure. Actually, there is no other technology that shapes social, cultural and economic change as much as ICT. The speed of these changes increases rapidly and transforms the world into an informational village where information and knowledge are provided and accessible for everyone. Place and time become irrelevant in a world where information can be transferred affordably and in a short time (Castells, 1989). In the economic sector well-developed networks play a decisive role: Manufacture, energy, service and health sectors rely on it and would come to a standstill if they have to work without

Table 1. Results of the ServQual questionnaires.

Question/City	Kuwait City			Manama			Doha			Abu dhabi			Dubai			Sharjah			Muscat			Gulf Cities		
	n = 5			n = 5			n = 4			n = 5			n = 4			n = 5			n = 6			n = 34		
	E	P	Q	E	P	Q	E	P	Q	E	P	Q	E	P	Q	E	P	Q	E	P	Q	E	P	Q
ICT vs. automobile infrastructure	4.80	3.40	-1.40	5.20	4.00	-1.20	4.75	4.00	-0.75	5.20	4.40	-0.80	5.75	4.67[2]	-1.08	5.80	4.80	-1.00	5.33	2.00	-3.33	**5.26**	**3.79**	**-1.47**
Knowledge-intensive facilities?	6.00	3.80	-2.20	6.60	4.80	-1.80	7.00	5.00	-2.00	6.80	5.80	-1.00	6.75	5.83[1]	-0.92	6.40	5.80	-0.60	6.33	4.00	-2.33	**6.53**	**4.90**	**-1.63**
Digital libraries?	5.40	2.20	-3.20	6.80	3.20	-3.60	6.75	4.50	-2.25	6.20	1.20	-5.00	5.75	2.88	-2.88	6.00	4.40	-1.60	5.67	2.25	-3.42	**6.06**	**2.88**	**-3.18**
Physical libraries?	2.60	1.80	-0.80	5.40	3.60	-1.80	5.00	4.50	-0.50	5.00	2.20	-2.80	4.63	2.75	-1.88	6.00	5.40	-0.60	6.33	3.17	-3.16	**5.04**	**3.32**	**-1.72**
Creative city?	5.60	1.80	-3.80	6.25[3]	4.50	-1.75	6.50	5.00	-1.50	6.40	3.20	-3.20	6.00	4.75	-1.25	6.20	4.80	-1.40	6.00	3.00	-3.00	**6.12**	**3.78**	**-2.34**
Meeting places?	4.60	4.20	-0.40	6.80	5.60	-1.20	5.75	5.00	-0.75	5.80	3.00	-2.80	6.50	5.25	-1.25	5.40	4.60	-0.80	6.67	3.50	-3.17	**5.94**	**4.38**	**-1.56**
Short distances?	3.60	4.00	0.40	5.00	6.00	1.00	4.33[2]	4.33[2]	0.00	5.20	3.40	-1.80	5.25	4.00	-1.25	5.00	5.20	0.20	5.33	2.17	-3.16	**4.84**	**4.09**	**-0.75**
Information market activities?	5.60	5.40	-0.20	6.40	6.00	-0.40	6.75	5.50	-1.25	6.40	4.20	-2.20	6.25	4.50	-1.75	5.40	3.40	-2.00	4.67	5.50	0.83	**5.85**	**4.94**	**-0.91**
Job polarization?	4.20	3.00	-1.20	3.20	3.20	0.00	5.00[1]	5.67[2]	0.67	4.40	5.30	0.90	4.00	4.25	0.25	4.20	3.60	-0.60	4.17	5.40[4]	1.23	**4.09**	**4.27**	**0.18**
Political willingness?	6.00	4.80	-1.20	6.40	4.20	-2.20	6.25	6.25	0.00	5.80	3.75[3]	-2.05	6.25	5.00[2]	-1.25	6.20	6.20	0.00	6.00	4.00	-2.00	**6.12**	**4.84**	**-1.28**
EGovernance?	6.00	2.80	-3.20	5.60	5.40	-0.20	6.75	5.00	-1.75	6.20	4.00[3]	-2.20	6.00	4.00	-2.00	6.00	4.20	-1.80	5.33	3.20[4]	-2.13	**5.94**	**4.06**	**-1.88**
Leisure facilities?	5.20	3.40	-1.80	5.40	4.40	-1.00	6.25	5.00	-1.25	5.80	4.80	-1.00	6.50	5.75	-0.75	6.00	4.80	-1.20	6.17	4.17	-2.00	**5.88**	**4.56**	**-1.32**
Shopping malls?	4.40	5.60	1.20	5.20	6.80	1.60	4.25	5.75	1.50	5.60	6.60	1.00	6.25	6.50	0.25	5.20	4.60	-0.60	4.33	6.17	1.84	**5.00**	**6.00**	**1.00**
Spectacular buildings?	4.00	2.20	-1.80	5.20	4.00	-1.20	4.50	6.00	1.50	5.40	6.00	0.60	5.50	6.50	1.00	5.40	5.60	0.20	4.67	2.83	-1.84	**4.94**	**4.59**	**-0.35**
Capital-intensive companies?	5.20	5.60	0.40	6.20	6.00	-0.20	5.00	5.75	0.75	5.80	6.40	0.60	6.25	5.00[2]	-1.25	5.20	4.80	-0.40	5.67	4.50	-1.17	**5.62**	**5.42**	**-0.20**
Global city?	3.80	2.10	-1.70	4.40	3.40	-1.00	6.00	5.75	-0.25	6.20	4.30	-1.90	6.25	5.50	-0.75	5.00	3.60	-1.40	3.50	1.83	-1.67	**4.91**	**3.62**	**-1.29**
Free flow of information?	6.60	4.20	-2.40	6.80	2.80	-4.00	6.75	4.50	-2.25	6.60	3.30	-3.30	6.50	4.00	-2.50	5.80	4.20	-1.60	6.67	1.83	-4.84	**6.53**	**3.46**	**-3.07**
Average for questions 1 to 17	**4.92**	**3.55**	**-1.37**	**5.70**	**4.58**	**-1.11**	**5.74**	**5.15**	**-0.59**	**5.81**	**4.23**	**-1.58**	**5.90**	**4.77**	**-1.13**	**5.60**	**4.71**	**-0.89**	**5.46**	**3.50**	**-1.96**	*5.57*	*4.29*	*-1.28*
Informational city? (q18)	-	**2.00**	-	-	**3.60**	-	-	**4.25**	-	-	**4.00**	-	-	**4.88**	-	-	**4.20**	-	-	**2.67**	-	-	*3.57*	-

it. ICT permits them to act and move globally. City-wide free WiFi provides us with an impression of an advanced digital city.

The experience of the importance of ICT and its infrastructure was evaluated with averagely 3.79 points, which is 1.47 points below the expectation. ICT is broached in politics, but is almost unnoticeable implemented (interview partner DO1). The politicians mostly concentrate on road traffic due to the relevance of cars for the inhabitants, and the resulting traffic problems around the city (interview partners DO1, KU5). A life without cars is, particularly in Muscat (experience value: 2.00), almost inconceivable for the natives. Local public transport is—if provided—only used by foreign workers (interview partner AD3). The governments still invest in the expansion of road networks. In the private domain ICT seems to take a central position since otherwise the ways to communicate are limited. Some of the cities, namely Doha and those in the UAE seem to focus more on ICT than the others. On the whole, the internet connections are actually still too slow (interview partners DU3, KU2, MA3) and free WiFi hotspot availability is described as a rarity (interview partner MU3). As in Figure 3, you can find free WiFi only at some special locations in the city.

Knowledge City

While knowledge has always contributed to economic success, its importance has strikingly increased with the emergence of the information society and the knowledge society (Castells, 1989). Nowadays "cities are taking a leading role as both knowledge-consumption and knowledge-production hubs" (Carrillo, Yigitcanlar, García, & Lönnqvist, 2014, p. xvii). The new awareness of knowledge in all sectors paves the way for the knowledge city (Carrillo, 2006). One basic aspect of knowledge cities conveys "the conglomeration of intense scientific, technological, academic, cultural and innovation activities

Figure 3. Free WiFi as an impression of an advanced digital city. Example: Muscat, Oman.

in urban spaces operating as engines of economic productivity" (Carrillo et al., 2014, p. 3). A second basic aspect of such cities includes well-educated and creative knowledge workers (Florida, 2005) and "creative communities" (Carrillo et al., 2014, p. 4) in order to establish knowledge-based urban development. Furthermore, the number of a city's libraries is an indicator "to assess the initial status of the city" (Ergazakis, Metaxiotis, & Psarras, 2006, p.75) as a knowledge city, since they provide access to information. The term "knowledge cities" includes more than the pure existence of knowledge institutions and of knowledge workers in the city. All of "these elements constitute the cities' core identity, and the way their citizens use knowledge to build their social infrastructure, their institutions and their future" (Carrillo et al., 2014, p. xvii).

People living in the Gulf region see the importance of knowledge-intensive companies and knowledge institutions in an informational city and give an average of 6.53 points. The average experience is with 4.90 still below that score. The region is already working on the establishment of those institutions, but still has a long way to go (interview partner SH5). Knowledge institutions are available, but knowledge-intensive companies are not present in Kuwait (interview partner KU4). Interviewees KU1 and MA3 voice misgivings whether the knowledge-intensity of knowledge institutions, even if they exist, is high enough to refer to them as 'knowledge-intensive institutions.' Many of the interviewees are of the opinion that the institutions and companies are not yet capable of building their own knowledge base. Neither in the institutions (interview partner MU1) nor in the companies (interview partner MU2) is research and development sufficient (Kosior et al., 2015). However, for the knowledge infrastructure the picture is more optimistic. You can find world-class domestic public universities (as, for instance, Qatar University in Doha and Sultan Qaboos University in Muscat) as well as branch campuses of international private universities as in Doha's Education City (Figure 4), Dubai's Academic City, Dubai's Knowledge Village or Sharjah's University City.

There is a huge gap of 3.18 points between the average expectation (6.06) and the average experience (2.88) in case of the existence of digital libraries. Digital libraries include e-resources as well as digital reference services (Mainka et al., 2013). Its content should be freely accessible for all inhabitants. Most

Figure 4. Entrance of Georgetown University in Doha's Education City. In the background Sidra, a hospital and research institution for mother and child.

of the interviewees are sure about the nonexistence of digital libraries. Some of them however include reference services without the corresponding full texts, which are offered to students by the universities, like Scopus and ACM in the term 'digital library' (interview partners DU1, KU4, SH3). The use of the term therefore differentiates from the intended one. However, Qatar National Library offers diverse full-text resources to its users for free (Henkel, Barth, Gremm, & Stock, 2018).

The institutions in Education City (Table 2) were goal-oriented selected in order to meet the challenges of Qatar's present and future industries and companies. We only give some examples: Texas A&M University prepares students for jobs in the oil and gas industry, Weill Cornell Medical College instructs students for working in healthcare-related institutions, Virginia Commonwealth University works for the creative industries, and University College London prepares the workforce for libraries and museums (Gremm, Barth, Fietkiewicz, & Stock, 2018). It is an open question, whether such "knowledge mega-projects" as the construction of Education City in Doha fit in with other more traditional knowledge institutions (as, for instance, Qatar University), knowledge-intensive companies in the country, and the general social values of the natives (Rizzo, 2017).

For our interviewees the existence of physical libraries (5.04) is not as important as that of digital libraries. From the average experience (3.32), it can be deducted that not much libraries are present in the Gulf region. However, for the interviewees the problem lies elsewhere. The culture in the Gulf region is not a knowledge-based one. People do not get encouraged to read as they do not need to acquire knowledge for their jobs. Nearly all of the respondents agree that citizens do not use the available libraries (interview partners AD1, KU5, MA3). People only visit libraries for entertainment (interview partner KU4) or if they are concerned with education, like teachers and students (interview partner SH5).

However, the field research revealed that impressive libraries with great potential to attract visitors are available in the cities. There are even important examples of world-class national libraries (as the Kuwait National Library and the Qatar National Library), academic libraries (e.g., the library of Qatar University) or special libraries as, for instance, the library for Arabic Poetry in Kuwait (Figure 5). However, public libraries and school libraries are weak in the entire region (Henkel, Barth, Gremm, & Stock, 2018).

Figure 5. Libraries are important building blocks of knowledge-intensive cities' infrastructure: Al-Babtain Central Library for Arabic Poetry in Kuwait City.

Table 2. Branch campuses in Doha's Education City.

University	Study Program	Industry	Typical Company
Virginia Commonwealth University in Qatar (1998)	B.A. Fashion Design, B.A. Graphic Design, B.A. Interior Design, B.A. Painting & Printmaking, B.A. Art History; M.A. Design	Art and Design	Al-Jameela Magazine, Al Jazeera, Arteligno Interior Designers, Katara Cultural Village, many start-ups
Weill Cornell Medical College in Qatar (2001)	Pre-Medical / Medical Education (M.D.)	Healthcare	Hamad Medical Corp., SIDRA
Texas A&M University at Qatar (2003)	B.S. Chemical Engineering, B.S. Electrical and Computer Engineering, B.S Mechanical Engineering, B.S. Petroleum Engineering, M.S. / M.Eng. Chemical Engineering	Chemical Engineering, Electrical and Computer Engineering, Mechanical Engineering, Petroleum Engineering	Qatar Petroleum, Maersk Oil Qatar, Qatar Shell, RasGas
Carnegie Mellon University in Qatar (2004)	Biological Sciences, Business Administration, Computational Biology, Computer Science, Information Systems (all Bachelor's degree)	Biology, Business Administration, Information and Communication	Vodafone, Microsoft, Qatar Petroleum Ooredoo, ictQatar, Qatar Airways
Georgetown University – School of Foreign Service in Qatar (2005)	B.S. Foreign Service (Culture and Politics, International History, International Politics, International Economics), M.S. Foreign Service, M.A. Arab Studies, M.A. German and European Studies, M.A. Latin American Studies, M.A. Global, International and Comparative History	Foreign Services, Politics	Government Al Jazeera, Qatar States Security, Qatar National Bank
Northwestern University in Qatar (2008)	B.S. Communication, B.S. Journalism	Communication, Media, Journalism	Al Jazeera, Qatar Shell
HEC Paris in Qatar (2011)	Executive MBA, MBA Strategic Business Unit Management	Management	All companies
Hamad Bin Khalifa University (2012)	B.S. Computer Engineering, M.A. Islamic Studies, M.S. Islamic Finance, M.A. Translation Studies, M.A. Audiovisual Translation, M.S. Data Science and Engineering, M.S. Cybersecurity, M.S. Public Health, M.S. Energy and Resources, Doctorate Law, M.S. / Ph.D. Biological and Biomedical Sciences, Ph.D. Computer Science and Engineering, M.S. / Ph.D. Sustainable Energy, M.S / Ph.D. Sustainable Environment	Islamic Studies, Islamic Finance, Translation, Computer Science, Energy, Law, Healthcare	Qatar National Bank, Qatar Telecom, Qatar Petroleum, Hamad Medical Corp., SIDRA, Qatar Biomedical Research Institute, Qatar Islamic Bank
University College London Qatar (2012)	M.A. Library and Information Studies, M.A. Museum and Gallery Practice	Librarianship, Museums	Qatar National Library, Qatar Museums

Creative City

In the twenty-first century a transformation of cities is happening. Cities have to find new ways to stabilize their economies for ensuring their economic power. Focusing on creativity seems like a good solution for today's societies (Chatterton, 2002). Also researchers increasingly consider creativity as a huge benefit for cities, raising the concept of the creative city (Landry & Bianchini, 1995). According to Landry (2000, p. 4) the concept "describes a new method of strategic urban planning and examines how people can think, plan and act creatively in the city." He thinks that a creative city should offer cultural resources in order to generate and encourage creativity. Nowadays industries focus on knowledge production driven by creativity (Landry & Bianchini, 1995). Thus the existence of creative professionals who work in these knowledge-intensive industries is obligatory. A city should aim at attracting this "creative class" (Florida, 2002) by being tolerant and by offering enough employment opportunities and a high quality of living. In return, they are rewarded with more creative economic outcome, which is beneficial for their economic power.

With an average of 6.12, the interviewees think that in order to become an informational city, the city has to be creative. However, they do not feel positive about the creativity in the Gulf region, resulting in an average score of 3.78. In most cities, there are large investments in the creative infrastructure. Newly built museums in Abu Dhabi, Manama and Doha (Figure 6) or the galleries in Sharjah are landmarks of a creative city. The average value that is given by the Kuwaiti interviewees is with 1.80 even much lower than the region's mean value. Interviewee KU3 is of the opinion that there is not enough technology in Kuwait. Even though people try to apply it, they still have many problems with it. Since the talent of the citizens in the Gulf region is still undeveloped, the few talents that are available come from abroad (interview partners DU3, MU3). The region takes countermeasures to this problem though by sending (male) citizens to study abroad to expand their knowledge and abilities (interview partner MU3). A high tolerance threshold does not exist at the Gulf. There are many taboo topics like politics and homosexuality and partly also religion. According to interviewee MU1, Muscat is more tolerant than the rest of the region.

Figure 6. Creative infrastructure: Museum of Islamic Art in Doha, Qatar.

The respondents in the Gulf region think with an average score of 5.49 that knowledge workers and creative people must be offered places to communicate. The availability of such "third places," i.e. places in addition to homes and working places (Oldenburg & Brissett, 1982; Oldenburg, 1999), in the Gulf region is rated with an average of 4.38. Especially in Abu Dhabi there seems to be a lack of meeting places. Here the average experience is only 3.00. The interviewees in Kuwait mentioned that the so called "Dewaniya" (or the "Majlis" in other Gulf countries) is a traditional meeting place like a tent or a house where people get together to cultivate contacts. Also malls, cafes, and restaurants are favored places to meet, but it is unknown if they are used to share and create knowledge (interview partners DU3, KU4). Interviewee SH3 states that meeting places become less important in advanced cities, but that in Sharjah these physical spaces are important since the city is still in the learning process.

Smart City

The smart city in the narrower sense is a green city including ICT, aiming at a sustainable economy and better quality of life (Chourabi et al., 2012; Hall et al., 2000). With Masdar City in Abu Dhabi (Figure 7), there is one experimental attempt in the entire region to construct a carbon-neutral and car-free smart city.

With an average of 4.34, the interviewees in the Gulf cities think that short distances should be realized in an informational city. Whether the citizens of the Gulf region can reach their destinations by short routes is estimated with an average score of 4.09. Especially the opinion of the interview partners in Manama stands out. They think with an average of 6.0 that Manama is characterized by short distances. In most of the cities at the Gulf, the offer of public transport is scarce. Due to our field research it can be said, that the investment in public transport has no visible effect in reality. Except for the metro in Dubai, which is accordingly a widely used means of transportation (interview partner DU2), and Doha's metro system (under construction) no railway or metro system exist. People resort to using cars which results in high traffic density and people being delayed (interview partner KU2). According to our interview

Figure 7. Wind tower in Masdar City, a "smart" and sustainable living lab.

partners in Muscat, a short urban structure does not exist in the capital area. Muscat expanded its urban area in recent years and is nowadays an urban conglomeration stretching 60 km rendering it impossible for Muscat to offer short distances. The long ways from one point of the capital area to another is also what we witnessed during our field research. In the past, Muscat had clusters and consequently short distances, but nowadays the companies and population are spread throughout the whole city (interview partner MU2).

Sectoral Mix and Labor Market

In a knowledge society, knowledge and capital-intensive industries and industries of information and communication technologies (ICT) are corresponding factors on the labor market (Sassen, 2001; Taylor, 2004; Florida, 2005). In this case, the term 'knowledge-intensive industries' means universities and science parks, 'capital-intensive industries' refers to banks and insurance companies whereas industries of ICT are computer, telecommunication and information service companies. Not least, creative industries, in terms of Florida's (2012) Bohemian Index, which measures the density of artists, writers and performers in a region, are a component of a city's labor market in the knowledge society. The workers, which are employed in the industries of the knowledge society, work rather with their brain, than with their hand.

The inhabitants of the Gulf cities answered the question concerning the necessity of companies with information market activities averagely with 5.85 points on the expectation and 4.94 on the experience side. The difference between the assigned scores of the single cities is conspicuous: Doha—having an own telecommunication company—gets 6.75 points for expectation and 5.50 points for experience and others like, for example, Sharjah—having no own telecommunication companies and depending on the neighboring regions Abu Dhabi and Dubai (interview partner SH4)—get only 3.40 points for the experience part. In this context, also some powerful telecommunication companies like Etisalat located in the UAE, Zain located in Kuwait City and Nawras and Omantel located in Muscat were named (interview partners KU1, MU1). Often, companies are located in centered regions like 'Internet City' and 'Media City' in Dubai (interview partner AD3). It seems to be characteristic for informational cities that especially companies of the information markets form clustered economic zones (Figure 8).

Figure 8. Clustered economic zones as a characteristic property of cities of the knowledge society. Internet City in Dubai as a prototypical example.

Nevertheless, the telecommunication market is seen as rather small (interview partner DO1). The interviewees agree with each other that their services are insufficient. Interview partner MU2 mentions that the providers are just specialized in sale and hence non-manufacturing companies. In contrast, a software branch does not exists is the Arab region at all (interview partners DO1, MA4).

The majority of the workforce in knowledge-intensive institutions consists of expats (Gremm, Barth, Fietkiewicz, & Stock, 2018). It is the plan of nationalization programs such as 'Emiratization,' Qatarization,' etc. to replace the foreign talents with natives (Wiseman, Alromi, & Alshumrani, 2014). However, apart from leadership positions, nationalization projects failed so far.

Averagely the interview partners answered the question regarding job polarization with 4.27 points on the experience side, which is 0.18 points higher than their expectation. Kuwaitis, Emirati and Omanis speak about a strict separation between highly and badly paid jobs, but also emphasize the importance of the existing middle class. Many of these jobs are indispensable (interview partners KU4, MU1, SH1). In contrast to most of the other natives, Omanis work beside the expats as common professions (interview partner MU1). The expat population in the Gulf cities—most of them live in the country only for work—is almost everywhere above 80 percent except in Muscat, where it is a bit lower. Nationalization programs like 'Qatarization' or 'Emiratization' are in progress to reduce the high number of expats (interview partners DO1, KU1). It is doubtful whether this question is appropriate for the Gulf region since most of the natives work in the (well-paid) public middle class service sector (Kosior et al., 2015).

Political Willingness and E-Government

Political willingness and e-Government are required conditions to establish an informational city. e-Government refers to the use of ICT and especially the inclusion of web-based services by the government. This form of government should be established primarily for the citizens since it provides a relief in governmental matters like the access to information, hold available by government (Awan, 2003).

The political willingness to perform changes exists in all of the analyzed cities (average experience rating: 4.84) and their future visions are supported by large sums of money. The willingness seems to be particularly high in Doha (6.25), Sharjah (6.20), and Dubai (5.00), however readiness to cooperate on a regional level only exist to a minor degree (interview partner DO1). The interviewees mention that the global trend of city development is recognized and that plans to change the future of the Gulf States exist. The interviewees in Dubai confirm that the city "has the willingness of contesting the world" (interview partner DU1) and therefore tries to improve their status. Other Emirati take the prototypical informational city Singapore as an example for their future urban development towards a knowledge society (interview partner AD2). The master plans for the future will be financed by today's oil revenues. However, in most cases the implementation proves difficult (interview partners KU3, MU5). Interviewee KU3 predicts that someday, when the oil reserves are empty, the government will be unemployed and the country will be empty since all expat people will leave.

When asked if Kuwait City, Manama, Doha, Abu Dhabi, Dubai, Sharjah and Muscat are characterized by e-governance, the interviewees give averagely 4.06 points. Thus, their experience is 1.88 points lower than their expectation of e-governance in an informational city. This difference is accounted for by the existence of e-government and e-participation, and by the lack of e-democracy (interview partner SH2). Information is available, but the citizens do not get the chance to participate in governmental matters, since all decisions are already made (interview partner KU1). Only some transactions can be done online, but one thing seems to be possible in all Gulf cities: the payment of parking tickets online (interview partners AD1, DU3, MA1). Interviewee KU3 talks about improvable services, such as the mail service of the government, while in Muscat up to 90 percent of the services can already be done online (interview partner MU3). Even though activity and participation on social media platforms like Facebook, Instagram and Twitter exists, it is not always sufficient (interview partners SH3, MU2).

Location Factors

In an informational city, another central aspect are soft location factors. The presence of several of these factors contributes to the attractiveness and the increase of the magnetic effect of a city. Soft location factors comprise cultural and leisure time facilities like museums, galleries, libraries, opera houses, symphony halls, theatres, and diverse events, architainment and attractive waterfronts as well as shopping opportunities (Hall, 1997a). In order to attract inhabitants, business companies and visitors, a city needs to offer a variety of those services to cover the different interests of the individual groups (Van den Berg, van der Meer, & Otgaar, 2007).

The presence of leisure time facilities in the Gulf region is rated 4.56 and is therefore not as high as the expectation (5.88). However, we visited attractive leisure parks as, for instance, Al Qasbah in Sharjah (Figure 9). The average experience value of the interviewees in Dubai is quite high. But not only do the residents think highly about the available leisure time facilities in Dubai but also interview partner SH4 confessed that residents of Sharjah go to Dubai for the consumption of leisure facilities.

The question about shopping malls is one of the two questions where the average expectation (5.00) lies below the average experience (6.00). All the interviewees decided unanimously that there are enough or even more than enough shopping malls in the Gulf region. These malls are not only used for the purpose of shopping but they serve as a social place for the people. Due to extremely high temperatures in summer, the climate in the malls is pleasant for the residents (interview partner MU4). Again it holds

Figure 9. Leisure time facilities as weak location factors. Al Qasbah with the Eye of the Emirates in Sharjah, U.A.E.

for Sharjah, that even though the emirate has enough shopping possibilities (interview partner SH3), people go to Dubai or Abu Dhabi instead (interview partner SH4).

The average expectation regarding the existence of architainment in the Gulf cities is 4.94 which is only 0.35 above the average experience. The average experiences in Muscat and especially in Kuwait are very low. Kuwait is accordingly not attractive regarding architainment and waterfronts. The cities' attractiveness is limited to light effects since there is no ancient substance (interview partner KU1). In Muscat the construction of buildings with more than eight floors is forbidden by Oman's building law (interview partner MU1). Other contemporary or sophisticated buildings do not exist, either. Muscat can score with its very nice nature made up of mountains, beaches, wadis (dry riverbeds), and also desert landscapes; however there are only few tourists (interview partner MU2). Unlike Muscat, Dubai attracts lots of tourists every year. Some of them even come to visit more than once (interview partner DU2). In Doha and Abu Dhabi the average experience is just a bit lower than in Dubai.

Spaces. First efforts researching world cities were made by 'The World City Hypothesis' (Friedmann, 1986), which deals with the city's placement in the world economy. Castells' (1989) city research is also included and defines cities by their placement in the space of flows: flows of money, power, and information compared to the former focus on space of places. According to Taylor (2004) world cities are never isolated but always interconnected. This position in the network of global regions is called 'cityness' (Taylor, Hoyler, & Verbruggen, 2010). Sassen (2001) choses the term 'global city' as a preferable expression for such significant metropolises in today's globalized world. However, we do not differ between the expressions 'world city' and 'global city' in this chapter and use the terms synonymously.

Figure 10. Bahrain's capital Manama with its Financial Harbour is one of the most important financial centers in the Middle East.

The Gulf cities seem to have a sufficient number of capital-intensive companies, since the expectation (5.62) and the experience (5.42) almost achieved the same valuation. Some of the cities have banking districts (interview partner MU2) where the British finance system prevails. Islamic banking—a system where no interests are charged—also co-exists, but since it is suspicious to many people, some cities, e.g. Muscat, refused to introduce this system for a long time (interview partner MU3). Manama (Figure 10) plays an important role as a financial center of the entire Middle East.

Regarding the aspect of informational cities and Gulf cities as global cities, the expectations were averagely higher (4.91) than the experiences (3.62). At it, interviewees in Doha (5.75), Dubai (5.50) and Abu Dhabi (4.30) are most confident of living in a global city. Especially in Dubai it was emphasized that the city is well-connected (interview partner DU2). Others, for instance in Sharjah, mention that their city is trying to become global (interview partner SH4). And still others, living in Kuwait, Manama or Muscat, do not classify their city as global. A global city has to be well connected and needs to attract people (interview partner KU1)—however, Muscat is called a "boring village" (MU1).

In the Gulf States, a strict policy of censorship is pursued (average experience value: 3.46): Voice over IP (VoIP) is officially banned. However, the Skype-website could be opened in all cities, but not in Muscat. The government tries to control every sector to operate their political power (interview partners DU2, MU2). Also, it can be assumed that the telecommunication companies do not want to compete with freely available software and that they want to generate more profit (KU4). However, a virtual private network (VPN) client is widely used to get access to Skype and other censored websites (interview partners MA3, KU4). The interviewees feel restricted in expressing their own opinion (interview partner MU4). The expected necessity of free flow of information in an informational city (average expectation value: 6.53) cannot be confirmed for the Gulf cities.

Ranking of the Gulf Cities

The final point of our research about the Gulf cities' cityness and informativeness is an overall ranking of Kuwait City, Manama, Doha, Abu Dhabi, Dubai, Sharjah, and Muscat (Table 3). In case of the interviewees' experiences (P), Doha receives the best score with 0.86 points above the average of the evaluated Gulf cities and is followed by Dubai and Sharjah. Abu Dhabi, Kuwait City and Muscat fall below the average. Regarding the difference (Q) between experience and expectation, Doha—where the interviewees' experiences do not deviate much from their expectations—comes first once more. Also Sharjah achieves again a good rank. The values of Kuwait City, Abu Dhabi and Muscat are once again below the average. Taking the last question on informational cities (Q18) into consideration the huge difference between the best (Dubai: 4.88) and the worst (Kuwait: 2.00) rated city is striking. To receive the final ranking of all cities, the positions of the rankings by P, Q and question 18 are aggregated: Doha is ranked first and hence has the highest standing as knowledge-intensive, informational city regarding the estimations of our interview partners. It is followed by Dubai and then Sharjah. The fourth and fifth positions are occupied by Manama and Abu Dhabi while Kuwait and Muscat come in last. Interview partner KU3 summarizes: "Kuwait was yesterday, Dubai is the present, and Doha is the future" (Gremm, Barth, & Stock, 2015). This result is in line with older findings by Ewers and Malecki (2010, p. 496) as they state, "Dubai (UAE) and Doha (Qatar), in particular, are pursuing a direct jump from resources to a post-industrial economy."

Table 3. Ranking of the Gulf cities by the ServQual results.

Ranking by P	P-Value	Ranking by Q	Q-Value	Ranking by Q18	Q18-Value	Ranking by Position	Position Value
Doha	5.15	Doha	-0.59	Dubai	4.88	Doha	4
Dubai	4.77	Sharjah	-0.89	Doha	4.25	Dubai	7
Sharjah	4.71	Manama	-1.11	Sharjah	4.20	Sharjah	8
Manama	4.58	Dubai	-1.13	Abu Dhabi	4.00	Manama	12
Abu Dhabi	4.23	Kuwait City	-1.37	Manama	3.60	Abu Dhabi	15
Kuwait City	3.55	Abu Dhabi	-1.58	Muscat	2.67	Kuwait City	18
Muscat	3.50	Muscat	-1.96	Kuwait City	2.00	Muscat	20
Gulf City	4.29	Gulf City	-1.28	Gulf City	3.57		

CONCLUSION

Receiving the status of an informational city seems for *Muscat* far away. Muscat has convincing shopping malls as a weak location factor, good information market activities and no problems with job polarization. All other dimensions from Table 1 exhibit large difference (Q) values and low perception (P) values.

For *Kuwait City*—a city concentrating rather on the development of automotive infrastructure than the development of ICT—an after-oil future with an empty cityscape is predicted. Gone are the times when Kuwait was attractive to foreigners due to its prosperity. Having impressive libraries that are almost not used by the citizens, meeting places where people rather talk about other things than knowledge-intensive topics and no architainment to attract people from different countries seem to be determining disadvantages in the struggle to become an informational city.

Abu Dhabi shows good perception values concerning capital-intensive companies, weak location factors (shopping malls as well as spectacular buildings) and knowledge-intensive facilities. All other items of our questionnaire do not exhibit strong indicators for being a city of the knowledge society.

The will to establish an informational city is existent in *Manama*, which is ranked exactly in the middle of all cities. There are good P-values concerning capital-intensive companies, information market activities, shopping malls, e-governance, short distances and meeting places.

Sharjah has (next to Doha) the second highest score for the dimension of a creative city. In Sharjah, the political willingness to create an informational city is high; there are good P-values for knowledge-intensive facilities, spectacular buildings, libraries, short distances, and capital-intensive companies.

Dubai, as the most known of the seven United Arab Emirates, focuses its strategy on ICT more than other cities in the region. The construction of an own metro system, the settlement of telecommunication providers and technology parks enabled a huge development during the last few years. Dubai is very weak concerning its libraries; however, all other dimensions of a prototypical city of the knowledge society are rather high. And, as expected, Dubai scores with its spectacular buildings.

According to our ServQual results, the willingness to become an informational city seems to be highest in *Doha*. The development of an ICT infrastructure is as high as in Dubai and also some telecommunication providers are located in the city. The plan is to educate and train nationals properly to pave the way to a global knowledge society. This proves that the willingness to pursue the status of an informational city definitely exists.

Out of the seven studied Arab cities, four of them (Muscat, Kuwait, Abu Dhabi, and Manama) cannot be classified as prototypical cities of the knowledge society for now. However, Sharjah, Dubai and especially Doha are able to unlock a potential for a transition towards a knowledge-intensive city. However, the present transformation is due to lots of local money, but foreign talents, many foreign universities and foreign companies. Without foreign help, the transformation process may stop. Can indeed "money buy development" (Ewers & Malecki, 2010, p. 504) towards a knowledge-intensive economy?

PROJECT DESCRIPTION AND THE CONTRIBUTION OF THIS ARTICLE

Since about 2010, the Information Science Department of Heinrich Heine University Düsseldorf, Germany, runs a research project on 'informational cities.' The research program started with the formulation of a theoretical framework. Since 2011, we have performed empirical studies in cities all over the world. Based upon these empirical findings, we adjusted and calibrated the conceptual framework. In our project on (more than 30) world cities, we analyzed the roles of e-government and m-government in informational cities, the use of open urban data in mobile applications, the new functions of physical and digital libraries, and the information behavior of people living or visiting smart cities in terms of their Twitter activities. For selected informational cities we studied citizen participation on city-level through citizen relationship management (CiRM) in the U.S. and in Germany, programs for information literacy instruction, creativity and entrepreneurship, labor markets and job polarization, 'spaces of flows' in terms of flows of capital, power, and information, and the culture-dependent use of social media. Additionally, we performed regional analyses and case studies on single cities in Tokyo, Yokohama, Kyoto and Osaka (Japan), Singapore, London (UK), Barcelona (Spain), Oulu (Finland), and New Songdo City (South Korea).

The main contribution of this article, being a regional study on cities on the Arabian Gulf, is to understand the Arabian way from an oil-based economy to a knowledge-intensive society. We visited and studied cities in Kuwait (Kuwait City), Bahrain (Manama), Qatar (Doha), the United Arab Emirates (Abu Dhabi, Dubai, Sharjah) and Oman (Muscat). This region is not only nowadays one of the wealthiest areas of the world, it has ambitious master plans to unlearn its reliance on oil and (by using the money from the oil industry) to construct ways towards knowledge economies. The implementation of the master plans seems to be very problematic; however, cities such as Doha (Qatar), Dubai and Sharjah (both in the United Arab Emirates) make good progress on their ways to knowledge-intensive economies.

ACKNOWLEDGMENT

Our interview partners were cited by using abbreviations, with KU for Kuwait City, MA for Manama, DO for Doha, AD for Abu Dhabi, DU for Dubai, SH for Sharjah and MU for Muscat and a following number for the interview order in each city.

Three research teams visited the Gulf States. In 2013, Adriana Kosior, Agnes Mainka, and Wolfgang G. Stock travelled to Dubai, Abu Dhabi, Doha, and Manama; in 2014, Julia Barth, Julia Gremm, Fee Hilbert, and Wolfgang G. Stock visited Kuwait City, Sharjah, and Muscat; in 2016, Julia Barth, Julia

Gremm and Wolfgang G. Stock traveled again to Qatar in order to conduct further interviews in Education City and Qatar University. We would like to thank all our interview partners for their time and for many valuable hints and suggestions.

This research received no specific grant from any funding agency in the public, commercial, or not-for-profit sectors.

REFERENCES

Abu Dhabi. (2008). *The Abu Dhabi Economic Vision 2030*. Abu Dhabi: The Government of Abu Dhabi.

Awan, M. (2003). E-Government: Assessment of GCC (Gulf Co-operating Council) countries and service provided. *Lecture Notes in Computer Science*, *2739*, 500–503. doi:10.1007/10929179_94

Bahrain. (2008). *Our Vision. From Regional Pioneer to Global Contender. The Economic Vision 2030 for Bahrain*. Manama: Academic Press.

Balkin, J. M. (2005). Digital speech and democratic culture: A theory of freedom of expression for the information society. In A. D. Moore (Ed.), *Information Ethics, Privacy, Property, and Power* (pp. 297–354). Seattle, WA: University of Washington Press.

Barth, J., Fietkiewicz, K. J., Gremm, J., Hartmann, S., Henkel, M., Ilhan, A., … Stock, W. G. (2017b). Informationswissenschaft in der Urbanistik. Teil 1: Konzeptioneller Forschungsrahmen und Methoden. *Information – Wissenschaft & Praxis, 68*(5-6), 365-377.

Barth, J., Fietkiewicz, K. J., Gremm, J., Hartmann, S., Henkel, M., Ilhan, A., … Stock, W. G. (2018). Informationswissenschaft in der Urbanistik. Teil 2: Erste empirische Ergebnisse zu smarten Städten. *Information – Wissenschaft & Praxis, 69*(1), 31-46.

Barth, J., Fietkiewicz, K. J., Gremm, J., Hartmann, S., Ilhan, A., Mainka, A., ... Stock, W. G. (2017a). Informational urbanism. A conceptual framework of smart cities. In *Proceedings of the 50th Hawaii International Conference on System Science* (pp. 2814-2823). Washington, DC: IEEE Computer Society. 10.24251/HICSS.2017.340

Carrillo, F. J. (2006). *Knowledge Cities: Approaches, Experiences, and Perspectives*. New York, NY: Butterworth Heinemann.

Carrillo, F. J., Yigitcanlar, T., García, B., & Lönnqvist, A. (2014). *Knowledge and the City: Concepts, Applications and Trends of Knowledge-based Urban Development*. New York, NY: Routledge.

Castells, M. (1989). *The Informational City: Information Technology, Economic Restructuring, and the Urban-regional Process*. Oxford, UK: Basil Blackwell.

Chatterton, P. (2002). Will the real creative city please stand up? *City*, *4*(3), 390–397. doi:10.1080/713657028

Chourabi, H., Nam, T., Walker, S., Gil-Garcia, J. R., Mellouli, S., Nahon, K., & Scholl, H. J. (2012). Understanding smart cities: An integrative framework. In *Proceedings of the 45th Annual Hawaii International Conference on System Sciences* (pp. 2289-2297). Washington, DC: IEEE Computer Society.

Dubai Strategic Plan. (2015). Highlights: Dubai Strategic Plan. Dubai, UAE: Academic Press.

Ergazakis, K., Metaxiotis, K., & Psarras, J. (2006). An emerging pattern of successful knowledge cities' main features. In F. J. Carrillo (Ed.), *Knowledge Cities: Approaches, Experiences, and Perspectives* (pp. 3–15). New York, NY: Butterworth Heinemann. doi:10.1016/B978-0-7506-7941-1.50004-X

Ewers, M. C., & Malecki, E. J. (2010). Leapfrogging into the knowledge economy: Assessing the economic development strategies of the Arab Gulf states. *Tijdschrift voor Economische en Sociale Geografie*, *101*(5), 494–508. doi:10.1111/j.1467-9663.2010.00628.x

Fietkiewicz, K. J., Mainka, M., & Stock, W. G. (2017). eGovernment in cities of the knowledge society. An empirical investigation of Smart Cities' governmental websites. *Government Information Quarterly*, *34*(1), 75–83. doi:10.1016/j.giq.2016.08.003

Fietkiewicz, K. J., & Pyka, S. (2014). Development of informational cities in Japan: A regional comparison. *International Journal of Knowledge Society Research*, *5*(1), 69–82. doi:10.4018/ijksr.2014010106

Fietkiewicz, K. J., & Stock, W. G. (2015). How "smart" are Japanese cities? An empirical investigation of infrastructures and governmental programs in Tokyo, Yokohama, Osaka and Kyoto. In *Proceedings of the 48th Hawaii International Conference on System Sciences* (pp. 2345-2354). Washington, DC: IEEE Computer Society.

Florida, R. (2002). The rise of the creative class: Cities without gays and rock bands are losing the economic development race. *The Washington Monthly*, *34*(5), 15–25.

Florida, R. (2005). *Cities and the Creative Class*. New York, NY: Routledge.

Florida, R. (2012). *The Rise of the Creative Class: Revisited*. New York, NY: Basic Books.

Friedmann, J. (1986). The world city hypothesis. *Development and Change*, *17*(1), 69–83. doi:10.1111/j.1467-7660.1986.tb00231.x

Friedmann, J. (1995). Where we stand: A decade of world city research. In P. Knox & P. Taylor (Eds.), *World Cities in a World-system* (pp. 21–47). Cambridge, UK: Cambridge University Press. doi:10.1017/CBO9780511522192.003

Giffinger, R., Fertner, C., Kramar, H., Kalasek, R., Pichler-Milanović, N., & Meijers, E. (2007). *Smart Cities: Ranking of European Medium-sized Cities*. Vienna, Austria: Centre of Regional Science.

Goos, M., & Manning, A. (2007). Lousy and lovely jobs: The rising polarization of work in Britain. *The Review of Economics and Statistics*, *89*(1), 118–133. doi:10.1162/rest.89.1.118

Gremm, J., Barth, J., Fietkiewicz, K. J., & Stock, W. G. (2018). *Transitioning Towards a Knowledge Society. Qatar as a Case Study*. Cham, Switzerland: Springer Nature. doi:10.1007/978-3-319-71195-9

Gremm, J., Barth, J., & Stock, W. G. (2015). Kuwait is the past, Dubai is the present, Doha is the future: Informational cities on the Arabian Gulf. *International Journal of Knowledge Society Research*, *6*(2), 51–64. doi:10.4018/IJKSR.2015040103

GSDP. (2008). *Qatar National Vision 2030*. Doha, Qatar: Qatar General Secretariat for Development Planning.

GSDP. (2011). *Qatar National Development Strategy 2011~2016*. Doha, Qatar: Qatar General Secretariat for Development Planning.

Hall, P. (1997a). Modelling the post-industrial city. *Futures*, *29*(4/5), 311–322. doi:10.1016/S0016-3287(97)00013-X

Hall, P. (1997b). *The First Megacity Lecture: Megacities, World Cities and Global Cities*. Rotterdam, The Netherlands: Stichting Megacities.

Hall, R. E., Bowerman, B., Braverman, J., Taylor, J., Todosow, H., & von Wimmersperg, U. (2000). *The vision of a smart city*. In *2nd International Life Extension Technology Workshop*, Paris, France.

Henkel, M., Barth, J., Gremm, J., & Stock, W. G. (2018). Qatar National Library as part of a country-wide knowledge infrastructure. In *Proceedings of the 2018 International Conference on Library and Information Science*. Taipei, Taiwan: International Business Academics Consortium.

Höselbarth, F. (2010). *The Education Revolution in the Gulf*. Hildesheim, Germany: Georg Olms.

Kosior, A., Barth, J., Gremm, J., Mainka, A., & Stock, W. G. (2015). Imported expertise in world-class infrastuctures. The problematic development of knowledge cities in the Gulf region. *Journal of Information Science Theory and Practice*, *3*(3), 17–44. doi:10.1633/JISTaP.2015.3.3.2

Kuwait. (2015). *Kuwait Mid-Range Development Plan 2015/2016 – 2019/2020*. Kuwait City: Academic Press.

Landry, C. (2000). *The Creative City: A Toolkit for Urban Innovation*. London, UK: Earthscan.

Landry, C., & Bianchini, F. (1995). *The Creative City*. London, UK: Demos.

Likert, R. (1932). A technique for the measurement of attitudes. *Archives de Psychologie*, *22*(140), 3–55.

Mainka, A. (2018). 21st *Century Cities*. Berlin: De Gryuter.

Mainka, A., Hartmann, S., Orszullok, L., Peters, I., Stallmann, A., & Stock, W. G. (2013). Public libraries in the knowledge society. Core services of libraries in informational world cities. *Libri*, *63*(4), 295–319. doi:10.1515/libri-2013-0024

Moon, M. J. (2002). The evolution of e-government among municipalities: Rhetoric or reality? *Public Administration Review*, *62*(4), 424–433. doi:10.1111/0033-3352.00196

Murugadas, D., Vieten, S., Nikolic, J., Fietkiewicz, K. J., & Stock, W. G. (2015). Creativity and entrepreneurship in informational metropolitan regions. *Journal of Economic and Social Development*, *2*(1), 14–24.

Nam, T., & Pardo, T. A. (2011a). Conceptualizing smart city with dimensions of technology, people, and institutions. In *Proceedings of the 12th Annual International Conference on Digital Government Research* (pp. 282-291). New York, NY: ACM. 10.1145/2037556.2037602

Nam, T., & Pardo, T. A. (2011b). Smart city as urban innovation: Focusing on management, policy, and context. In E. Estevez, & M. Janssen (Eds.), *5th International Conference on Theory and Practice of Electronic Governance* (pp. 185-194). New York, NY: ACM. 10.1145/2072069.2072100

Nefiodow, L. A. (1991). *Der fünfte Kondratieff: Strategien zum Strukturwandel in Wirtschaft und Gesellschaft.* Wiesbaden, Germany: Gabler, FAZ.

Oldenburg, R. (1999). *The Great Good Place*. New York, NY: Marlowe & Company.

Oldenburg, R., & Brissett, D. (1982). The third place. *Qualitative Sociology*, *5*(4), 265–284. doi:10.1007/BF00986754

Parasuraman, A., Zeithaml, V. A., & Berry, L. L. (1988). SERVQUAL: A multiple-item scale for measuring consumer perceptions of service quality. *Journal of Retailing*, *64*(1), 12–40.

Rizzo, A. (2017). Why knowledge megaprojects will fail to transform Gulf countries in post-carbon economies: The case of Qatar. *Journal of Urban Technology*, *24*(3), 85–98. doi:10.1080/10630732.2017.1311569

Sassen, S. (2001). *The Global City: New York, London, Tokyo* (2nd ed.). Princeton, NJ: Princeton University Press. doi:10.1515/9781400847488

Sharjah. (2018). *Sharjah Economy.* Retrieved from https://www.sharjah.com/v/economy/

Stallmeyer, J. C. (2009). Landscapes of informational urbanism. *Journal of Landscape Architecture*, *6*(3), 34–39.

Stallmeyer, J. C. (2011). *Building Bangalore. Architecture and Urban Transformation in India's Silicon Valley*. London: Routledge.

Stock, W. G. (2011). Informational cities: Analysis and construction of cities in the knowledge society. *Journal of the American Society for Information Science and Technology*, *62*(5), 963–986. doi:10.1002/asi.21506

Stock, W. G. (2015). Informational urbanism. *Journal of Systemics, Cybernetics and Informatics*, *13*(6), 62–69.

Stock, W. G., & Stock, M. (2013). *Handbook of Information Science*. Berlin: De Gruyter Saur. doi:10.1515/9783110235005

Stryjakiewicz, T. (2010). Location factors of the creative and knowledge-intensive industries in European metropolitan regions. *Geografický Časopis. The Geographical Journal*, *62*(1), 3–19.

Tanfeedh. (2017). *The National Program for Enhancing Economic Diversification. TANFEEDH Handbook.* Muscat, Oman: National Program for Enhancing Economic Diversification.

Taylor, P. J. (2004). *World City Network: A Global Urban Analysis*. London, UK: Routledge.

Taylor, P. J., Hoyler, M., & Verbruggen, R. (2010). External urban relational process: Introducing central flow theory to complement central place theory. *Urban Studies (Edinburgh, Scotland)*, *47*(13), 2803–2818. doi:10.1177/0042098010377367

Thite, M. (2011). Smart cities: Implications of urban planning for human resource development. *Human Resource Development International*, *14*(5), 623–631. doi:10.1080/13678868.2011.618349

UAE. (2014). *Vision 2021*. Abu Dhabi: United in Ambition and Determination.

Van den Berg, L., van der Meer, J., & Otgaar, A. H. J. (2007). The attractive city: Catalyst of sustainable urban development. In *XVI Congreso de Estudios Vascos* (pp. 485-491). Bilbao, Spain: Academic Press.

Wiseman, A. W., Alromi, N. H., & Alshumrani, S. (2014). Challenges to creating an Arabian Gulf knowledge economy. In A. W. Wiseman, N. H. Alromi, & S. Alshumrani (Eds.), *Education for a Knowledge Society in Arabian Gulf Countries* (pp. 1–33). Bingley, UK: Emerald. doi:10.1108/S1479-367920140000024002

ADDITIONAL READING

Gremm, J., Barth, J., Fietkiewicz, K. J., & Stock, W. G. (2018). Transitioning Towards a Knowledge Society. Qatar as a Case Study. Cham: Springer Nature. doi:10.1007/978-3-319-71195-9

Kosior, A., Barth, J., Gremm, J., Mainka, A., & Stock, W. G. (2015). Imported expertise in world-class infrastuctures. The problematic development of knowledge cities in the Gulf region. *Journal of Information Science Theory and Practice*, *3*(3), 17–44. doi:10.1633/JISTaP.2015.3.3.2

Chapter 6
The National Plan of E-Government:
How the UAE Government Modernized Citizen Services

Badreya Al-Jenaibi
United Arab Emirates University, UAE

ABSTRACT

This chapter explores the readiness of federal authority employees and their current skills, adoption, and knowledge about e-government. It also examines the new technologies and e-technologies in GCC and UAE. The researcher has collected information and opinions from different areas in the country creating a survey analysis in this research. The study aims are to discover the experts among the employees, explore the developments of e-governments in the UAE federal agencies, gain information about the strategies of the services, and gain knowledge about training and challenges that face the employees who work in federal positions. Research questions include, Are federal organizations employees prepared to adopt and use the e-government? What are the struggles that face them?

INTRODUCTION

The concept of e-government has been introduced in many countries with the corporation of ICT to transform several systems to make them more reachable, effective, and efficient solutions for the government (Garson, 2006, Digital Government 2018). With financial crisis facing many countries, they transformed their agencies and departments using a "CITIZEN-CENTRIC"(Abhichandani, T. (2008, p. 1) model. Garson (2006) stated that it is expected that the quality of citizens' life would be enhanced and they will avail themselves more of government services. Miller and Walling (2013) added that the government wants to efficiently and effectively improve and deliver services to their citizens.

DOI: 10.4018/978-1-5225-7347-0.ch006

On the other side, Bertot, Jaegar, and McClure (2008) wrote that evidence shows it is quite difficult to achieve these goals and focusing on the needs of the citizens who are disconnected from the initiatives of the government. In this tough situation, the government needs to take an integrated approach to deliver the needs required by their citizens (Chan, 2017). Moon (2002) agreed that e-government programs have been conducted around the world and achieved success with them. It is not an easy task to transform the services of the government so the existing systems are completely automated and updated. This study focuses on the growth of e-government and its needed function within the UAE government. In this study, an analysis is also conducted about the current situation of e-government in the UAE and the administration's objective toward the transformation. It is a detailed approach that will address strategies related to technology and the awareness of the need of transitioning into e-government. It will also discuss why the global government needs e-government at this time in human history.

Smart or electronic government refers to the delivery of national or local authorized information and services via the Internet or other digital means to citizens and businesses or other governmental agencies (Garson, 2006). It facilitates in many ways better service delivery to citizens, empowerment of the people through access to information without the bureaucracy, and improved productivity and cost savings in doing business with suppliers and customers of the government (Al-Khouri, 2012). Radan, (2012) indicated that the smart government refers also to the improvements or streamlining in internal processes through changes to the risk approach taken, underpinning law or regulations, procedures, processes, forms, channels used, and job functions, that, in turn, improve information and service provision.

The UAE is one of the countries planning to embrace the concept of electronic governance. Mohammad Abdullah Al Gergawi, UAE Minister of Cabinet Affairs said: "The UAE government has embraced innovation and set high benchmarks in government efficiency and trust. We are happy that the Government Summit partnership with WEF has led to a tangible and positive outcome as the Smart Toolbox, which takes trust in government as a unifying theme" (Weinzierl, 2014, p. 1). Recently, Vice President and Prime Minister and Ruler of Dubai, His Highness Sheikh Mohammed bin Rashid Al Maktoum was briefed about the distinctive features and services in pending in the next period and steps for complementing transformation to the smart government by mid-2015 as per the strategic plan announced during the recent government summit. He said that, "The aim is to make smart services part of the daily life for all populations, and more accessible and direct. The more services that are smart, the happier people become" (Jivani, 2014, p. 2). The UAE has become one of the leading countries in this field and some of its citizens are still not familiar with this term of "smart" governance.

Research Aims

This topic was chosen according to its importance in developing the UAE and improving the efficiencies of the employees' performance in using the new services of e-Government. Furthermore, it is one of the latest achievements in the United Arab Emirates. In this research, the researcher will focus on the employees and the available experiences in UAE governmental arrangements used for e-government services, the main resources for e-government initiatives in the work place, discussions about the e-government development level, the training courses available in e-government, the challenges for e-government, and, finally, the importance of e-government and its rules and regulations.

The researcher has collected information and opinions from different areas in the country creating a survey analysis in this research. The study aims are to discover the experts among the employees, explore the developments of e-governments in the UAE federal agencies, gain information about the strategies of the services, and gain knowledge about training and challenges that face the employees who work in federal positions. Research questions focus on: Are federal organizations employees prepared to adopt and use the e-government? What are the struggles that face them?

Research Hypotheses are:

- Can e-government increase personal security for its people?
- What is the extent of providing security for the citizen when it is used for e-Government?
- Does providing e-government increase use of online services?
- Are UAE employees more confident and ready to use e-government than other nationalities?
- Is there a relationship between the cost of e-government and providing e-government services?
- What is the cost of developing e-government services?
- Is there a relationship between the IT skills and background among staff and their readiness for implementing e-Government?

LITERATURE REVIEW

Meaning of E-Government?

E-Government refers to “Electronic Government”; it’s a pioneering initiative that provides online electronic service from the government departments to their people. The e-Government project originally launched in the year 2000 by the order of His Highness Sheikh Mohammed Bin Rashid Al Maktoum, UAE Vice President & Prime Minister and Ruler of Dubai. In the year 2013, the name changed to the Dubai Smart Government. Currently, the e-government provides over 2000 services online throughout the UAE (Dubai Smart Government, 2014). Burn & Robins (2003) wrote that “E-Government can be defined as the use of Information and communication technology (ICT) to provide and improve government services by enabling electronic transactions and interaction between citizens, businesses, and other arms of government” (p. 25). E-Government is also designed “to provide public services, to improve managerial effectiveness and to promote democratic values; as well as a regulatory framework that facilitates information intensive initiatives and fosters the knowledge society”(Ramon, 2012, p. 17.).

Castells (2009) and Varian et al., (2005) agreed that Information and Communications Technologies (ICT) have affected the ways in which people, governments, and businesses interact with each other. The rapid diffusion of the Internet, mobile telephones, and broadband networks demonstrate how pervasive this technology has become (Al-Jenaibi, 2010). Today, ICT is considered one of the fundamental building blocks of modern societies and digital economies (Yaghi, Al-Jenaibi, 2017). So, the e-government is an important service and should be available in all countries to all citizens to accomplish their transactions in a fast and easier way.

Federal E-Government Tactical Structure

The UAE set a strategic framework for 2012–2014 for the transition in these three years:

The UAE vision 2021 envisages development of a knowledge-based economy that will be diverse and flexible led by skilled professional Emiratis. The vision contains four important components with detailed objectives related to national identity, economy, education, and health. It seeks to make the UAE a land of ambitious and confident people who hold on to their heritage; a strong federation; a competitive economy led by creative and knowledgeable Emiratis; and finally a high quality of life in a generous and sustainable environment (Al-Khouri, 2012, p.136).

This framework was developed in tandem to the other six reverences in the strategy: UAE Strategy 2011–2013, UAE Government ICT Strategy, Services Development Strategy, Federal government budget, Current Situation Analysis, and Benchmarking (Al-Jenaibi, 2011). Al-Khouri (2012) argued that cooperation is necessary to achieve the goal of the 38 initiatives set by the government, four of which covered the e-government area:

1. Strengthen the regulatory framework and governance mechanisms for e-government in the country.
2. Provide infrastructure support of information systems in the United Arab Emirates.
3. Launch and provide e-government applications and services.
4. Develop effective mechanisms for performance management.

The E-Government Accomplishments and Growths in the UAE

Al Suwaidi & Rajan (2013) discussed the failure factors of e-government in developing countries and regions. Lack of funding was the first failure factor and after that there were capacity deficits, poor change management, lack of a thoroughly developed strategy, poor technical infrastructure & lack of interoperability, privacy & security constraints, dominance of politics & self-interests, legalization & policy issues, inefficient leadership roles, and lack of e-service adoption. In the other hand, there were success factors, mainly good strategy & vision, followed by requisite competencies, effective change management, internal political desire, effective design, privacy & security, encouraging take-up of services, adequate technical infrastructure & interoperability, effective project management, and external pressure.

According to Radan (2012), the United Nations survey in 2012 stated that the UAE was among the emerging leaders transitioning in e-government development —in 2010 the UAE placed 49th and in 2012 it moved to 28th. The plans of the government services is improving and increasing year by year, making government information available to people electronically by an open data system with a format that is friendly and easy to use. Richard Kerby, interregional adviser on e-government at the United Nations, discussed the importance of government online websites to keep the people using and trusting their leadership, stating that goals include "making the website easy to navigate, having a multimedia design that will keep users interested, and displaying the information in multiple languages, so that foreigners, not only the local population, are able to make use of the services." (Al-Khaleej Times, 2012, p. 2)

With the continuity of the development of the e-Government in the UAE, His Highness Sheikh Mohammed bin Rashid Al Maktoum, Vice President and Prime Minister of the UAE and Ruler of Dubai announced the promotion of the "Mobile Government" that makes services easier to access by the people by their mobiles at any time, which will cost approximately 200 million dirhams. His Highness focused on delivering government services to citizens with the best communication methods to achieve a high quality life for UAE nationals as one of the visions of 2021:

The Mobile Government Forum looks forward to raising awareness among government entities to take advantage of mobile phone services by adopting innovation and the best technologies in the field to offer the highest level of services comparable to the private sector by sharing experiences with other outstanding entities in the world (ENA, 2013).

Development of E-Government in the UAE

When e-government and developed countries are kept into dual consideration, they provide ICTs to progress the living standards of their citizens. Many of the developing countries include Qatar, UAE, and Jordan, which are planning to apply proper strategy and infrastructure related to e-government. the focus is on several features like simplicity, integrity, and efficiency with the reduction in cost of services provided by government.

Initially, initiatives were not successful (Al Khouri, 2012). There are a few reasons for the failure, likely current infrastructure is not compatible economically and socially. Al-Shafi (2012) argued that the reasons for the failure arose after investigation, and included an absence of trust, skills, strategy, and interest from the citizens. This clearly shows that not only does the government need to take interest, but the people should be involved as well.

The people who make policies should look beyond these issues and adopt different models of e-government being used in developed countries to overcome the problem (Al-Shafi & Weerakkody, 2008). Al-Khouri (2013) stated that the government of United Arab Emirates is working on improving the services of public sector organizations so that they can achieve greater standards in the electronic amenities. E-government services rose in UAE from 2008 to 2012 according to the report of United Nations at the same time the development index increased from 0.571 to 0.734. Not only does the UAE come in at 28th position among all the developed countries, but they are 5th in Asia (Human Development Report 2014). It clearly shows that they are making efforts to lead in Asia.

Rapid development has been underway in UAE over the past 40 years, increasing alongside discovering oil and working for the progress of uniting the country together with seven emirates (World Economic Forum, 2007). Throughout these years they faced many issues related to this transformation, like cultural incompatibility, unskilled staff, and conflict with the citizens. Linnefell (2014) stated that this long progressive movement also started regarding implementing the e-government. This helps in seeking the attention of the policy-makers and authorities who understand that there is a need for different technologies regarding communication, electronics, and information. Mourtada & Al-Khatib (2014) call attention to the UAE government always being one step ahead in accepting the latest technologies so they can improve efficiency. Through these technologies, the government have improved their services and standards in sectors such as education, health, and judicial services.

The GCAA STRATEGIC PLAN (2012–2013) published that there were different guidelines related to the strategic build for e-government in UAE, which are as follows:

- Federal and local government should cooperate with one another and maintain healthy relations
- Keep updating the policies along with the progress of decision-making
- Develop human resources who should be well trained
- All rules and regulations should be updated and reviewed
- Ministries should be authorized with their duties and managing the activities along with public and joint policy-makers.

The UAE has gone through several different stages of development to implement e-government. There are 77 million Internet users in the region and most of them are social network users depending on digital communication (Al-Khouri, 2013). The government is very intense about adopting e-government services, as they want to compete with other nations without depending on the oil, products, and services provided by them (Al-Jenaibi, 2017). Flanagan (2011) agreed that, in this advanced time, the UAE is among the nations who has hold top position in an Information Technology world. They are among those nations whose governments invest in, adopt, and achieve the ICT which needs to be implemented in private and public sectors. The UAE has developed a framework for strategies for federal e-government for the year 2012–2014 fulfilling several initiatives over the three years. This framework had two aims:

- Firstly, UAE has to achieve the position as one of the best countries with ICT implementation
- Secondly, the focus remained on the citizens to develop an innovative government. (UAE government, 2014)

Profits of E-Government in the Community

E-government is beneficial for the improvement of the services provided to the people. According to a Commonwealth of Australia (2003) survey, it is found that 80% of the people responded to the benefits they achieved from e-government. It makes life easier for the citizens as they can easily retrieve the information they want to discover (Nixon, Rawal & Rajesh, 2011). E-government is favorable for the citizens, businesses, and the commonwealth, the advantages of which are discussed in the detail below:

- **Citizens:** Shareef, Kumar & Dwivedi (2011) wrote that the features provided by the e-government to the citizens are as follows:
- **Convenience:** Citizens can easily access the services of the government from their homes and offices without rushing to reach an outside office with limited office hours (Bwalya & Healy, 2010). It also facilitates public services and facilities according to their venue and schedules (Rorissa, Demissie, & Pardo, 2011). Citizens can easily access their services with their mobile devices or through the Internet.
- **Upgraded Customer Service:** E-government enhances and deploys the resources from the back end to process the customer service on the front end (Rokhman, 2011). Electronic self-service is introduced, which can interact with the customer according to their routine. The simple tasks that require effort and more time have been reduced, saving energy (Sipior, Ward & Connolly, 2011). When customers require assistance for any difficult tasks, the system can provide virtual and skilled assistance (Bwalya & Healy, 2010).
- **Increase in the Accessibility of the Information:** Rokhman (2011) agreed that e-government has made progress in allowing citizens access to the information of the government.
- **Businesses:** the following are benefits provided to businesses by e-government:
- **Reduction in the Cost of Conducting Business:** There is a decrease in the cost of conducting business regarding all interactions involving with the government (Hussein, Mohamed, Rahman Ahlan & Mahmud, 2011).
- **Accessibility is Improved:** E-government facilitates greater accessibility of information to businesses in a more effective and efficient manner (Yonazi, Sol & Boonstra, 2010). Business is more successful when the proper plan is made. Business plans are created from the data provided and

predictions are made for the future. The business more likely achieves success if governmental data is accessible to the businesses (Bhattacharya, Gulla & Gupta, 2012).

- **Commonwealth:** According to Backus (2001) the picture of the commonwealth can be enhanced with the help of e-government. E-government provides benefits with a rise of efficiency and standards discussed in detail below:
- **Efficiency Increases:** Productivity increases when the e-government helps the commonwealth. It also assists employees in highlighting the serious activities and all routines can be managed and handled electronically.
- **Rise in the Standards:** Because all the transactions and interactions are conducted online, it reduces paperwork and automates the tasks of the citizens (Beynon-Davies, 2007). The citizens can more easily receive responses from the government through electronic applications. It provides facility to the commonwealth as the nation can easily manage their workforce and work on problems faced by the citizens during the processing of data.

Necessity of E-Government in the Government

E-government is defined as utilizing information technology that is supportive toward the operations of the government and provides the services of the government to the citizens (Castells, 2009). It is needed as for the following reasons:

1. **Effectiveness:** Related to the productivity, e-government also associates efficiency with a reduction of costs. Reduction in cost is one of the main factors for ICT used by the government. According to Simaanya (2014), the following are the ways through which efficiency can be achieved:
    a. Paperwork is replaced by online applications and all manual work is computerized
    b. Files are managed efficiently and less workforce is used
    c. Data is shared with the government
    d. Most of the newsletters and publications are distributed online
2. **Steering Customers:** The government is comparable to a business who focuses on customers and works on strategies; by switching to an e-government (Flak, Nordheim, & Munkvoid, 2008), software is created that creates focus groups and appropriate services are provided to them.
3. **Consequences of Policies:** After implementing e-government and policies, different results will be achieved:
    a. Tax is collected easily as the information is shared by the agencies
    b. Health information is beneficial and demand of health services is decreased
    c. Information must be shared among local and federal governments, assisting inter-agency issues (Cournède, Goujard & Pina, 2013).
4. **Achieves the Needs of the Economy:** E-government reduces fraud within agency services; resolution reestablishes trust. Different online tools are used for improving the productivity of government interactions, and this will better support small businesses as well (Hung, Chang & Yu, 2006).). Information related to the government can be accessible via portals and the ICT tools used by the government provide more efficiency and stability than the private sector.
5. **Attractive With the Citizen:** E-government helps in establishing trust between the government and its citizens. Through this process, corruption is reduced and accountability is promoted. Users of the services can easily consult and give feedback through contact via email links, surveys, blogs, and

questionnaires (Heeks, 2005). Citizens are involved in making policies and constitutions, and they participate actively because their votes matter (Simaanya (2014). The voice of every individual is heard (Smart Dubai Initiative, 2017).). E-government is used as a tool that simplifies the administration processes, makes processes more transparent, and delivers services in a more efficient and effective way to better unite the government.

Readiness of UAE Organizations of E-Government

Lootah & Geray (2006) believed that over the last two years, we can easily observe that e-government has grown significantly. At this stage, we need to conduct research on the issues faced by the organizations and study the effectiveness of e-government. Nowadays, tools of e-readiness used in certain organizations fail to address developing problems (Niaros, Kostakis, & Drechsler, 2017). These tools of e-readiness also don't focus on the considerations of e-government. Most of these tools assess e-services, availability, and its use for the ICT. Research is important to conduct which tools will offer the best framework for adopting e-government and provide the most assistance to the ICT managers in improving readiness in the public sector.

Al-Khouri (2012) agreed that identity management should be introduced in the UAE for the personal identity of every individual. It is divided into three sections: Identity cards issued to individuals, a structure of public keys, and the united management of identity. E-government services are dependent on the authentication of user. A passcode can be provided to the users for authentication, as is seen in other countries like Singapore and the UK. In some countries, a token system has been introduced, similar to smart cards. A national ID program introduced in UAE provides the same functionality. These ID cards have authentication keys used for signing different documents. E-government welcomes an authentication architecture providing flexibility. It can also be considered a digital signature, with their legal weight equivalent to a written one.

The infrastructure of the public key provides authentication and non-repudiation. It covers the major issues like privacy, identity, and trust. There is also a feature provided by ID cards called the Federated Identity Management (FIM). It provides a single-sign-on, which means citizens can sign in and easily avail themselves of various services (Dubai eGovernment, 2013). The federal and local government provides the services to their citizens with their websites. Users are authorized to use databases and facilitate different functions like validating certificates or authenticating applications and all these tasks are easily conducted with an ID card.

E-Government and the Uprising and Changes in the UAE Administration and Organization

According to the government's point of view, e-government implementation is delivering an undeniably necessary job. Flanagan (2011), stated that the e-government will assist the function of government and the allocation of legislative information and services. The main purpose of e-government is to potentially support an improved portfolio of community services with a disciplined and cost worthy approach. The estimated advantages of e-government comprise of improved services, efficiency, enhanced accessibility of local services, and more simplicity and responsibility for government workers.

Jivani (2014) argued that E-government is about how we use ICT to present accessibility to legislative information and provide communal services to society and business followers. However, the users have not discovered how to take the most advantage of the power of this technology. Presently there is an issue related to its stability from its applications and restrictions to perceptibly defend the enormous investments in the ICT scheme for the past years (Mourtada & Alkhatib, 2014). The standard civic opportunities regarding the government's endeavors are structured according to the capability of the administration to effectively progress its public's quality of life. Administrations require that their strategy, convention, and system permit civilian contribution and deal with the desires of progressing regarding to service provisions. The life cycle of service delivery needs to be engineered and planned in order to fulfill the public's belief in improved public security and excellence of life. The responsibility of the government strategy is to form a much better public-centric and determined government.

Government policies should enable major governmental modifications which:

(1) Progress development in services, (2) decrease redundant costs and workload on organizations, (3) fortify learning and education organizations, (4) promote excellence in administration performance, (5) promote advancement and new functions, (6) Promote market circumstances and build a business atmosphere for the purpose of encouraging productive financial market.

Sponsors of e-government identify prospects for society to participate in a better role in community strategy. They also pressure the system's capabilities to link them, rapidly and straightforwardly, to full services—no need to wait your turn, no queues, 24/7 service. Deducting the cost is the main issue to address making assessments to go online. Developed e-government has the ability to deduct expenses by as much as 90%, through updated infrastructures and integrated organizations that decrease governmental load and development time sequences, and enhance responsiveness (Dubai eGovernment, 2013). Moreover, evaluating conventional over-the-counter accommodations, online facilities can decrease extensive substantial expenses as e-governmen does not require a physical location, public access, electricity, desk services, and so on.

Certainly, ICT presents the prospective for growth and struggle in the community sector, particularly in the region of consumer service and generally in the organizational quality plan. Such antagonism aggravates decreasing the expenses of government services by computerization and automation, as well as sustaining stress on firms to enhance performance and transform from an conventional approach (Al-Khouri, 2013). The private zone has always challenged the public zone and cooperation is necessary a mechanism for enhanced quality and for continued effectual resources consumption. Improved automation in the public zone encourages more levels of stability among the public and private sectors.

Al-Khouri (2012) stated that government organizations and public zone organization specifically are giving more consideration to basic potentials and contracting out other sustainable tasks to be conveyed by the private zone. For that purpose, ICT has participated as an essential role in serving the government to accomplish outstanding output achievements. Conversely, and regardless of high expenses and the pervasive implementation of complicated ICT infrastructure, a lot of other countries persist in hanging back from committing to the main procedures of financial development and efficiency. Government speculations in ICT up until now have been extremely concentrated on governmental validations, financial prudence, and service transformation without focusing on building a public-oriented scheme that encourages and supports civilian contribution.

The main insufficiency in such attempts is that they have been considered and implemented from a government point of view instead of depending on public requirements and assumptions (Heeks, 2005). Such a constricted vision of e-government needs to report the ICT successions with a dubious eye. If steps are implemented to deal with other side of society and authority without any support, e-government may generate backing, especially if there is any net increase in leverage ICT to decrease and reform organizational systems and service conveying systems. It was recommended by other researchers that governments take on new methods that entrench a logical transformation supporting the design and estimation of public zone organizations. This is visualized to have significant suggestive abilities for improvising the services provided by public organization, and severe outcomes for the public value related to the provided service. In Middle East, government is considered as the way to produce a more improved economy (Serageldin, 2011). It is considered to possess a fundamental role in organizing and routing the method of change and transformation, which will enhance public reliance. Developing confidence in e-government is a complicated issue. Appropriate literature illustrates that there are vast apprehensions about the capabilities of digital networks to unenthusiastically impact public confidentiality and safety.

Reassuring Local to Implement E-Government

Trust is most likely to be one of the most significant features in accomplishing e-government policies. To achieve the desired objectives for e-government of producing and providing excellence in quality and assimilated public services, the people should trust the implicit atmosphere (Seres, 2010). Without believing in the government, people will not contribute in the success of e-government. A literature review and experiential studies of e-government determines the decisive factors for the implementation of e-government from both a citizen's and government's point of view is trust and protection. Tassabehji & Elliman (2006) added that an experiential study identifies that the level of trust is a steady process that occurs over time instead of a cumulative procedure. There are a number of overlying and reliable factors that have the capability to effect trust-building. These are categorized into two main groups: pre-interactional and interactional aspect.

For the adopting e-government services successfully, the public must have an interest to take on e-government services, which includes the purpose of collecting and giving information by the means of online channels. According to the UAE government (2014), currently, the people are looking forward to online governmental services and demanding services that match those of the private sector in each phase of quality, availability, and quantity. Such demands and requirements made by the citizens exert more pressure on governments to produce services with high quality and provide an efficient and effective system. The difficulty occurs when the public takes on multiple roles during government interactions (Serageldin, 2011). Single role-based characteristics are diminishingly appropriate in present government dealings. This forces governments to gain qualities that are citizen-centric and deliver services and resources modified to the authentic service and resource requirements of the users whether they are citizens, inhabitants, government workers, business associates, or a combination of any of the above.

The UAE is one of the highest empowering governments in implementing developing ICT in its private and governmental areas. The policy of seven steps was implemented by the government, consisting of benchmarks that included e-government practices adopted internationally in Canada, USA, Southern Europe, Singapore, the European Union, and GCC countries. The result of this attempt was the current description and prioritization of the plan and the main focus regions. The growth approach has been implemented in the three major dimensions of e-Services, e-Readiness, and the ICT atmosphere.

The e-Service dimension is related to speeding up the velocity of e-Transformation within government associations and the terms of better quality electronic services by modern delivery channels such as the Internet, fixed and mobile phones, and kiosks, as well as the conventional service centers.

Karanasios (2011) believed that E-Readiness is concerned with intensifying the potential of federal organizations in terms of ICT, institution arrangement, HR potential and capabilities, and their eagerness for e-Transformation. The ICT atmosphere aspect includes organizational aspects like policies and the legislation required for maintaining the execution of e-government initiatives. This has produced the progress of five strategic objectives.

The Substructure of National Identity Management Made by UAE

The purpose of this infrastructure is to improve safety in the homeland and build a particular structure to manage identities and protect the transactions of e-government.

A federated identity is the preferred way of connecting a person's electronic identity and characteristics, saved in multiple identity management systems (Meena, 2009). Such structures would permit everyone to use the identical username, password, or other individual credentials to log in to the networks of more than one venture for the purpose to carry out transactions.

Al-Khouri (2012) wrote that being a quintessential part of strategy, smart identity cards are provided to all citizens. These national identity cards of the UAE are considered greatly developed and smart cards with sufficient protection. The card is presented with credentials saved directly and securely in the smart chip. It therefore allows the creation of a person's identity on-site and virtually, allowing protected and trusted dealings (Karanasios, (2011). The multi-feature validation, which provides both match-on-card4 and match-off-card5 features, facilitates authentication and confirmation of any specified identity. Then the cardholder is allowed to use all the identity-based services. There are several features of a UAE ID card which are as follows:

1. It can be recognized on site
2. It can be recognized remotely
3. It can get verified through different applications (Al-Khouri, 2012)

It needs verification by PIN number, authenticating biometrically and through digital signatures. This system eliminates identifying users through numerous systems. To meet the conditions of successful e-government, this system should be simplified and easy to access. At the present time, the UAE is implementing the identity management system and providing smart cards in public regional areas. Some of the systems are already deployed to handle the card, which includes most e-Gate services located at airports (Al-Khouri, 2008). It provides to card holders the capability to easily immigrate with the authentication of biometrics. People living in Dubai can access and sign in to online local government applications and use different e-services. An Abu-Dhabi portal presents different services, like displaying individual profile to the Abu Dhabi Police.

Al-Khouri (2012), added that It has been forecast that all e-government services will be accessed by signing up for the UAE identity card and PIN. The national identity card can be used by both local and federal authorities. The structure of the UAE federated identity management system ensures trustworthy and protected access from different locations, and therefore presents superior mobility. The main strategy to implement e-government in UAE is to provide services to the public through various applications and mediums. The main aim of national identity card is establish a building block to deploy e-government and e-services in the country successfully. In a period of rising digital connectivity, governments are concentrating on communication with their people inside the virtual world. The governments recognize that conventional physical systems are now inadequate and there is an obvious necessity to build new facilities to recognize electronic identities.

The government of the UAE is determined, as an element of its national development policy, to possess a secure, distinctive, and tamper-proof digital reorganization for their citizens. Such identity management systems possessed by the national government would achieve higher levels of faith, assurance, and support contribution (UAE government, 2014). This identity management system can eliminate all duplicate databases used to identify users. The system developed by the government does not require any procedures to process the transactions or need to establish relationships with the customer.

The UAE e-government proposals will flourish when people can go beyond the physical limitations to accomplish their transactions. A citizen should have the ability to use his identity card to process his e-government and e-commerce transactions through websites. To the scope that the UAE federated characteristics permit government organization to present restricted access to information or other resources, e-government has the prospective to allow new stages of association between several organizations. An identity management system can be re-engineered and important resources can be expanded with the use of multi-feature systems. Private and government systems are merging to cover the wider aspects. The government and e-commerce applications should be in regional spheres.

E-Government: The Power of Technology

E-Government in its simplest form is about the use of ICT to provide access to governmental information and deliver public services to citizens and business partners. However, practitioners have still not yet figured out how to exploit its full benefits. There is an equilibrium problem between e-government applications and limitations arising from the difficulty to tangibly justify the gigantic investments necessary in ICT systems.

The average public expectations concerning governments' efforts are shaped according to the ability of the government to successfully improve citizens' quality of life (Karanasios, 2011). Governments need to ensure that their policies, regulations, and systems enable citizen participation and address the needs of improving the delivery of services. The service delivery lifecycle needs to be reengineered and redesigned so as to meet citizen's expectations of enhanced social security and overall satisfaction.

Government policies should enable governments to undertake radical organizational changes that:

1. Foster growth in services
2. Reduce unnecessary costs and regulatory burdens on firms
3. Strengthen education and training systems
4. Encourage good management practices
5. Foster innovation and new applications
6. Foster market conditions and create a business environment that promotes productive economy (Al-Khouri, A. 2012).

Questions and Answers About e-Government in the UAE

What are the current employee experiences among the UAE governmental authorities using e-government services?

According to the UAE e-government strategy (2014), it aims to establish an effective and reliable information technology infrastructure and provide electronic services to customers through multiple innovative electronic channels in line with the UAE Vision 2021 and the UAE Government Strategy 2012–2014. Therefore, based on the study done by researchers, there is a wide range of experiences that the governmental employees might have gained and earned from their work so far in this field. Some employees might have global experience gained by working with international institutes or organizations like the European Union or United Nations (Al-Khouri, 2012). Researchers define the international experience as the experience gained by working with different sectors in different countries. So, international experience could be gained after working in different countries. On the other hand, the local experience is gained by working with local authorities or sectors. Some researchers reclassify a local experience into more detail, like experience with local government. So in this type, experience is only related to the work done with a governmental utility in the Emirate, Province, or State. Another specialized experience is related to working in designing the e-government. Other types could be those from working in the private sector that provides services to the e-government. Researchers also consider using the governmental e-services as experience in the e-government. Therefore, all UAE citizens who have used any e-governmental services are also targeted in the research. In addition, all students and researchers who have conducted any kind of studies in this field are also targeted and considered. In order to include all sides and assure that the research is comprehensive, researchers included the consultation in the e-government as part of the experience.

What are the main resources for e-government initiatives in the workplace?

Researchers like Goldkuhl (2007) have a focused interest in the e-government initiative resources in governmental workplaces. Researchers can examine the e-government strategy of utilizing both centralization and decentralization to help government departments optimally employ their resources and provide online services to customers in a cost-effective manner for all parties (Karlsson,Holgersson, Söderström & Hedström, 2012). Researchers like Al-Jenaibi (2012). consider that federal ministries, local governments, private sectors, and technological innovations are the possible initiative resources at the governmental workplaces.

Is the e-government development level excellent?

The development of the government is part of a comprehensive plan of efforts for several government agencies and aims to develop government services in the UAE (Al-Khouri, 2013) . As-Saber, Hossain, and Srivastava (2007) believed that in order to ensure correct development involving integration and coordination between these efforts, the researchers measure the development process according to several levels, the international level, the national level, the federal departmental level and the local level, including the municipalities and local authorities.

What are the challenges for the e-government?

Researchers have discussed many challenges that employees of the governmental sector face. Goldkuhl (2006) believed that the main goal of e-government is to be able to provide an increased bunch of public services to the citizens in cost-effective manner. The first major challenge is finances that include the cost of developing the e-services in addition to the cost of providing the services. According to Holgersson, Söderström, Karlsson, and Hedström (2010), Another major challenge type is the issues related to the internet and IT. They focus on the related issues to the employees as well as the citizens. The third major challenge type is technical and may include the verity of spoken languages in the UAE. Also, the ease and the hardness of using those services by the citizens were studied by the researchers.

What is the importance of the e-government and its rules and regulations?

According to the UAE federal e-government Strategy of the United Arab Emirates 2012–2014,

The federal government surpassed other Arab countries in issuing the first two laws that support e-Transformation: i.e. the law on Electronic Commerce and Transactions and the Prevention of Information Technology Crimes. The first law relates to the use of electronic documents and e-Commerce and aims at providing a safe environment and legal protection against fraud and other crimes, while the second law is designed to combat IT crimes and precisely identifies the crimes punishable by law and the penalties for each crime. On the other hand, there is a need to develop legislations and laws in certain areas such as freedom of information and data privacy. (p. 16)

So from this quote, the researchers feel certain the UAE federal government has issued enough laws to regulate the e-government services. Therefore, the researchers concentrated on this important issue of the e-government. In this part, the knowledge level of the employees regarding the government laws, rules, and organizations was questioned.

Method

In most surveys, data is collected for a wide range of variables, which may be qualitative or quantitative. In this research, the quantitative method of collecting data was used. Quantitative research explains phenomena by collecting numerical data that is analyzed using mathematically based methods, which has many advantages . First, it can generalize research findings when the data is based on random samples of sufficient size. Second, it is useful for studying large numbers of people. Also, data collection using quantitative methods is relatively quick. Moreover, the researcher may construct a situation that eliminates the confounding influence of many variables, allowing one to more credibly establish cause-and-effect relationships. Finally, data analysis is relatively less time-consuming (using statistical software). The software used is SPSS Statistics. The researcher distributed 500 surveys and received 407 from the following authorities in the country: Red Crescent, SCA (Securities & Commodities Authority), IA Insurance Authority, Tanmia, General Authority of Youth and Sport, Awqaf, Etisalat, UAE Identity, Customs, Federal Electricity and Water Authority, General Civil Aviation Authority, Gpssa, and NTA transportation.

According to Kraut et al. (2004), the survey has advantages that made the researcher choose this method—it is a cost-effective and efficient means of gathering information estimating the characteristics of a target population without interviewing them. Fan and Yan, (2010) argued that other methods such as face-to-face interviews need more time and may need recording devices that are expensive, require time and access to the interviewee, results can't be generalized, and visibility can lead to interviewer influence, when the interviewer inadvertently guides the interviewee in a specific direction (Opdenakker, 2006). The survey had two types of questions: close-ended questions and open-ended questions. The survey includes many questions for the employees of the institutions about the kind of training they received from the country in using e-government services, as well as their rating and opinion about applying e-government in the UAE, and the challenges they faced in skills, technologies, and cost of these electronic services. The researcher used the IBM SPSS Statistics 21.0 for analyzing the data of the survey.

Hesse-Biber (2010) argues that one of the best research practices is to employ qualitative and quantitative methods, which improves the availability and accessibility of information included in the research. Therefore, the researcher has integrated both methods of survey and interviews in this study. Those methods facilitate the perception of the research to its readers (Frechtling 1997). It also help in the clarification of the context, philosophy and aim of the text. The survey provides measurable quantities; it is related to numbers and statistics. On the other hand, interviews neither present any assessment or evaluation for the hypotheses nor deal with any relation among variables (Mertens, 2009). In addition, the outputs of the survey are more credible and transparent as they were obtained from the people who actually use the smart public services. Differently, the interviews were more comprehensive, as they were carried out with members of the federal authorities. The reason why federal authorities were involved in the study is that the government compelled them to follow the Smart Government system; all federal agencies should apply that system by the end of 2015. The study is carried out on only those who use or belong to the Smart Government system at the time, in order to reach the best results possible. The researcher attempts to closely investigate the merits and demerits of, and the restrictions on the Smart Government through hearing from the people who use to it and those who benefit from it. With interviews, the researcher is able to collect data and reliable information from the inside of the system, so that the ethos of the system could be clarified and properly introduced in the study. Clarke (2009) maintains Hesse-Biber's argument that making use of both qualitative and quantitative methods makes it easier for the researcher to collect data and communicate with a wide range of participants. As for this study, the participants would be the users of public smart services in the United Arab Emirates, whether employees or clients. While in an interview, the research is able to examine and interpret the interviewee's emotions and reactions. That is what is missing in the distance surveys, i.e. the researcher cannot ever figure out the emotions and reactions of the individual who answers the survey (Wyse, et al, 2016.(However, one advantage of surveys is that they make it easier for participants to share information and opinions, being unseen and maybe even anonymous. In order to assess and evaluate the situation of the system, the researcher interviewed a number of managers, as they are the ones responsible for applying the strategy and policies of Smart Government. Here are some of the federal authorities that some managers of which were interviewed: Emirate's Red Crescent, Insurance Authority, Emirate's Identity Authority, Federal Authority for Government Human Resources, Federal Electricity and Water Authority, Telecommunication Regulatory Authority, General Authority of Islamic Affairs and Endowments, Federal Transport Authority, and the Federal Customs Authority. Those authorities are a part of a total of eighteen federal authority in the United Arab Emirates. The researcher was willing to contact all the authorities, however it was not possible and many managers had not cooperate due to being very busy working. To help the researcher gather information, the researcher assigned five extensively trained students of the University of United Arab Emirates who were paid to obtain information and question interviewees.

Among the interviews, ten were carried out on the phone while eight were conducted in person. The interviewees consisted of 2 females and 16 male, their ages were 38-55 years old. Three of them are master's degree holders, while fifteen are bachelor's degree holders. Interviews took place at the cities of Abu Dhabi and Dubai, where most of the UAE federal authorities' offices are located. Each interview lasted for 15-20 minutes. None of the interviews was recorded, yet all the information was written down. The data was collected in March and April of 2017. There were not any differences in the course of all of the interviews; all revolved around the purposes, merits, obstacles, suggestions, and recommendations related to the smart government system. The original language of the interviews is Arabic and then they were translated into English. The analysis was based on certain points of discussion and coded identities of managers. The researcher and his team were not able to reach 4 of the 18 federal authority in United Arab Emirates due to lack of sufficient number of clients and services.

Qualitative Data Analysis

The analysis was based on the following points: purposes, merits, obstacles, comments, suggestions, recommendations related to the smart government system application. The identity of each manager was coded with a number, as it is better for the study that all managers remain anonymous.

Purpose of the Application of Smart Government in UAE

Interviewees agreed on that each authority exerts its utmost effort to integrate technology into their work and services. "For example, the Red Crescent Authority provides a charitable service in which donations are gathered online; donations are not exclusively made at the authority's place, but from anywhere" (Interviewe #1, 2017). Then, interviewee #5 comments that the authority tries to convince and induce the public to make use of the technology it offers, which will make it easier for them to perform whatever they want from any place at any time. Every single authority must use the Smart Government system and its applications according to schedules set by the government to be followed. In addition, every authority must assign a team working on the implementation and application of the system as well as following-up its progress, in order to present the greatest benefit for the public.

Here are some interviewee's comments on how far the authorities have changed and improved:

1. Federal authorities is taking a huge leap towards technology, as this system is most affective in terms of saving time and number of employees in customer service, so it is considered a major improvement.
2. It is most beneficent for the clients. Being online, it offers the clients great comfort in exchange for rapid response and effective service.
3. Smart Government system is a cutting-edge technology that will be exploited in the advancement of many states including United Arab Emirates. It is a great benefit for the public as well as the authorities, and it was also a great success with the clients in solving their issues. United Arab Emirates proved its superiority regarding the application of the system when it surpassed many other states applying the same system.
4. It is a great deal for the authority and its clients; less time, less paper work, less effort, less congestion, etc.

The interviewees considered that this system to turn the public from mere receivers into participants, and the researcher believes that it is a method that inspires for more positivity to come. The system does not only serve customers, but the whole community with all its groups. This technology will lead the first steps towards a new era beginning with applying Smart Government's technological methods on the field of education. Then follows other fields and other technologies, because obviously we cannot dispense with technology nowadays. It indeed helps the government with many of its work. The government is doing its best to serve the public, so "it had to impose the Smart Government system on all federal authorities and they all have to follow" (Interviewee #9, 2017) .The customers are already using the Smart Government applications so much that they have become more experienced in them more than the employees themselves have. Interviewee #11 further commented, "the community is changing; you can see kids with iPads and they know how to access and download programs." Four managers believed that technology has to satisfy the requirements of society, and in turn, society has to cope up with technology, and it should be easy with the presence of all these smart devices. Therefore, it was the perfect time to apply this technology. "If we did not apply as many technologies as we can, we would not be able to cope up with all these new technologies appearing every day" (Interviewee #18, 2017). Internet has been a part of most of the people's lifestyle, so most of the clients are used to speed access to whatever they want, so why not introduce to them the Smart Government system. It was agreed upon that federal authority is devoted to please their customers and that it easily done with the new technology applied.

Assessment of the Application of Smart Government in UAE

Each authority had different outcomes for the assessment The authorities have advertised their new smart applications in various ways, whether online or on the ground. The authorities "should also benefit from those who has an experience in using smart applications" (Interviewee #16, 2017). In addition, there is shortness of knowledge, expertise and personnel. The reason is that most of employees are new to this kind of technology, and they cannot professionally deal with those applications. There are also many clients who do not favor using applications, but they prefer to deal with customer service in person (Interviewee #10, 2017). Interviewee #7 maintained the opinion of Interviewee #10 adding that all parties should share responsibility for learning and using this technology.

One challenge that faces the application of Smart government technology is the employees aged 50 or over does not know how to use it or even refuse to do so; However, there is a training available for them, and some of those employees accepted to take a training. Another challenge is that the principles and philosophy of Smart Government is not yet clear to the government itself. The researcher received positive feedback from all the interviewees; they all find Smart Government beneficial and user-friendly (even if not, there is a training available. They also expressed their desire for the application of more similar systems. Here are the main points discussed in the interview:

- Absence of clear purposes The decentralized management structure
- Agreement on the proposals regarding Smart Government (Interviewee #14,2017).
- Limited number of experts in Smart Government
- Lack of knowledge and awareness by the public and the employees
- Unclear perception of Smart Government by the government itself (Interviewee #1, 2017).
- Lack of knowledge of technology and smart applications by the government
- Lack of instructions regarding the use of Smart Government
- Lack of knowledge regarding the services offered by Smart Government (Interviewee #17,2017)

Eleven interviewees believed in the necessity of a cooperation arising between the private and public sectors for the sake of achieving advancement in Smart Government. Furthermore, All Interviewees agreed on the application of Smart Government; however, the government should settle the challenges against its application first.

Positive Feedback on Smart Government

According to the employees, there are many merits of Smart Government. Yet, managers tended to criticize the absence of centralized management to return to. Smart technologies are widely accepted and used in United Arab Emirates because of their speed and effectiveness, and the Smart Government technology is not an exception. Also, there is improved progress at workplace and speed. The employees of each authority cooperate and do their best to satisfy the requirements of the government regarding Smart Government system. In addition, "private and public sectors also cooperate not only for the advancement of Smart Government but also because it brings them an interest" (Interviewee #3, 2017). Smart government should widens its network to include more areas, thus increase the number of users who benefit from the services and privileges it offers. Smart Government is not exclusive to official communications and applications, but it contains every single information or data in the authority. Interviewee #4 even suggested that the authority dedicates a workshop for the employees qualified to use these applications, or even train and qualify some employees to work at these workshops. The choices and services offered to the customer online "does not take any effort that is worth of mention compared to the procedures followed in the authority's place" (Interviewee #5, 2017)). Interviewees #15, #17 and #18 agreed on the following features of Smart Government:

- Has a clear work plan
- Its applications are quickly-downloaded, easily-accessed, and user-friendly
- It is considered a portal and a digital library for all official documents, data, and information.

Suggestions for Spreading Smart Government Made by the Employees

1. To create a centralized management represented by a department that could be returned to if needed.
2. To clarify the principles, policies, and strategies of Smart Government
3. To impose regulations for Smart Government
4. To unite the government bodies and institutions by a system
5. To train employees to be qualified for using Smart Government
6. To clarify the vision, purposes, and rules of using Smart Government in federal authorities
7. To cooperate with the organizations related to the public sector
8. "To develop a definite plan that suits the system's purposes and method of implantation" (Interviewee #5,2017).
9. To present adequate information according to a work schedule
10. To bring the employees up to date on the latest software
11. To train the government employees
12. To set and define the cost of the Smart Government
13. To Create a body responsible for referring all data and information to Abu Dhabi Government
14. To provide the employees working on field with smart devices containing Smart Government applications

15. To study the experiences of other states and use it in improving the Smart Government system in United Arab Emirates

Descriptive Results (Demographic Information)

Table 1 showed the demographic information for the research sample: their gender, age, nationality, location, and level of education. Result shows that more than half of respondents were male (55%),45% of them were female. Among respondents, 59% were between 26 to 35 years old, followed by 20–25 years (24%), and above 35 (17%). In terms of education level, more than half the respondents hold a bachelor's degree as a highest level of education (60%), followed by high school graduates (29%).

Table 1. Respondents' profile

	Frequency	Percent
Gender		
Female	184	45.2
Male	223	54.8
Age		
20–25	97	23.8
26–35	241	59.2
above 35	69	17.0
Nationality		
United Arab Emirates	295	72.5
Arab	101	24.8
others	11	2.7
Emirate		
Abu-Dhabi	144	35.4
Dubai	102	25.1
Sharjah	24	5.9
Ajman	35	8.6
Uma-AlQuwain	32	7.9
Fujurah	19	4.7
Ras-Alkhimah	51	12.5
Education Level		
High school	117	28.7
Diploma	19	4.7
Bachelor	247	60.7
Master	22	5.4
PhD	2	.5

The length of work experience varies among employees, from 1–4 years (48.2%), 5–9 years (38.8%), and the rest more than 9 years. Employees were asked how many workers are under their supervision, and more than 63% of them have 1–4 employees, 19.4% have 4–9, and 17% have more than 10 employees. From the respondents, 43.7% work at an office, 31.9% participate in field work, and 24.3% work in both the office and in the field.

Descriptive Results (Variables)

In the following Table 2, descriptive analyses for two sub-variables (challenges of e-government and readiness for adopting e-government) together with eight constructs have been presented. Constructs measuring "challenges of e-government" use the five-point scale of agreement (1: Not a challenge to 5: Very important challenges), and constructs of "readiness for adopting of e-government" were measured using a five-point scale of agreement (1: strongly agree to 5: strongly disagree). Two variables and its constructs, together with questions measuring the variables, are presented in Table 3. All items are sorted based on highest to lowest mean.

Table 2. Details of questions measuring variables

Variables (Constructs)	Questions
Challenges of e-Government	
Skills and background about e-government	IT skills and background among staff IT skills and background among locals Staff recognize seriousness of giving personal information Desire of fewer staff in the use of e-government services Diversity of languages in the UAE
Cost of e-government	The cost of the development of e-government services The cost of providing e-government services The cost of government in the development of e-government laws Special funds only for e-government applications Internet availability for employees
Challenges of e-government	The provision of e-government services easy to use by staff How difficult is it to use e-government applications The extent of providing security for the citizen using e-government
Readiness for Adopting e-Government	
Readiness of employees	I am confidence about using e-government and its style I am happy that my organization applies the e-government system I am ready to use the e-government I have no fear or doubt regarding e-government I have knowledge and ideas about the e-government I have an idea about the e-government process. I have had sufficient training about the e-government system I can handle e-government problems that arise in my agency
Readiness of organization	The current atmosphere in the organization is suitable for E-government The organizational structure of the institution is ready formatted for the application of e-government Directors of this institution are ready to implement e-government methods This institution's staff is ready to implement e-government methods Current working conditions are suitable for applying the e-government methods Rules and procedures for working in this institution are suitable for the e-government methods
Benefits	E-government will improve employee performance E-government will improve my performance at my current job
Cooperation	Cooperation between my organization and private organizations will develop more employees jobs Cooperation between my organization and private organizations is useful The organization will improve quality for everyone if there is cooperation with the private sector
Care about customers	I do care about our customers The auditors and enterprise customers are the first priority for me

Challenges of E-Government

In terms of skills and background, a majority of respondents don't know whether the use of e-government services is challenging (mean: 2.79) based on the scale defined as (1) Not challenge; (2) Simple challenge; (3) I don't know; (4) Important challenge; (5) Very important challenge. In terms of cost of e-government, a majority of respondents don't know about Internet availability for employees (mean: 2.74). In terms of challenges of e-government, a majority of respondents don't know whether the provision of e-government services is easy to use by staff (mean: 2.46).

Readiness of E-Government

Respondents were asked questions whether they feel employees are ready for adapting to e-government. Among questions measuring the readiness of employee, the lowest mean was for "I am happy that my organization applies the e-government system" with a mean: 1.57 (between strongly agree to agree) and the highest mean was for "I had sufficient training about the e-government system" (mean: 2.28; (2) agree).

Respondents were asked questions about whether their organization was ready for adapting e-government. Among questions measuring the readiness of organization, the lowest mean was for "Directors of this institution are ready to implement e-government methods" (mean: 1.92; agree). The mean of all items ranged around 2, which shows a majority of respondents agree that their organization is ready for adapting e-government.

Respondents were asked about the benefits of e-government, and the majority of them agree that e-government will improve their current job and employee performance (mean: 1.7).

Respondents were asked about Cooperation between organizations about e-goverment, and the majority of them agree that their organization has Cooperation with the private sector (mean: 1.8).

The majority of respondents do care about their customers (mean: 1.76), and the auditors and enterprise customers are the first priority for respondents (mean: 1.71). The majority of respondents agree or are neutral about the idea that e-government could cause issues (mean: 2.65). It has been also found that the majority of respondents agree about the idea of most people have the ability to use the Internet (mean: 2.17).

Overall Level

Table 3 shows the overall mean for all constructs. The overall mean ranged between 1.77 to 2.603 and shows the majority of respondents believe they are ready for adopting to e-government and e-government is simple challenging.

TESTING HYPOTHESIS

Data Preparation

Before testing the hypothesis, data was screened and cleaned. Missing data for ordinal-scale questions have been replaced by the mean. In the next step, reliability of the measurement was tested using Cronbach Alpha. In order to test the reliability of the measurement (testing the quality of internal consistency

Table 3. Overall mean

Constructs	Mean	Std. Deviation	Variance
IT Skill	2.60	0.86	0.74
Challenges	2.43	0.99	0.98
Cost	2.29	0.93	0.86
Readiness of Organization	1.98	0.69	0.48
Readiness of Employee	1.84	0.56	0.31
Cooperation	1.80	0.69	0.48
Care of customer	1.78	0.67	0.45
Benefits	1.77	0.76	0.58

of items measuring one construct), Cronbach's coefficient alpha, the most commonly used in statistics, was determined. The Cronbach alpha coefficient is recommended to be above 0.7 (Pallant, 2011). Table 2 shows the reliability of each construct. Results show that the value of Cronbach's Alpha for all constructs ranged from 0.7 to 0.85, which indicates high inter-correlation between items in the constructs. This result shows that the questions statistically measure its constructs, and items in the constructs have high internal consistency.

Factor Analyses

Exploratory factor analysis (EFA) is generally used to discover the factor structure of a measure and to examine its internal reliability. EFA is often recommended when researchers have no hypotheses about the nature of the underlying factor structure of their measure. In this study, a factor analysis was tested to identify how many factors (constructs) are defined under "challenges to adopt e-government".

The 13 questions measuring e-government challenges were subjected to run exploratory factor analysis using the SPSS version 22. The Kaiser-Meyer-Olkin (KMO) Test was 0.826, exceeding the recommended value of .6 (Kaiser, 1974) and Bartlett's Test of Sphericity (Fields, 2005) reached statistical significance, supporting the factorability of the items. Maximum likelihood components analysis with Promax rotation method revealed three components with eigenvalues exceeding 1, explaining 38.23%, 11.51%, and 9.48% of the variance respectively. Hence, it was decided to retain these three components for further investigation.

There was a weak negative correlation between the three factors: [factor 1 & 2: ($r = .51$); factor 2 & 3: ($r = .49$); and factor 3 & 1 ($r = .61$)]. This confirms these three factors are measuring different things, since there was no strong correlation among factors. As a result, e-government challenges are generated under three factors with high reliability. The factors are: Skills and background about e-government (.874); Cost of e-government (.906); and Cost of government in the development of e- government laws (.804).

Table 4. Factor analysis for e-government challenges

Questions	Factor		
	1	2	3
Staff realize the seriousness of giving personal information	.902		
IT skills and background among locals	.874		
Desire of fewer staff in the use of e-government services	.569		
Diversity of languages in the UAE	.473		
IT skills and background among staff	.403		
The cost of providing e-government services		.906	
The cost of government in the development of e-government laws		.804	
Special funds only for e-government applications		.603	
The cost of the development of e-government services		.312	
Internet availability for employees		.297	
How difficult is it to use e-government applications			1.045
The extent of providing security for the citizen using e-government			.566
The provision of e-government services easy to use by staff			.560
Reliability (Cronbach's Alpha)	0.79	.740	0.8
% of Variance	38.23	11.51	9.48
Cumulative %	38.23	49.74	59.22
KMO		826	
Bartlett's Test of Sphericity		0.00	
Correlation among factors			
Factor 1	1.000	.51	.61
Factor 2	.51	1.000	.49
Factor 3	.61	.49	1.000

Note: factor 1: Skills and background about e-government; factor 2: Cost of e-Government; factor 3: Challenges of e-government

TESTING THE HYPOTHESIS

Correlation

A Spearman-rho correlation was run to test the correlation between variables.

The correlation was found to be statistically significant between IT Skills and Cost [r = 0.471, p = .00 < 0.05]; Challenges and IT Skills [r = 0.440, p = .00 < 0.05]; and Challenges and Cost [r = 0.595, p = .00 < 0.05]. These findings suggest that respondents believe the more challenges in "IT skills", the higher the challenges in "Cost" and "Challenges to adopt e-government".

The correlation test result shows that there is a significant positive correlation between "Readiness of employees" and "Readiness of organization" [r = 0.54, p = .00 <0.05]; "Benefits" and "Readiness of organization" [r = 0.38, p = .00 < 0.05]; "Benefits" and "Readiness of employees" [r = 0.49, p = .00 < 0.05]; "Cooperation" and "Readiness of employees" [r = 0.41, p = .00 < 0.05]; "Cooperation" and "Readiness of organization" [r = 0.24, p = .00 < 0.05]; "Care about customers" and "Readiness of employees" [r = 0.33, p = .00 < 0.05]; "Care about customers" and "Readiness of organization" [r = 0.32, p = .00 < 0.05]; "Cooperation" and "Benefits" [r = 0.31, p = .00 < 0.05]; "Care about customers" and "Benefits" [r = 0.38, p = .00 < 0.05]; "Care about customers" and "Cooperation" [r = 0.67, p = .00 < 0.05].

The result indicates that the more employees are ready to adopt E-government, the more the organization is ready for adopting e-government. Benefits, care about customers, and cooperating with private sectors are also found to have a positive correlation with readiness of adopting e-government by employees/organization, and shows the highest level of benefits, care about customers, and cooperating with private sectors, with the greater amount of employee and organization readiness to adopt e-government.

CONCLUSION

As a summary, this study shows that the more employees agree with the benefits of e-government, care about their customers, and cooperate with the private sector, the more they feel they are ready for adopting e-government. It was also found that the more employees felt challenged by e-government's cost and IT skills, the more they felt their organization will be challenged when adopting e-government. In this technological era, government is paying attention to communicating with their citizens virtually as well directly. With the passage of time, e-governance will gradually become the priority of most countries. The government requires improvement and the best solution suggested is e-government. The UAE is one of the nations who have started work on procuring e-government because it is not only beneficial for citizens but also for businesses and the commonwealth. E-government provides various benefits to the community, like efficiency, focusing on the need of citizens, and improvement of policies. The government of the UAE is also introducing an identity management system that only requires a single sign-in once before the citizen can easily access services. Along with the improvements in the technological sector, the UAE is now ranked high among the countries progressing in information technology. There is a framework of readiness as well. In the end, we would like to conclude that e-government should be implemented in every aspect of the public sector by the government as it is beneficial for every habitant in the country, whether related local or central government, or the citizens. In conclusion, the UAE has become one of the most developed countries embracing the latest technology in the world. E-government is one of the most important goals and an achievement that leads to success in the society of authorities and companies around the country. The country is working hard to improve awareness about how to use the new technologies and the electronic government.

Discussion

In conclusion, the researcher confirmed the importance of developing e-government in the UAE under the UAE Federal e -Government Strategic Framework, the UAE E-Government Failure & Success Factors Rank Model, and the achievements and development of the current state of e-government. The survey method was used in this research among 407 governmental employees in Dubai & Abu Dhabi in federal offices currently applying e-government. The achievement of the UAE placing among world leaders proved that e-government will make the country more developed. The suggestions that have arisen from this survey about e-government in the UAE include make more campaigns to alert people about the importance of using the e-government services, describing how it will abbreviate their time and guarantee their services. On the other hand, for employees the government has to provide more training and workshops for them to gain more professional, high quality of work. The UAE is a country developing year after year, and the e-government services will eventually be applied to all institutions in the country.

The most important challenges the government faces are getting customer satisfaction for the services provided by providing intelligent and easy applications to use. The smart government as a modern concept is embodied in the UAE. Federal employees believes that the goal of the government and smart application is the optimal use of information technology in order to provide government services to citizens, residents, visitors, and the business sector, government departments, and their employees, across multiple electronic channels in order to facilitate their transactions and their lives. His Highness Sheikh Mohammed bin Rashid Al Maktoum (2013) believes that the transformation of the government and all government services to e-government relies on smart mobile phone applications and working around the clock with government agencies. He gave the UAE agencies, specifically in Dubai, 24 months to make the shift and apply e-government in every public place. Still, the most important current requirements that the federal authorities need from the government is to develop a set of instructions, manuals, and policies consolidated within the framework of the transition to smart government. These guidelines are considered a road map for smart government to ensure the provision of easy-to-use services that meet the needs of customers and their expectations as to the unified reference services from the federal authorities. It is necessary to find and create an introductory guide for smart services, planning the implementation of smart services, policy implementation of smart services, and security of the development of smart services and standards of design interfacing Smart Services applications.

A majority of respondents, when asked about the benefits of e-government, agreed that E-Government will improve current jobs and employee performance (mean: 1.7). Accordingly, it is an important necessity in our communities to contribute to the development of e-government. However, this development must deal with prudence and intelligence, and people must raise the ceiling on education, knowledge, and awareness of how to use these smart services. These smart services have already contributed in the revitalization of the economic mobility, even in terms of foreign investments in the country, by supplying them with all the information they required. People urgently need smart services in places such as markets, houses, and universities because they are becoming the perfect solution to connect with the world and find out the latest news. If people find smart services anywhere, it reflects the presence of development and progress, and helps in the mental development of individuals to think about themselves because the new and useful is innovative and keeps pace with the times.

Respondents were asked questions regarding whether the employees are ready for adapting e-government. Among questions measuring the readiness of employees, the lowest mean was for "I am happy that my organization is applying the e-government system" (mean: 1.57;between strongly agree to agree), and the highest mean was for "I had sufficient training about the e-government system" (mean: 2.28; (2) agree). Overall the mean ranged between 1.77 to 2.603 and shows the majority of respondents believe they are ready for adopting e-government and e-government is simple challenging. Subsequently, the United Arab Emirates has achieved great development in the field of modern technology, and created a sophisticated infrastructure in this area. As indicated by Radan (2012), the United Nations survey in 2012 showed that the UAE was among the emerging leaders in e-government development; in 2010 the UAE placed in 49th position and in 2012 it moved to the 28th position.

There are changes in the works for the future government actions, so smart work systems effectively include the new system to provide the best results. Mourtada & Al-khatib (2014) discussed that the UAE government is always a step ahead when accepting the latest technologies so they can improve efficiency.

Through these technologies, the government have improved their services and standards in many sectors, such as education, health, and judicial offices. In addition, the old system will continue to work in conjunction with modern intelligent systems to provide all categories of customers' requirements until the full transition to a smart system of government.

According the employees' survey on smart government challenges, the results indicate that e-government challenges are generated under three factors with high reliability. The factors are: Skills and background about e-government (.874); cost of e-government (.906); and cost of government in the development of e-government laws (.804). Moreover, the correlation was found to be statistically different between IT skill and Cost [$r = 0.471$, $p = .00 < 0.05$]; Challenges and IT skill [$r = 0.440$, $p = .00 < 0.05$]; as well as Challenges and Cost [$r = 0.595$, $p = .00 < 0.05$]. These findings suggest that respondents believe that the more challenges there are in "IT skills", the more challenges they have in "cost" and "challenges to adopt e-government".

In connecting the results with the literature review, M. Al-Suwaidi & A. Rajan (2013) discussed the failure factors of the e-government in developing countries and regions. Lack of funding was the most prominent failure factor, after which there were capacity deficit, poor change management, and lack of thoroughly developed strategy. As a result, there are some challenges that should be overcome regarding this initiative—the first of transparency and corporate governance, the second of readiness of other government agencies, such as the quasi-governmental companies, the corporation between public and private sector, and the administrative obstacles and transformation of the existing administrative system by the smart system of government .Also the link between actors represents a fourth hurdle, especially relating to the proceedings, to provide a unified system to terminate the transactions, such as ports, customs, and Free Zones which are linked to all through Dubai Trade. Consequently, the social concept is still the main obstacle, as a government employee must be creative and highly productive, which corresponds to some resistance to change. Al-Khouri (2012\) stated that government policies should enable governments to undertake radical organizational changes such as (1) foster growth in services, (2) reducing unnecessary costs and regulatory burdens on firms, (3) strengthening education and training systems, (4) encouraging good management practices similarly suggested by the employees in the collected survey.

The smart employee is the foundation that can provide intelligent service to the needs of customers and traders to suit different circumstances. Employees are the main tools of shifting the old government system to the smart government system, but there is also an importance of reviewing the legislation and federal laws for harmonization and restructuring procedures and achieving the greater coordination between ministries and federal agencies. The aim is to link services with one another through the formation of joint and specialized working groups. The formation of a technical support team uniform working around the clock seven days a week is important to ensure the highest level of efficiency of operations.

The study proves the remarkable advantages of Smart Government to all its parties: clients, authorities, and the society. Clients will pay much less for efficient services. In addition, they can perform whatsoever online without having to go out. Which means that clients pay less in exchange for a higher quality services compared to those provided at the authority place. These outcomes second the opinion of Melhem (2013) that Smart Government is both timesaving and effort-saving. Moreover, Smart Government proved its effectiveness in the experience of the United Arab Emirates. Smart Government resulted in increase in job opportunities. Smart Government is much easier to transfer information, data, documents through the Internet rather than all paper work and walking around. Thus, Dohyoon (2012) thinks that the clever government improve the communication between persons and government, it makes use of advancements and build on prior achievements of governments. It could be said that all studies reached the same findings: Smart Government facilitate communication between the public and the authorities and it is a great way for both parties to get things done with the least effort made. All transaction are done online, which means that more is accomplished in less time. It is certain that not all clients benefit from the Smart Government system for so many reasons, such as their inability to deal with technology. Yet, the government is the ultimate beneficiary of this technology. Smart Government also guarantees providing a high quality service with the least possible errors. Biygautane (2013) suggested that UAE should use technology as a method to develop the services of the public sector, because it facilitates the communication between the government and the public. The clients themselves agreed that technology is better for them, as it presents improved services.

The more government expand the technology of Smart Government and make it more user-friendly, the more clients will resort to using it. Smart Government links the people with the government through technology. Most of the clients already use smart applications on their devices. That is because these users are used to smart applications and trust the Smart Government applications. Managers noticed that Smart Government has so much effect that clients that they started to use it even inside the authority itself. It has become a kind of social experience not just a service provider. The researchers refer to other significances of Smart Government, which are it promotes the position of UAE government as a technology-dependent government, and it makes people more conscious about the benefit of smart applications. Using devices that supports smart applications, people can reach the government services from any place and at any time. Thus, it is obvious that the clever devices having many shapes. The managers considered that Smart Government is a huge transformation for the methods of the government. The government now is more creative and innovative and it is taking transactions to a completely new level. The study shows that the ultimate purpose of Smart Government is the development and advancement of the public sector and customer service. Smart Government has an indispensable role in the public sector; it serves a great help in facilitating the processes of communication and service provision. Clever government supports information transfer, shared society profitability, non-dividend transfer of written papers, as well reviews. Smart Government need much attention and effort from the government in training employees, advertising the new technology, preparing the applications for use, etc. However, It will save the government much time, effort, and cost in the future. In the age of technology, every state exerts its utmost effort to meet the requirement of the people, so the states constantly introduce new systems based on up to date technologies. Yet, many clients sustain from using this technology because of many reasons, among which: they are not used to this kind of technology, they are not convinced with it, they do not trust its efficiency, or the absence of good internet services. Addressing these problems help increase the number of users of smart application, thus improve the public sector's quality of service. The Public are not only receivers of the services provided to them through Smart Government, but they are central participants to the development of this technology. The customer's suggestions, complaints, and feedback are crucial for the development of Smart Government. But there are some problems the public cannot help in solving, they include: Absence of clear purposes, the decentralized management structure, lack of information, and unclear philosophy, principles, policies, and strategies. However, Smart Government is a client-oriented system that aims at ensuring the best possible quality of the provided services. In conclusion, the limitation of this study is due to insufficient data and inadequate publications related to this subject. This study is limited to the application of Smart Government and its users in the United Arab Emirates. All published studies about Smart Government are limited. For example, Al-Khouri's studies conducted in the years of 2012 and 2013 discusses only the National density services ignoring all other applications of Smart Government. This study evaluates and assesses the use of Smart Government applications, which is studied for the first time. It also contains 18 interviews conducted with managers from federal authorities, the results of 1500 survey conducted with users of smart government applications of different authorities, and suggestions and recommendations for new strategy and policy. This study will be useful for scholars, students, and even the authorities, as it provides them with the public's feedback about the online services. Further studies may target the willingness of the employees in both private and public sectors to use Smart Government system. Moreover, there is a dire need to conduct further studies that examine the employees' feedback about Smart Government.

Suggestions

- Training for the employees and individuals on how to use e-government applications in smart governments.
- Providing free WIFI for customers in public places.
- Providing e-government educational courses and workshops for employees.
- Given the official documents are needed in every federal agencies, there is a need to connect the passport and nationality ID with smart government for official services use.
- Facilitate services such as hospital appointments, the payment of traffic fines, and the exchange of information between government agencies.

Research Limitation

Smart Government system was the fruit of many changes and transformations. So, it requires many things that include: a huge solid infrastructure that is qualified for communications, a qualified trained staff, regulations, etc. The government has to raise the public awareness of the benefits of this technology and induce them to use it by providing appropriate facilities. It has also to provide free Wi-Fi in public places, so clients can reach Smart Government applications from any place at any time. Finally, it has to convince the public of this online services and make them trust it so that they resort to using it when needed.

It has been the aim of this paper to cover the UAE e-government case and present a roadmap about the electronic phenomena and the practice of federal authority agency employees of the e-government. As a concept study, this paper aimed at sparking interest and inspiring scholarly and practitioner discourse in this area of study inside the community of electronic government research and practice,. This paper is the first paper in the UAE that covers the topic of readiness of federal agencies employees of using the e-government in the UAE. Also, most of the federal authorities in the UAE have not been studied before. A huge number of surveys were collected in a small country like the UAE, which made the results valid and unique.

Of the distributed surveys, 407 were returned. The survey was distributed among employees, who are more difficult than customers to collect data from as they are busy and not available all the time. Also they are in different agencies and emirates. However, the study did not cover other local places or the private sector. Further research is needed about e-government in the private sector and comparative studies must be engaged between private and public sectors.

REFERENCES

Abhichandani, T. (2008). *Evaluation of E-Government Initiatives for Citizen-Centric Delivery: Analysis of Online Public Transit Information Services*. VDM Verlag Publishing.

Al-Jenaibi, B. (2010). Differences between gender treatments in the Work Force. *Cross-Cultural Communication*, *6*(2), 63–74.

Al-Jenaibi, B. (2011). The practice of Public Relations Departments in increasing social support in the diverse workplaces of the United Arab Emirates. *Cross-Cultural Communication*, *7*(3), 41–54.

Al-Jenaibi, B. (2012). The scope and impact of workplace diversity in the United Arab Emirates–A preliminary study. *Malaysia Journal of Society and Space*, *8*(1), 1–14.

Al-Jenaibi, B. (2017). E-governance: A Study of Public Agencies in the United Arab Emirates, South. *Asian Journal of Management*, *24*(1), 2017.

Al-khaleej Times. (2012). *UAE among emerging e-Government leaders*. Retrieved on June 8,2015, from http://www.khaleejtimes.com/mobile/inside.asp?xfile=/data/todayevent/2012/September/todayevent_September18.xml&section=todayevent

Al-Khouri, A. M (2012). E-Government Strategies The Case of the United Arab Emirates (UAE). *European Journal of ePractice,* 135-140.

Al-Khouri, A. M. (2013). *Technological and Mobility Trends in E-Government*. Retrieved from www.id.gov.ae/assets/zbOAvc6B.pdf.aspx

Al-Khouri, A. M., & Bal, J. (2007). Electronic Government in the GCC Countries. *The International Journal of Social Sciences (Islamabad)*, *1*(2), 83–98.

Al-Ohali, Y., Al-Oraij, A. A., & Shahzad, B. (2011, July). KSU News Portal: A Case Study. In The 2011 international conference on Internet Computing (ICOMP'11) (pp. 18-21). Academic Press.

Al-Shafi, S. (2012). Free Wireless Internet park Services: An Investigation of Technology Adoption in Qatar from a Citizens Perspective. *Journal of Cases on Information Technology*, *10*(3), 21–34. doi:10.4018/jcit.2008070103

Al-Shafi, S., & Weerakkody, V. (2008). The Use Of Wireless Internet Parks To Facilitate Adoption And Diffusion Of E-Government Services: An Empirical Study In Qatar. *Proceedings of the 14th Americas Conference on Information Systems (AMCIS 2008).*

Alwagait, E., Shahzad, B., & Alim, S. (2014). Impact of social media usage on students' academic performance in Saudi Arabia. *Computers in Human Behavior.*

As-Saber, S., Hossain, K., & Srivastava, A. (2007). Technology, society and e-government: In search of an eclectic framework. *Electronic Government. International Journal (Toronto, Ont.)*, *4*(2), 156–178.

Australia. (2003). *An Over Review of the E-government Benefits Study*. Retrieved April 20, 2015 from: www.finance.gov.au/agimo-archive/__data/assets/file/.../benefits.pdf

Backus, M. (2001). *E-Governance and Developing Countries, Introduction and examples, Research published on April 2001*. Retrieved May 12, 2015 from: from www.iicd.org/files/report3. doc

Beynon-Davies, P. (2007). Models for e-government, Transforming Government: People. *Process Policy, 1*(1), 7–28.

Bhattacharya, D., Gulla, U., & Gupta, M. P. (2012). E-service quality model for Indian government portals: Citizens' perspective. *Journal of Enterprise Information Management, 25*(3), 246–271. doi:10.1108/17410391211224408

Burn, J. & Robins, G. (2003). Moving towards e-government: a case study of organizational change processes. *Logistics Information Management, 3*(2), 25-35.

Bwalya, K. J. (2012). *E-Government in Emerging Economies: Adoption, E-Participation, and Legal Frameworks*. Hershey, PA: IGI Global Publishing.

Bwalya, K. J., & Healy, M. (2010). Harnessing e-government adoption in the SADC region: a conceptual underpinning. *Electronic Journal of E-Government, 8*(1), 23-32.

Castells, M. (2009). *The Rise of the Network Society: The Information Age - Economy, Society, and Culture* (2nd ed.; Vol. 1). Chichester, UK: John Wiley & Sons. doi:10.1002/9781444319514

Chan, K. (2017). *What Is A 'Smart City'? Expatriate Lifestyle*. Retrieved on March, 17, 2018 from https://www.expatriatelifestyle.com/life-and-style/What-Is-A-Smart-City

Clarke, P. (2009). Understanding the experience of stroke: A mixed-method research agenda. *The Gerontologist*. PMID:19386828

Colesca, S. E. (2009). Understanding Trust in e-Government, Inzinerine Ekonomika –. *The Engineering Economist*, (3), 7–15.

Cournède, B., Goujard, A., & Pina, Á. (2013). *How to Achieve Growth-and Equity-friendly Fiscal Consolidation?* Academic Press.

Digital Government. (2018). *Tech and data: can 'digital government' be radically smarter?* Retrieved on March, 17, 2018 from https://apolitical.co/flagship/big-data-and-digital-government/

Dubai eGovernment. (2013). *eGovernment to 24/7 smart government*. Retrieved on May, 22, 2015, from the website: http://www.dsg.gov.ae/SiteCollectionImages/Content/DeG%20Documents/June-2013-en.pdf

Fan, W., & Yan, Z. (2010). Factors affecting response rates of the web survey: A systematic review. *Computers in Human Behavior, 26*(2), 132–139. doi:10.1016/j.chb.2009.10.015

Fields, A. (2005). *Discovering statistics using SPSS*. Beverly Hills, CA: Sage Publications.

Flak, L., Nordheim, S., & Munkvoid, B. (2008). Analyzing Stakeholder Diversity in G2G Efforts: Combining Descriptive Stakeholder Theory and Dialectic Process Theory. *E-Service Journal, 6*(2), 6.

Flanagan, B. (2011). *UAE is most high-tech country in Mena region*. Retrieved on June, 17, 2015 from: http://www.thenational.ae/business/technology/uae-is-most-high-tech-country-in-mena-region

Frechtling, J. A. (1997). *User-friendly handbook for mixed method evaluations*. Diane Publishing.

Garson, D. G. (2006). *Public Information Technology and E-Governance*. Sudbury, MA: Jones and Bartlett Publishers.

Goldkuhl, G. (2006, October). What does it mean to serve the citizen?–Towards a practical theory on public e-services founded in socio-instrumental pragmatism. *Proceedings of the International Workshop on E-services in Public Administration*, 27-47.

Goldkuhl, G. (2007). What does it mean to serve the citizen in e-services?-Towards a practical theory founded in socio-instrumental pragmatism. *International Journal of Public Information Systems*, *3*(3), 135–159.

Heeks, R. (2005). *Implementing and Managing E-Government: An International Text*. Sage Publishers.

Hesse-Biber, S. N. (2010). *Mixed methods research: Merging theory with practice*. Guilford Press.

Hirschfeld, B. (2012). *Global Thesis Update: Technology and the Arab Spring*. Retrieved on April 11, 2015 from http://worldperspectivesprogram.org/2012/04/12/global-thesis-update-technology-and-the-arab-spring/

Holgersson, J., Söderström, E., Karlsson, F., & Hedström, K. (2010). Towards a roadmap for user involvement in e-government service development. In Electronic Government (pp. 251-262). Springer Berlin Heidelberg. doi:10.1007/978-3-642-14799-9_22

Human Development Report. (2014). Retrieved on March 25, 2015, from:www.pnud.org.br/arquivos/rdh2014.pdf

Hung, S.-Y., Chang, C.-M., & Yu, T.-Y. (2006). *Determinants of user acceptance of the ... agent: a research note*. European and Mediterranean Conference on Information Systems (EMCIS), Alicante, Spain.

Hussein, R., Mohamed, N., Rahman Ahlan, A., & Mahmud, M. (2011). E-government application: an integrated model on G2C adoption of online tax. *Transforming Government: People, Process and Policy*, *5*(3), 225–248.

Jivani, M. (2014). *UAE government launches world's first store for smart government application*. Retrieved on June 8,2015, from http://techview.me/2014/03/uae-government-launches-worlds-first-store-smart-government-application/

Kaiser, H. F. (1974). An index of factorial simplicity. *Psychometrika*, *39*(1), 31–36. doi:10.1007/BF02291575

Karanasios, S. (2011). New and Emergent ICTs and Climate Change in Developing Countries. Manchester, UK: Centre for Development Informatics, Institute for Development Policy and Management, SED, International Development Research Centre. Retrieved from http://www.niccd.org/KaranasiosClimateChangeEmergentICTs.pdf

Karlsson, F., Holgersson, J., Söderström, E., & Hedström, K. (2012). Exploring user participation approaches in public e-service development. *Government Information Quarterly*, *29*(2), 158–168. doi:10.1016/j.giq.2011.07.009

Kraut, R., Olson, J., Banaji, M., Bruckman, A., Cohen, J., & Couper, M. (2004). Psychological research online: Report of Board of Scientific Affairs' Advisory Group on the Conduct of Research on the Internet. *The American Psychologist*, *59*(2), 105–117. doi:10.1037/0003-066X.59.2.105 PMID:14992637

Linnefell, W. (2014). *E-government Policy Formation*. Retrieved on February 22, 2014 from: www.ejeg.com/issue/download.html?idArticle=348

Lootah, R., & Geray, O. (2006). *Dubai eGovernment Case Study*. Retrieved on February 23, 2014 from: http://www.oecd.org/dataoecd/4/40/36986277.pdf

Meena, R. S. (2009). Environmental-informatics---A solution for long term environmental research. *Global Journal of Enterprise Information System, 1*.

Mertens, D. M. (2014). *Research and evaluation in education and psychology: Integrating diversity with quantitative, qualitative, and mixed methods*. Sage Publications.

Miller, W., & Walling, J. (2013). *Government in the twenty-first century: New Avenues of Study. Taking Sides*. New York, NY: McGraw Hill.

Moon, M. J. (2002). The Evolution of E-Government among Municipalities: Rhetoric or Reality? *Public Administration Review*, *62*(4), 424–433. doi:10.1111/0033-3352.00196

Mourtada, R., & Alkhatib, F. (2014). *2014 UAE Social Media Outlook: Increasing Connectivity Between Government and Criticizes GCAA strategic Plan*. Retrieved on February 22, 2015 from: https://www.gcaa.gov.ae/

Niaros, V., Kostakis, V., & Drechsler, W. (2017). Making (in) the smart city: The emergence of maker spaces. *Telematics and Informatics*, *34*(7), 1143–1152. doi:10.1016/j.tele.2017.05.004

Nixon, P. G., & Rawal, R. (2011). *Enabling Democracy? e-Government, Inclusion and Citizenship*. The Hague University.

Opdenakker, R. (2006, September). *Advantages and Disadvantages of Four Interview Techniques in Qualitative Research*. Retrieved June 9,2014, from Forum Qualitative Sozialforschung website: http://www.qualitative-research.net/index.php/fqs/article/view/175/391

Pallant, J. (2011). *SPSS Survival Manual* (4th ed.). Crows Nest, Australia: Allen & Unwin.

Radan, S. (2012). UAE among emerging e-Government leaders. *Khaleej Times*. Retrieved on June, 15, 2015 from http://www.khaleejtimes.com/

Ramon, G. (2012). *Enacting electronic government success: An integrative study of government-wide websites, organizational capabilities, and institutions*. Springer.

Rashid. (2013). *HH launches Mobile Government, ICT Fund offers AED 200m in support of the initiative: Emirates News Agency*. Retrieved on May 22, 2014, from http://www.wam.org.ae/servlet/Satellite?c=WamLocEnews&cid=1290004835681&pagename=WAM%2FWAM_E_Layout&parent=Collection&parentid=1135099399983&rendermode=preview-admin-1135099398363

Reddick, C. G. (2010a). *Politics, Democracy and eGovernment: Participation and Service Delivery*. Hershey, PA: Information Science Reference. doi:10.4018/978-1-61520-933-0

Reddick, C. G. (2010b). *Citizens and E-Government: Evaluating Policy and Management.* Hershey, PA: Information Science Reference. doi:10.4018/978-1-61520-931-6

Rokhman, A. (2011). e-Government adoption in developing countries; the case of Indonesia. *Journal of Emerging Trends in Computing and Information Sciences*, *2*(5), 228–236.

Rorissa, A., Demissie, D., & Pardo, T. (2011). Benchmarking e-government: A comparison of frameworks for computing e-government index and ranking. *Government Information Quarterly*, *28*(3), 354–362. doi:10.1016/j.giq.2010.09.006

Serageldin, I. (2011). *Science and the Arab spring. Issues in Science and Technology*. Retrieved on May 16, 2015, from http://www.issues.org/27.4/p_serageldin.html

Seres, A. (2010). Three Database Management Systems (DBMS) compared. *Open Source Science Journal*, *2*, 65–82.

Shahzad, B., & Alwagait, E. (2014). Best and the Worst Times to Tweet: An Experimental Study. *WSEAS, 15th International Conference on Mathematics and Computers in Business and Economics (MCBE'14), Proceedings of 15th International Conference on Mathematics and Computers in Business and Economics.*

Shareef, M. A., Kumar, V., Kumar, U., & Dwivedi, Y. K. (2011). e-Government Adoption Model (GAM): Differing service maturity levels. *Government Information Quarterly*, *28*(1), 17–35. doi:10.1016/j.giq.2010.05.006

Simaanya, M. (2014). *Employee Management System* (Doctoral dissertation). University of Zambia.

Sipior, J. C., Ward, B. T., & Connolly, R. (2011). The digital divide and t-government in the United States: Using the technology acceptance model to understand usage. *European Journal of Information Systems*, *20*(3), 308–328. doi:10.1057/ejis.2010.64

Smart Dubai Initiative. (2017). *Dubai Initiative*. Retrieved on March, 17, 2018 from http://www.smart-dubai.ae

Tassabehji, R., & Elliman, T. (2006). *Generating Citizen Trust in e-Government using a Trust Verification United Nations Survey (2014). E-government for the Future we Want.* Retrieved on July, 28, 2015, from http://www2.unpan.org/egovkb/documents/2010/E_Gov_2010_Complete.pdf

UAE government. (2014). *EGovernment in the United Arab Emirates 2014. Retrieved from* government.ae/.../eGov...E.../22b5703a-d8d5-4fd4-ad1f-5d2c2bcde8cc

Varian, H. R., Farrell, J., & Shapiro, C. (2005). *The Economics of Information Technology*. New York: Cambridge University Press.

Weinzierl, A. (2014). *Global Agenda Council Launches Guide to Good Government and Trust-Building.* Retrieved on August 25,2015, from www.oecd.org/governance/WEF_NR_FutureOfGovernmentSmart-Toolb

World Economic Forum. (2007). *United Arab Emirates and the World: Scenarios to 2025.* Retrieved on September, 8, 2015, from www.weforum.org/pdf/scenarios/UAE.pdf

Wyse, S. V., Perry, G. L., O'Connell, D. M., Holland, P. S., Wright, M. J., Hosted, C. L., ... Curran, T. J. (2016). *A quantitative assessment of shoot flammability for 60 tree and shrub species supports rankings based on expert opinion*. Academic Press.

Yaghi & Al-Jenaibi. (2017). Happiness, Morality, Rationality, and Challenges in Implementing Smart Government Policy. Public Integrity, 1-16.

Yonazi, J., Sol, H., & Boonstra, A. (2010). Exploring issues underlying citizen adoption of e-government initiatives in developing countries: The case of Tanzania. In *Proceedings of the 10th European conference on e-government: National center for taxation studies University of Limerick, Ireland* (p. 425). Academic Press.

Chapter 7
Exploring Knowledge Activities for Bringing Saudi Arabia and the BRICS Together

Abdulkader Alfantookh
SRMG, Saudi Arabia

Saad Haj Bakry
King Saud University, Saudi Arabia

ABSTRACT

This chapter explores the promotion of knowledge cooperation between the Kingdom of Saudi Arabia (KSA) and the BRICS (Brazil, Russia, India, China, and South Africa). The knowledge activities considered include knowledge creation, diffusion, utilization, and sharing. The chapter introduces a knowledge assessment framework based on UNESCO knowledge society paradigm, and on available international indicators. The framework gives an index of "28 indicators." It provides a criterion for comparing the measured values of the indicators among countries to enable identifying prioritized knowledge cooperation opportunities that bring countries together. The framework is used to investigate these opportunities between the KSA and the BRICS. It has been found that knowledge activities in KSA enjoy high enablement levels from which the BRICS can gain, while knowledge activities in the BRICS emphasize knowledge productivity from which KSA can gain. Bridging the gap between the two sides will be useful to both, and to building global knowledge societies that lead to world peace and prosperity.

INTRODUCTION

The countries identified as the BRICS, Brazil, Russia, India, China and South Africa, have grouped together, in the early years of the 21st century, to become an important economic block that enjoy advanced development, and that also represent an emerging powerful global economic power (BRICS, 2017). Identifying cooperation opportunities between the BRICS countries and the GCC countries, of Bahrain, Kuwait, Oman, Qatar, Saudi Arabia, and the UAE, which also represent an important global economic block, would be beneficial to both (ITU, 2005). This chapter is concerned with assessing knowledge

DOI: 10.4018/978-1-5225-7347-0.ch007

cooperation opportunities between the Kingdom of Saudi Arabia (KSA), the largest economic power of the GCC block, and the BRICS.

This introductory section aims at providing the necessary background that enables addressing the problem considered. The given background has three main parts:

- The first highlights the international drive toward knowledge as a source for development.
- The second identifies international views on assessing knowledge issues in different countries.
- The third introduces the work presented in this chapter.

The International Drive Toward Knowledge for Development

At the turn of the century, that is in the year 2000, world leaders gathered in the United Nations (UN) and issued the Millennium Development Goals (MDG) that set development directions toward better world. Among these development directions were: education, global partnership, and the availability of new technology, especially the information and communication technology (ITU, 2005).

In the year 2003, the world leaders met again in Geneva in what is known as the World Summit on the Information Society (WSIS), which emphasized knowledge activation as a mean for development, through the widespread use of information and communication technology (ICT) (WSIS, 2005). In 2005, the UN Educational Scientific and Cultural Organization (UNESCO) started replacing the term information society with that of knowledge society emphasizing the need to create knowledge and to utilize it for supporting development (UNESCO, 2005).

In 2010, UNESCO introduced a paradigm for the knowledge society. The paradigm identifies knowledge activities as consisting of knowledge creation, knowledge preservation, knowledge diffusion and knowledge utilization (UNESCO, 2010). In 2013, UNESCO stressed the need for spreading the drive toward building global knowledge societies in order to promote human development alongside technological innovation. It considered that the core aspiration for peaceful and sustainable knowledge societies must acknowledge the interest of all stakeholders. If every country plans and drives its ways toward building its own knowledge society according to these principles, global knowledge societies seeking peace and development can be reached for the benefits of all (UNESCO, 2013).

International Assessment of Knowledge Issues

Various international indices and international databases provide indicators, with periodically updated data, on different knowledge related issues. Four important sources of such indicators have been used in this chapter. They are as follows: the "Global Innovation Index: GII" (Dutta et al, 2016) the "Global Competitiveness Index: GCI" (Schwab et al, 2016); the "International Telecommunication Union: ITU Measures of the Information Society" (ITU, 2016); and the "World-Bank Database" (World Bank, 2017). It should be noted that indicators and data provided by these sources are usually of mixed origins and not necessarily originated by the source concerned. For example, "education related issues" in all of these sources are mostly taken from UNESCO (2017), which may in turn depend on national data produced by the countries concerned.

The knowledge issues addressed by the indicators presented by the sources given above are of diverse nature. GII and GCI include "81" and "119" indicators respectively covering a wide range of issues of: political, regulatory, educational, business, technology, and creativity nature. ITU indicators are concerned with issues associated with "ICT and society". The indicators of the "World-Bank Database" are concerned with the various national and international development issues. Reviewing these indicators need to be performed for selecting core indicators that are directly related to knowledge activities in the society, and that can provide essential tools for knowledge development and cooperation toward building global knowledge societies that promote peace and foster development.

The Presented Work

The work presented in this chapter has three main integrated purposes, as explained in the following.

- It is concerned with the development of a framework that provides basic tools for assessing the state of knowledge in a country, for the purpose of supporting knowledge development, and knowledge cooperation toward building global knowledge societies. The targeted framework is based on UNESCO knowledge society paradigm, and on available international indicators, with comparison criterion for identifying knowledge cooperation opportunities.
- The framework is used to investigate the current state of knowledge development in the KSA and the BRICS; and to derive knowledge cooperation opportunities between KSA and the BRICS. Table 1 identifies these countries in terms of their: population; GDP per capita; and GDP growth.
- Remarks and recommendations are finally concluded for putting knowledge cooperation opportunities into practice and for promoting the drive toward global knowledge societies for the benefit of all.

Table 1. Identifying the countries concerned relative to the world.

Country	Population	GDP per Capita [US $]	GDP Average (5 Years) Annual Growth (%)
KSA	32,300,000	20,029	4
Brazil	198,700,000	11,320	2.5
Russia	143,178,000	14,090	1.3
India	1,236,787,000	1,503	5
China	1,350,695,000	6,040	7.5
South Africa	52,275,000	7,314	1.9
The Wold	7,043,105,591	10,438	2.3

ASSESSMENT FRAMEWORK

The framework developed for the assessment of the state of knowledge in the KSA and the BRICS involves the following three main stages.

- The first stage is concerned with the development of the structure of the assessment framework, considering UNESCO knowledge society paradigm.
- The second stage is associated with selecting available international indicators to assess the various structured components of the framework.
- The third stage is related to providing a criterion, based on the outcome of the assessment of indicators, for identifying knowledge cooperation opportunities among the assessed countries.

These stages are explained in the following.

Assessment Framework Structure

The assessment framework is based on UNESCO knowledge society paradigm (UNESCO, 2010). The paradigm identifies four main dimensions for the knowledge activities: knowledge diffusion; knowledge creation; knowledge preservation; and knowledge utilization. The framework structure reconsiders these activities according to the following.

Knowledge diffusion is considered to be associated with the following four main "education" issues:

- Enablement of knowledge diffusion through education, including: expenditure on education; and quality of education.
- General education, which involves: enrolment in primary education; and enrolment in secondary education.
- Tertiary education, considering: enrolment on tertiary education; and science and engineering in this education.
- Education outcome and training support, which involves: school life expectancy and staff training.

Knowledge creation is considered to be concerned with the following main "research and development: R&D" issues.

- Enablement of knowledge creation that involves R&D: expenditure; quality; collaboration; and number of researchers.
- Outcome of knowledge creation, which includes: publication of research articles; in addition to the registration of national and international patents.

Knowledge preservation is interpreted as access to the available modern tools that not only support knowledge preservation, but also activate knowledge sharing, which is an essential feature for global knowledge societies. These tools are those of the ICT. Access to these tools is viewed to involve access to: fixed telephone; mobile telephone; computers; the Internet; in addition to international bandwidth.

Knowledge utilization is interpreted as knowledge impact or performance related to knowledge. It is considered to be concerned with the following four main issues:

- Business activities, including: the generation of new businesses; and the growth of the "Gross Domestic Product: GDP".
- Investment, including: "Foreign Direct Investment: FDI" in both directions: inflows and outflows.
- Technology, including: high-to-medium technology output; and high-technology exports.
- Health and employment, including: life expectancy; and employment.

Figure 1 gives a general view of the framework structure.

Assessment Indicators

The assessment indicators associated with the above framework structure are given in the following Tables in term of: indicator definition, indicator measurement, the international source from which the indicator is selected, and the latest data for the countries concerned.

- Table 2 gives the indicators of the of knowledge diffusion.
- Table 3 provides the indicators of knowledge creation.
- Table 4 identifies the indicators of ICT access that support knowledge preservation and sharing.
- Table 5 gives the indicators of knowledge impact which is related to knowledge utilizations.

Collectively, the indicators of each of the above Tables provide a sub-index associated with its knowledge dimension; and these sub-indices collectively provide an overall index associated with the framework as a whole. In addition, Tables (2, 3 and 5) have sub-groups within their groups; each sub-group also provides a sub-sub-index, with these sub-sub-indices producing collectively the sub-index of the group. The sub-indices and the sub-sub-indices are also identified by the Tables.

Figure 1. The structure of the assessment framework

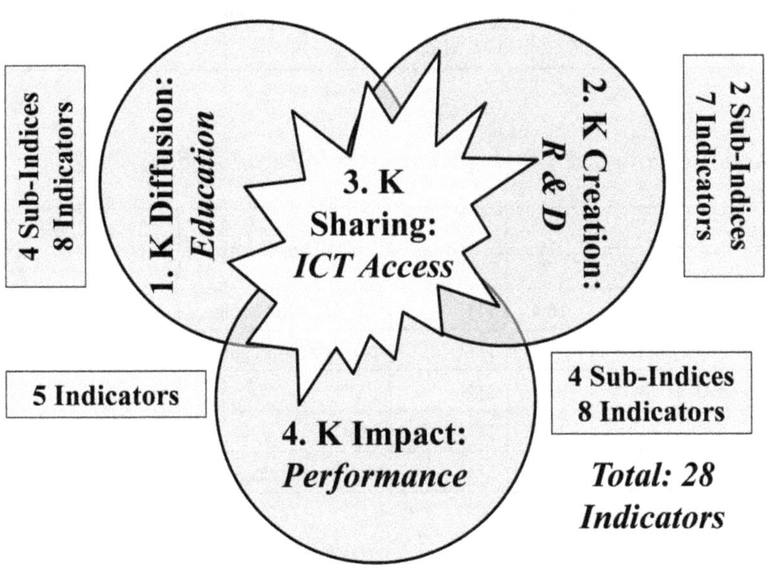

Table 2. Knowledge diffusion indicators: Education (Sub-Index)

The BRICS					SA	Source	Measurement	Indicator	Sub-Sub-Index
S	C	I	R	B					
6	NA	3.2	4.1	5.8	5.1	GII	(%) of GDP	Expenditure	Enablement
2.2	4.0	4.2	3.5	2.7	4.1	GCI	Levels: 1-7	Quality	
85.0	99.9	93.3	96.2	98.5	96.5	GCI	(%) of Age Group	Primary Enrolment	General Education
101.9	89.0	68.5	95.3	99.4	116.2	GCI	(%) of Age Group	Secondary Enrolment	
19.2	26.7	24.8	76.1	NA	50.9	GCI	(%) of Age Group	Tertiary Enrolment	Tertiary Education
NA	NA	NA	28.1	12	29	GII	(%) of Graduates	Science & Engineering Graduates	
NA	13.1	11.7	14.0	14.2	15.6	GII	(Years)	School Life Expectancy	Outcome & Training
4.9	4.3	3.9	3.8	4.3	4.1	GCI	Levels: 1-7	Staff Training	

Table 3. Knowledge creation indicators: R&D (Sub-Index)

The BRICS					SA	Source	Measurement	Indicator	Sub-Sub-Index
S	C	I	R	B					
0.8	2.0	0.8	1.1	1.2	1	GII	(%) of GDP	Expenditure	Enablement
4.7	4.3	4.0	4.0	4.0	4.2	GCI	Levels: 1-7	Quality of R&D Institutions	
4.5	4.4	3.9	3.6	3.8	4.2	GCI	Levels: 1-7	University-Industry Collaboration	
736.6	1392.8	NA	2602.6	1202.8	47.4	GII	Per Million Population	Researchers' Headcounts	
15.8	15.8	9.8	10.7	14.9	9.3	GII	Per Billion GDP	Published Articles	Achievements
1.1	43.66	2.0	11.5	2.1	0.4	GII	Per Billion GDP	National Patents	
0.5	1.5	0.3	0.4	0.3	0.33	GII	Per Billion GDP	International Patents	

Table 4. Knowledge preservation & sharing: Access to ICT (Sub-Index)

The BRICS					SA	Source	Measurement	Indicator
S	C	I	R	B				
9.2	19.3	2.3	28.5	22.3	16.4	ITU	No. per 100 People	Access to fixed telephone
147.5	80.8	70.8	152.8	125	176.5	ITU	No. per 100 People	Access to mobile telephone
25.8	43.8	11.9	69.7	48.8	72.6	ITU	(%)	Households with a computer
39.4	43.9	13.0	67.2	42.4	72.7	ITU	(%)	Households with Internet access
3720	4230	6782	41246	42946	46682	ITU	(bps) per User	International Internet Bandwidth

Table 5. Knowledge utilization: Knowledge impact / Performance (Sub-Index)

The BRICS					SA	Source	Measurement	Indicator	Sub-Sub-Index
S	C	I	R	B					
6.5	NA	0.1	4.3	2.2	0.47	GII	New Firms per 1000 People (Age: 15-64)	New Businesses	Activities
1.9	7.7	5.00	1.3	2.5	4.0	WB-Data	(5) Years Average (%)	Annual GDP Growth	
1.5	3.0	1.7	2.6	3.4	1.7	GII	(%) of GDP	FDI Inflows	Investment
1.1	1.4	0.5	2.4	0.4	0.62	GII	(%) of GDP	FDI Outflows	
28.8	43.6	31.8	23.9	39.6	30.4	GII	(%) of Manufactured Output	High-Medium-Tech Output	Technology
1.9	27.8	2.8	1.5	3.3	0.1	GII	(%) of Total Exports	High-Tech Exports	
56.1	75.2	66.2	70.5	73.6	75.5	GCI	(Years)	Life Expectancy	Life
75	95.5	96.4	94.5	93.9	94.4	WB-Data	(%) of Labor Force	Employment	

In addition to obtaining the indicators by the values concerned with their definitions, they are also given relative values related, in percent (%), to the value of the best country in the world in the indicator concerned. Assuming the value of the indicator for the country concerned to be "*X*", and the value for the best country to be "X_{best}", the relative percentage value "*x*" is calculated as follows.

$$x = (X/ X_{best}) * 100$$

This has two main advantages.

- It reflects comparison associated with a reachable value, since a country has been able to reach.
- It enables merging groups of indicators into sub-sub-indices, and these into sub-indices, so that assessment can be viewed at different levels.

Cooperation Opportunities Criterion

The values of the indicators obtained for KSA, and for each of the countries of the BRICS, provide the state of knowledge of the country concerned, according to the view of the framework. Identifying a knowledge cooperation opportunity, associated with any specific indicator, between KSA and any country of the BRICS, can be based on the difference between the value reached by KSA and that reached by a BRICS country. The importance of the opportunity will depend on the difference gap between KSA and the BRICS country concerned. Figure 2 classifies the difference gaps and assigns priorities to the cooperation opportunities that may result; and it also states if KSA is in a position to "give" experience to the BRICS in in the issue of the indicator concerned, or it is in a position to take. The difference gaps have been identified according to the following.

Figure 2. Identifying knowledge cooperation opportunities between KSA and the countries of the BRICS

<table>
<tr><td colspan="11">Identifying Cooperation Opportunities</td></tr>
<tr><td colspan="5">(KSA Gives Experience)</td><td>KSA</td><td colspan="5">(KSA Takes Experience)</td></tr>
<tr><td colspan="11">$y = (X_{BRICSCountry} / X_{KSA}) * 100$</td></tr>
<tr><td colspan="11">25 33 40 50 66 150 200 250 300 400</td></tr>
<tr><td>High</td><td>Above Ave</td><td>Ave.</td><td>Below Ave</td><td>Low</td><td>Close Range</td><td>Low</td><td>Below Ave</td><td>Ave</td><td>Above Ave</td><td>High</td></tr>
<tr><td>1</td><td>2</td><td>3</td><td>4</td><td>5</td><td rowspan="2">Cooperation Opportunities</td><td>5</td><td>4</td><td>3</td><td>2</td><td>1</td></tr>
<tr><td colspan="5">Cooperation Priority</td><td colspan="5">Cooperation Priority</td></tr>
</table>

If "$X_{BRICSCountry}$" represents the value of an indicator for a BRICS country, and if "X_{KSA}" represents the value of that indicator for KSA, then a measure of the gap, *"y"*, can be given as the following ratio of the two values.

$$y = (X_{BRICSCountry}/X_{KSA}) * 100$$

If the two values are equal, then then the gap measure will be "y = 100". If this measure "y < 100", then KSA can "give" experience to the BRICS country in the issue concerned; and if the measure "y > 100", then KSA can "take" experience from the BRICS. Five knowledge cooperation priorities have been defined, according to the following.

- If *"y = 350"* or more, then the cooperation priority is *"1"* and KSA takes experience; and if *"y = 29" or less,* then the cooperation priority is also *"1"*, but KSA gives.
- If *"350 > y >= 300"*, then the cooperation priority is *"2"* and KSA takes experience; and if it is *"33 >= y > 29"*, then the cooperation priority is also *"2"*, but KSA gives.
- If *"300 > y >= 250"*, then the cooperation priority is *"3"* and KSA takes experience; and if it is *"40 >= y > 33"*, then the cooperation priority is also *"3"*, but KSA gives.
- If *"250 > y >= 200"*, then the cooperation priority is *"4"* and KSA takes experience; and if it is *"50 >= y > 40"*, then the cooperation priority is also *"4"*, but KSA gives.
- If *"200 > y >= 150"*, then the cooperation priority is *"5"* and KSA takes experience; and if it is *"66 >= y > 50"*, then the cooperation priority is also *"5"*, but KSA gives.

Smaller gaps are considered to be insignificant to exchange experience.

INVESTIGATIONS

This section investigates and produces results on the following.

- The current knowledge development state of KSA and the BRICS, relative to the countries scoring world best in the indicators considered. These countries and their latest scores in these indicators are given in Table 6.
- Knowledge cooperation opportunities between KSA and the BRICS, with priorities according to the cooperation criterion.

KSA and the BRICS Knowledge Development State

The knowledge development states of KSA and the BRICS are illustrated in the following Figures, in values relative to the countries scoring best in the indicators concerned,

Table 6. Best countries by indicator according to latest data

Knowledge preservation / sharing *(ICT)*			**Knowledge diffusion**		
Best country		**Indicator**	**Best country**		**Indicator**
X_{best}	**Country**		X_{best}	**Country**	
123.8	Monaco	Access to fixed telephone	12.98	Lesotho (S Africa)	Expenditure
304	Macao China))	Access to mobile telephone	6.00	Switzerland	Quality
94.3	Luxemburg	Households with a computer	100	Singapore	Primary Enrolment
96.4	Qatar	Households with Internet access	135.5	Australia	Secondary Enrolment
4244992	Luxemburg	International Internet Bandwidth	114	Greece	Tertiary Enrolment
Knowledge utilization: *Impact / Performance*			53.22	Thailand	Science & Engineering Graduates
Best country		Indicator	19.86	Australia	School Life Expectancy
X_{best}	Country		5.7	Switzerland	Staff Training
22.51	Cyprus	New Businesses	Knowledge creation		
14.2	Paraguay	Annual GDP Growth	Best country		Best country
38.65	Hong-Kong	FDI Inflows	X_{best}	Country	
655.23	Mauritius (Africa)	FDI Outflows	4.36	S. Korea	Expenditure
70	Singapore	High-Medium-Tech Output	6.4	Switzerland	Quality of R&D Institutions
27.81	China	High-Tech Exports	6.00	Finland	University-Industry Collaboration
83.5	Hong-Kong	Life Expectancy	10679	Finland	Researchers' Headcounts
			67.5	Iceland	Published Articles
99.5	Qatar	Employment	43.66	China	National Patents
			23.7	Barbados (Caribbean)	International Patents

- Figure 3 shows the state at the overall index level that combines all "28 indicators".
- Figure 4 gives the state at the sub-indices level of: knowledge diffusion through education, which has "8 indicators"; knowledge creation through R&D, which has "7 indicators"; knowledge preservation and sharing through ICT access, which has "5 indicators"; and knowledge utilization through impact and performance, which has "8 indicators".
- Figure 5 illustrates the state at the knowledge diffusion through education sub-sub-indices level of: enablement, which has "2 indicators"; general education, which has "2 indicators"; tertiary education, which has "2 indicators"; and outcome and training, which has "2 indicators".
- Figure 6 gives the state at the knowledge creation through R&D sub-sub-indices level of: enablement, which has "4 indicators"; and achievements, which has "3 indicators".
- Figure 7 gives the state at the knowledge preservation and sharing through ICT access "5 indicators".
- Figure 8 shows the state at the knowledge utilization impact and performance sub-sub-indices level of: activities, which has "2 indicators"; investment, which has "2 indicators"; technology, which has "2 indicators"; and life issues, which has "2 indicators".

These Figures visualize two main issues.

- The state of KSA and the BRICS, relative to the states of the countries scoring world best in the indicators considered, seems to have large differences according to the issues considered. In this respect, it has been observed that KSA and the BRICS are relatively in higher state in some issues like general education, relatively average in some other issues like education enablement and R&D enablement, and relatively low in issues like investment.

Figure 3. The overall proposed knowledge index for SA & the BRICS(relative to best countries)

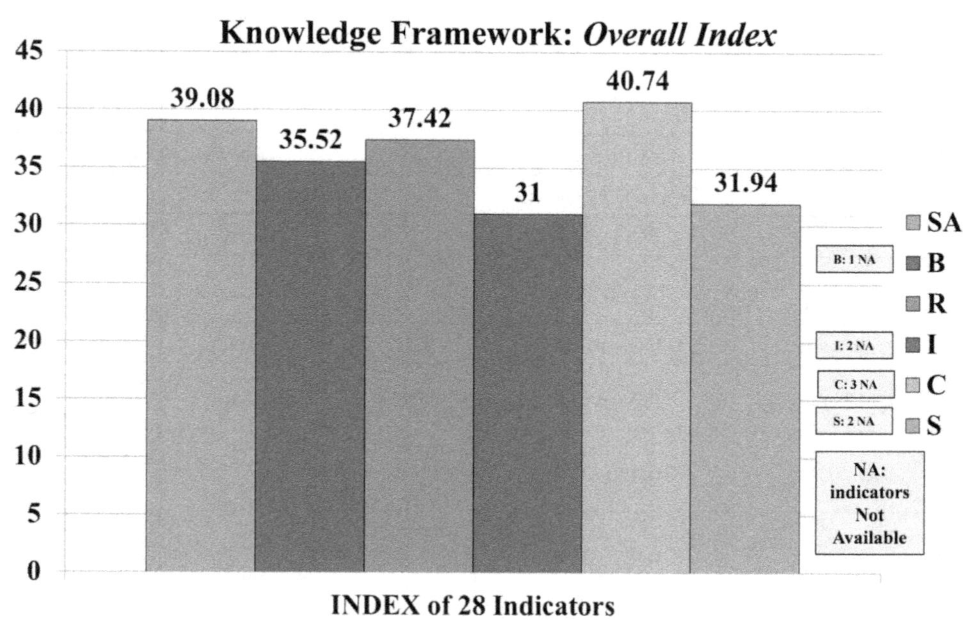

Figure 4. The proposed knowledge sub-indices for SA & the BRICS (relative to best countries)

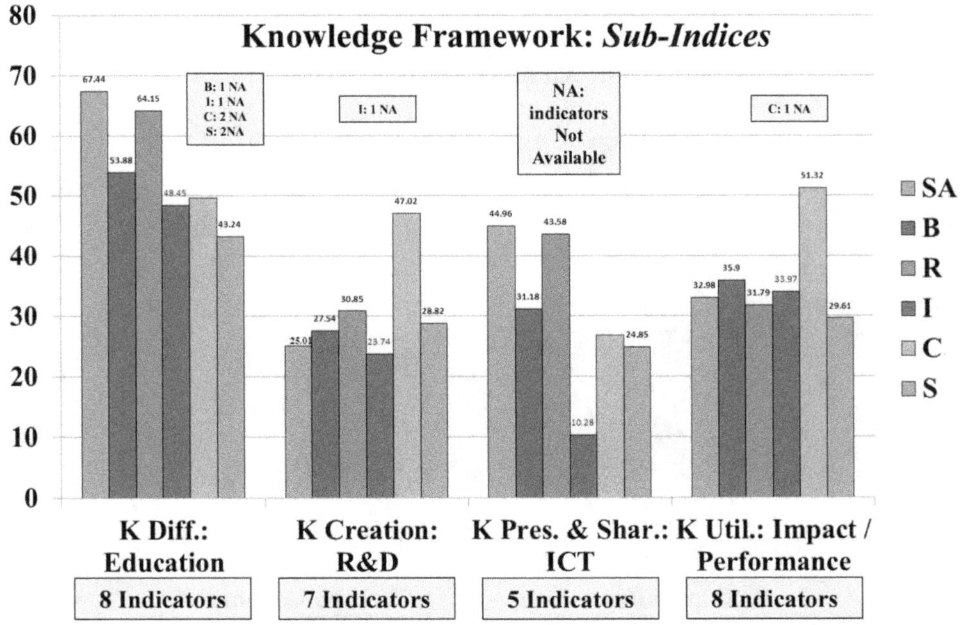

Figure 5. The proposed knowledge diffusion (education) sub-sub-index for SA & the BRICS (relative to best countries)

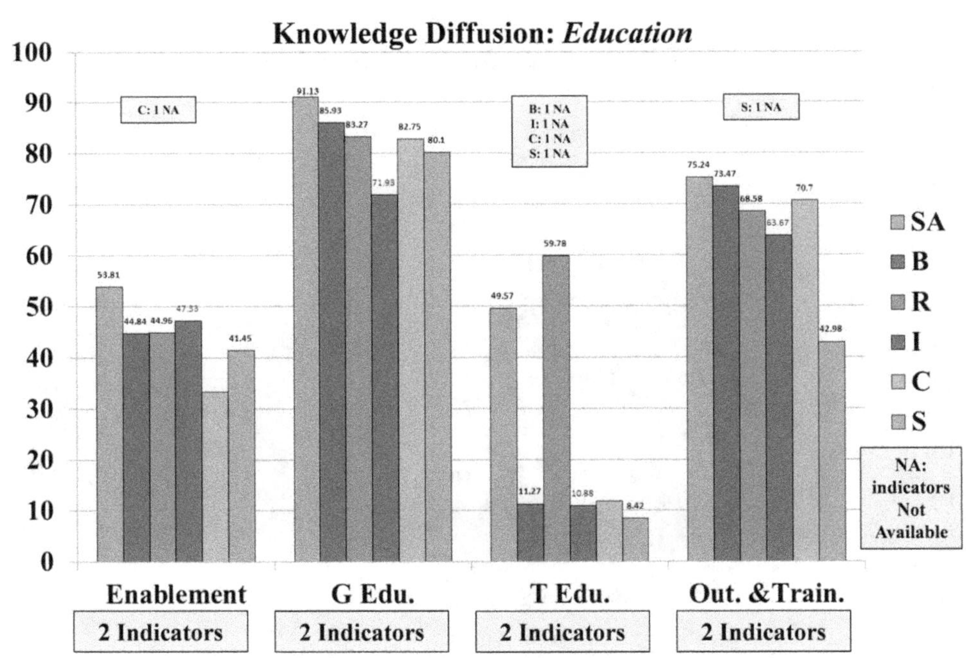

Figure 6. The proposed knowledge creation (R&D) sub-sub-index for SA & the BRICS= (relative to best countries)

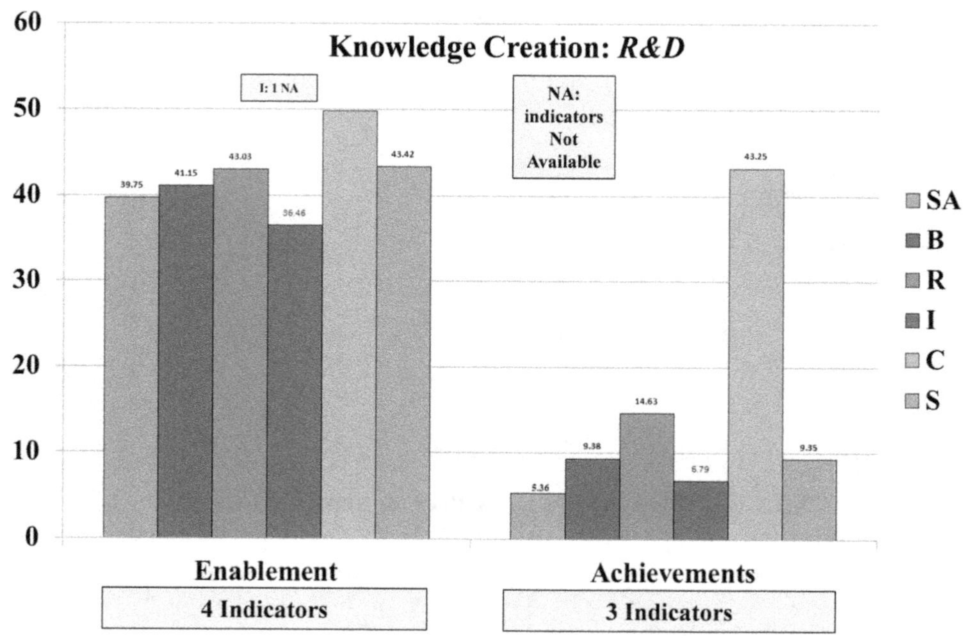

Figure 7. The proposed knowledge preservation & sharing (ICT) sub-sub-index for SA & the BRICS (relative to best countries)

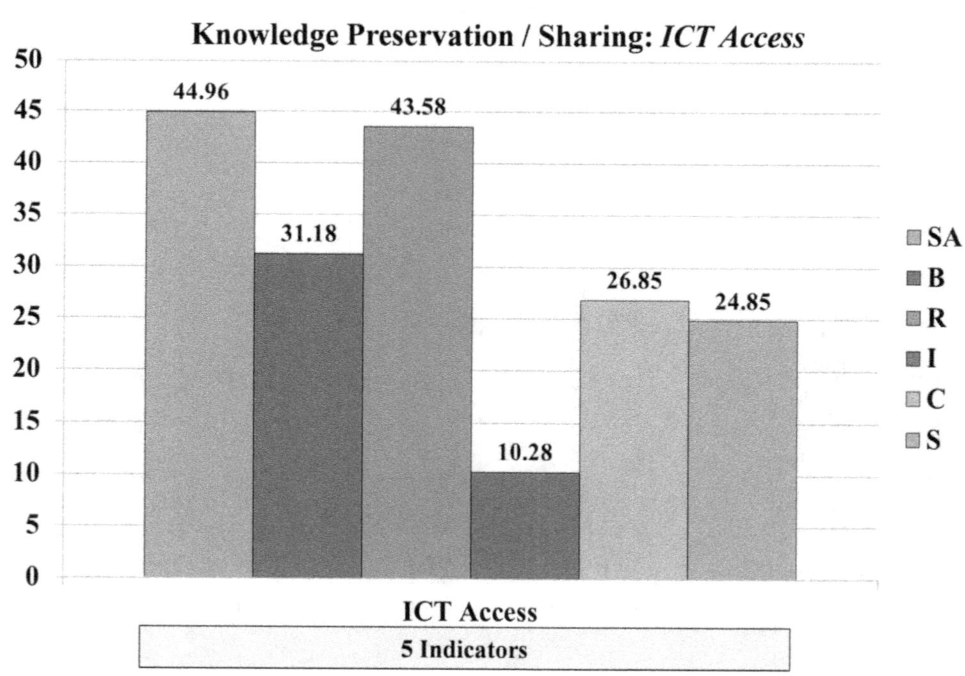

Figure 8. The proposed knowledge utilization (Impact & Performance) sub-sub-index for SA & the BRICS (relative to best countries)

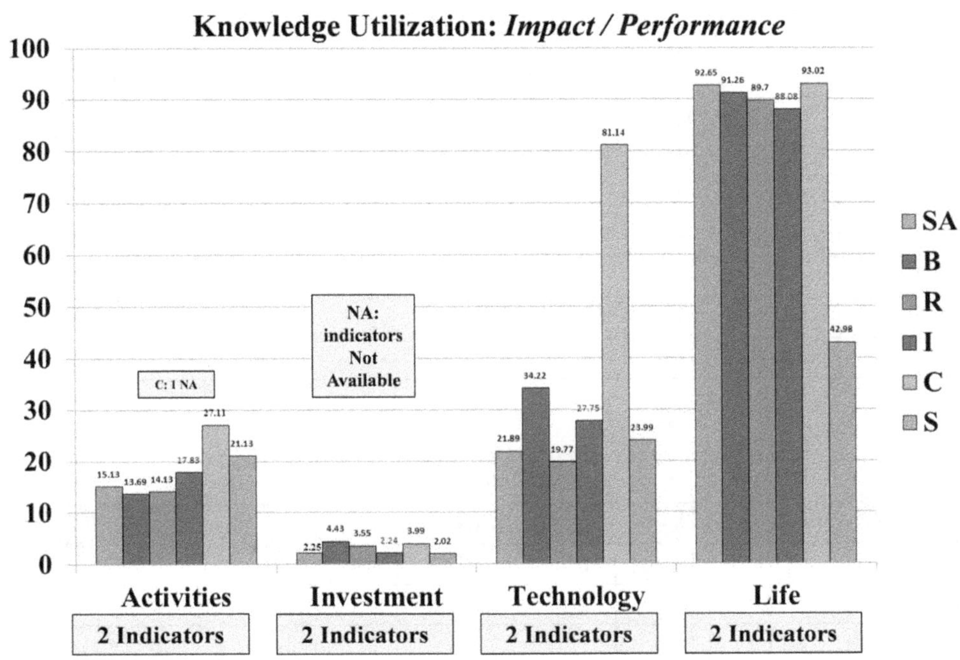

- The state of KSA and the BRICS, with regards to each other's also seems to differ. For example, KSA is higher than all BRICS in knowledge diffusion, around average among them in R&D enablement, and lower than all in R& D achievements.

KSA and the BRICS Knowledge Cooperation Opportunities

The cooperation opportunity criterion, associated with the framework, has been applied to identifying knowledge cooperation opportunities between KSA and the BRICS, taking into account the priority levels considered. The results obtained for all "28 indicators" of the framework are given in the following Tables.

- Table 7 shows that the indicators of knowledge diffusion through education provide KSA-the BRICS with knowledge cooperation opportunities in "5" out of "8" issues expressed by the indicators. The priorities of these opportunities are in the range "3 to 5". In all these, KSA is in a position to give experience to the BRICS.
- Table 8 illustrates that the indicators of knowledge creation, through R&D, provide KSA-the BRICS with knowledge cooperation opportunities in "5" out of "7" issues expressed by the indicators. The priorities of these opportunities are in the range "1 to 5". In all these, KSA is in a position to receive experience from the BRICS.
- Table 9 shows that the indicators of knowledge preservation and sharing, through ICT access, provide KSA-the BRICS with knowledge cooperation opportunities in "5" out of "5" issues expressed by the indicators. The priorities of these opportunities are in the range "1 to 5". In all these, KSA is in a position to give experience.

Table 7. Cooperation opportunities by priority: Knowledge diffusion (Education)

KSA Takes					KSA Gives					KSA	Indicators
S	C	I	R	B	S	C	I	R	B		
						NA	5			5.1	Expenditure
					5				5	4.1	Quality
										96.5	Primary Enrolment
							5			116.2	Secondary Enrolment
					3	5	4		NA	50.9	Tertiary Enrolment
					NA	NA	NA		4	29	Science & Engineering Graduates
					NA					15.6	School Life Expectancy
										4.1	Staff Training

Table 8. Cooperation opportunities by priority: Knowledge creation (R&D)

KSA Takes					KSA Gives					KSA	Indicator
S	C	I	R	B	S	C	I	R	B		
	4									1	Expenditure
										4.2	Quality of R&D Institutions
										4.2	University-Industry Collaboration
1	1		1	1			NA			47.4	Researchers' Headcounts
5	5			5						9.3	Published Articles
3	1	1	1	1						0.4	National Patents
5	1									0.33	International Patents

Table 9. Cooperation opportunities by priority: Knowledge Preservation / Sharing (ICT)

KSA Takes					KSA Gives					KSA	Indicator
S	C	I	R	B	S	C	I	R	B		
					5		1	5		16.4	Access to fixed telephone
						4	4			176.5	Access to mobile telephone
					3	5	1			72.6	Households with a computer
					5	5	1		5	72.7	Households with Internet access
					1	1	1			46682	International Internet Bandwidth

- Table 10 illustrates that the indicators of knowledge utilization impact and performance, provide KSA-the BRICS with knowledge cooperation opportunities in "6" out of "8" issues expressed by the indicators. The priorities of these opportunities are in the range "1 to 5". In most of these issues, KSA is in a position to take, and in some, it is in a position to give.

The results obtained shows that KSA can deliver some knowledge related experience to the BRICS, and that in the mean-time, it also needs other knowledge related experience from the BRICS. Knowledge collaboration will be beneficial to both in building their knowledge societies.

REMARKS AND RECOMMENDATIONS

The use of the knowledge framework presented in this chapter enabled the investigation of essential knowledge society issues. The results obtained showed that KSA scored higher than all the BRICS in two sub-indices of the framework: the knowledge diffusion through "education" sub-index; and the knowledge preservation and sharing through the "ICT" sub-index. Both of these are associated with the enablement of the knowledge society. In the knowledge creation through "R&D" sub-index, KSA scored close to the lowest of the BRICS "India"; this is despite the fact that KSA scored relatively higher in R&D enablement. In addition, in knowledge utilization impact and performance, the scores were mixed, but mostly in favor of the BRICS. As a whole, knowledge cooperation opportunities between the two sides exist, providing benefits to both. For bringing KSA and the BRICS together, concluded knowledge opportunities are explored in the following, considering those between KSA and each of the BRICS countries.

KSA-Brazil Opportunities

- Opportunities in which KSA can give experience to Brazil are associated with the following issues: science and engineering graduates (at priority 4); and quality of the education system, Internet access, economic growth, and investment outflow (at priority 5).

Table 10. Cooperation opportunities by priority: knowledge utilization (impact, performance)

KSA Takes					KSA Gives					KSA	Indicators
S	C	I	R	B	S	C	I	R	B		
1			1	1		NA	1			0.47	New Businesses
	5							3	5	4.0	Annual GDP Growth
	5		5	4						1.7	FDI Inflows
5	4		5						5	0.62	FDI Outflows
			1	1						30.4	High-Medium-Tech Output
1	1	1	1	1						0.1	High-Tech Exports
										75.5	Life Expectancy
										94.4	Employment

- Opportunities in which KSA can take experience from Brazil are related to the following issues: researcher's headcounts, national patents, new businesses, high-medium-tech out-put, and high-techy exports (at priority 1); investment inflows (at priority 4); and research publications (at priority 5).

KSA-Russia Opportunities

- Opportunities in which KSA can give experience to Russia are concerned with the following issues: economic growth (at priority 3); and fixed telephone access (at priority 5).
- Opportunities in which KSA can take experience from Russia are related to the issues of: researcher's headcounts, national patents, new businesses, high-medium-tech out-put, and high-tech exports (at priority 1); and investment inflows and outflows (at priority 5).

KSA-India Opportunities

- Opportunities in which KSA can give experience to India are associated with the following issues: fixed telephone access, computer access, Internet access and bandwidth, and new businesses (at priority 1); tertiary education enrollment (at priority 4); and education expenditure and secondary education enrollment (at priority 5).
- Opportunities in which KSA can take experience from India are related to the issues of: national patents, and high-techy exports (at priority 1).

KSA-China Opportunities

- Opportunities in which KSA can give experience to China are concerned with the issues of: Internet bandwidth (at priority 1); access to mobile phone (at priority 4); and tertiary education enrollment, computer access and Internet access (at priority 5).
- Opportunities in which KSA can take experience from China are related to the following issues: researcher's headcount, national patents, international patents, and high-techy exports (at priority 1); R&D expenditure, investment outflow (at priority 4); and research publication, economic growth and investment inflows (at priority 5).

KSA-South Africa Opportunities

- Opportunities in which KSA can give experience to South Africa are concerned with the following issues: Internet bandwidth (at priority 1); tertiary education enrollment, and computer access (at priority 3); quality of the education system, and access to fixed phone (at priority 4); and Internet access (at priority 5).
- Opportunities in which KSA can take experience from South Africa include: researcher's headcount, new businesses, and high-techy exports (at priority 1); national patents (at priority 3); and research publications, international patents and investment outflows (at priority 5).

It is generally noticeable that KSA enjoys higher level of knowledge society enablement from which the BRICS can gain; while the BRICS enjoy higher level of knowledge productivity from which KSA can gain. Bridging the two sides together will be beneficial to both, and will contribute to building global knowledge societies that support cooperation, promote peace and enhance development.

REFERENCES

BRICS. (2017). *BRICS Information Sharing and Exchange Platform*. BRICS Business Council. Retrieved from http://www.brics-info.org/

Dutta, S., & Lavin, B. (Eds.). (2016). The Global Innovation Index (GII) 2016: The Human Factor in Innovation. Cornell University, INSEAD and WIPO.

GCC. (2017). *The Cooperation Council of the Arab States of the Gulf, Secretariat General*. Retrieved from http://www.gcc-sg.org/eng/

ITU. (2016). Measuring the Information Society Report: 2016. International Telecommunication Union.

Schwab, K., & Sala-i-Martin, X. (Eds.). (2016). *The Global Competitiveness Report*. World Economic Forum.

UNESCO. (2005). *United Nations Education Scientific and Cultural Organization*. Paris: Towards Knowledge Societies.

UNESCO. (2010). United Nations Education Scientific and Cultural Organization, Toward Inclusive Knowledge Societies: A review of UNESCO's action in implementing the WSIS outcomes, Paris, 2010. UNESCO.

UNESCO. (2013). *United Nations Education Scientific and Cultural Organization*. Paris: Renewing the Knowledge Societies Vision for Peace and Sustainable Development.

UNESCO. (2017). *UNESCO Institute of Statistics: UIS*. UNESCO.

World Bank. (2017). *World-Bank Data*. Retrieved from http://data.worldbank.org/indicator/NY.GDP.MKTP.KD.ZG

World Telecommunication Development Report. (2005). Geneva: Access Indicators for the Information Society.

WSIS. (2005). *World Summit on the Information Society: Declaration of Principles: Building the Information Society: a global challenge in the new Millennium, Geneva 2003-Tunis 2005*. WSIS.

Section 3

Streamline and Emerging Technologies for Technology-Driven Innovation: Case Studies

Chapter 8
Classification of Traffic Events in Mexico City Using Machine Learning and Volunteered Geographic Information

Magdalena Saldana-Perez
Instituto Politecnico Nacional, Mexico

Miguel Torres-Ruiz
Instituto Politecnico Nacional, Mexico

Marco Moreno-Ibarra
Instituto Politecnico Nacional, Mexico

ABSTRACT

Volunteer geographic information and user-generated content represents a source of updated information about what people perceive from their environment. Its analysis generates the opportunity to develop processes to study and solve social problems that affect the people's lives, merging technology and real data. One of the problems in urban areas is the traffic. Every day at big cities people lose time, money, and life quality when they get stuck in traffic jams; another urban problem derived from traffic is air pollution. In the present approach, a traffic event classification methodology is implemented to analyze VGI and internet information related to traffic events with a view to identify the main traffic problems in a city and to visualize the congested roads. The methodology uses different computing tools and algorithms to achieve the goal. To obtain the data, a social media and RSS channels are consulted. The extracted data texts are classified into seven possible traffic events, and geolocalized. In the classification, a machine learning algorithm is applied.

DOI: 10.4018/978-1-5225-7347-0.ch008

INTRODUCTION

Internet is a huge source of information. Many web sites, social media, and repositories, let programmers, analysts, and users access to information about what happens in real world. Most of the internet content is created by users. People are not aware about how much data they have been producing during the last years (Li, 2015).

Social media represent a source of information about everything that concerns people such as education, technology, health, politics, and environment, among others. One advantage of social media is that makes possible to locate people or events letting users share their location (Zhao, 2015).

Internet and social media are used to identify and solve social problems; for example, are used by social movements to ask governments for justice and information about their management. Also, are used to communicate ideas and to connect people, to share safety measures when natural disasters occur, and to propagate information about certain situations and conditions that affect people's daily activities, such as traffic, air pollution, weather, rain, among others (Adamko, 2015).

One of the purposes of people when participating in social media is to be useful for others Han et al. (2017). When a person reports something that he perceives in his environment, for example a flooding, helps others to search for alternatives in order to avoid the area and to inform authorities about the problem, looking for a response.

In social media people act as sensors, the information they provide is the sensor data; in their posts citizens communicate the events and problems they perceive on roads through short texts, photos, and images; what is more, if the user allows it, many social medias are able to obtain the persons coordinates, which helps to identify the described events. The number of publications and posts related to a certain event on internet let analysts and citizens to understand the problems magnitude, and to formulate hypotheses and solutions.

Due to the huge development and growth of cities, the problems on their roads such as traffic, air pollution, among others, have considerably increased, affecting almost all the activities developed in a city.

Every day bottlenecks, traffic jams, and car accidents affect the economical, educational, and social activities in a city; the time a person spends in traffic events also affects his health. Social media has open the opportunity to have information about the real situation on the roads, letting people who have access to this information, change their routes or itineraries when problems are reported.

The data management to obtain information about the factors that affect the social dynamics, can be achieved by different computational and statistical methods. For example, classifying messages related to a specific event, or resubmitting the messages posted by trusty sources in order to make them accessible to more people (Dou, 2013).

Some researches interested in traffic at urban areas make use of data generated by vehicles with GPS implemented device (Zhang, 2011), such data are used to predict traffic in specific study areas. Alternate to the GPS generated data, cans the social media information increase the predictions accuracy? Or the ambiguity in certain social media posts decreases the predictions accuracy? Few researching works consider it.

In the present approach we propose the integration of a social media and a web RSS service to extract posts related to traffic events in a city, such as car accidents, protest marches, traffic jams, and others, that make difficult the displacement of people and vehicles through the studied area. After the extraction of the posts, they are textually processed, and geolocated to identify the geographic place where the event is reported. Finally, a machine learning algorithm is used to classify the processed data into six different kinds of traffic problems detected on the area.

A first version of the present approach is presented in the Traffic analysis based on short texts from social media (Saldana-Perez & Moreno-Ibarra, 2016).

The paper is organized as follows; in section two we show some related works about volunteered information; in section three we depict our methodology for the data integration and its geolocation; in section four we show the obtained results. Finally, in section five we point out our conclusions, and section six presents the discussions.

RELATED WORK

In this section we describe some works and concepts related to social media, to user generated content (UGC), and to the Volunteered Geographic Information (VGI).

The explosion and acceptance of social media, the technological advances, and the people's interest in generating content has increased updated data about what happens in real world that can be used by researchers to analyze the urban dynamics, and to generate possible solutions to many urban problems (Yang et al., 2016). Every image, text, photograph, design or concept created by an individual on internet whose motivation is to communicate an idea is part of the user generated content.

Volunteered Geographic Information (VGI) includes all the information generated by users on internet with coordinates or references to geographic places that make possible the geolocation of such information (Goodchild, 2007; Lin, 2013). The main principle of VGI is that it can be produced by people with or without advanced knowledge in the area to which the information bellows. This aspect opens the opportunity to more people to participate and increases the internet contents (Jiang, 2015; Cinnamon, 2013; Goodchild, 2015; Ricker, 2014; Vidal-Filho, 2013; Resch, 2015, Utomo & Ardiyanto, 2018).

Approximately 70% of smartphone users and 73.4% of internet users have a relation with VGI data. Although the high demand of this kind of information, most of the users consult or download it, few of them are producers (Havlik, 2013).

VGI has resulted useful when natural hazards or social problems appear, the data obtained from its analysis has let researchers and aid communities perform solutions, such as in the Haití earthquake (2013) when VGI was used to map affected areas and evacuation routes, in the Mexico earthquake (2017) when VGI helped to find missing people and to identify dangerous buildings.

Social media provide internet users the resources to generate VGI. One of the main characteristics of the social media information is that it is compound of few elements that in most of the cases provide concise ideas, and that can have geographical information such as coordinates or geographical elements described on their texts. This also, has increased the generation of multiple social media services such as Foursquare, Facebook, Twitter, Instagram, among others (Yilmaz & Hero, 2018).

In the present approach the social media Twitter is consulted to extract posts about the traffic situation in the study area. Twitter is a space on internet where people has the opportunity to create, get, and share information about many different topics such as health, environment, politics, sports, and more (Park et al., 2016). One of the main characteristics of Twitters posts is that they reflect the users ideas and feelings, or information about a specific topic. Users have the option to make their tweets public or not, and they can decide to share their coordinates or not in the tweets metadata (Pernici, 2018). According to (Leetaru et al. 2013), just one percent of the tweets produced every day have the coordinates of their place of publication.

In the case of tweets without coordinates, there are options to geolocate them; some researches use geographic information systems to geo-locate tweets, some others use specialized internet services such as the geo-location service of Google Maps (Bakerman et al. 2018); also there exists some researches that implement gazetteers to geolocate tweets in specific areas (Zheng & Sun, 2018, Dredze et al., 2013; Huang & Carley, 2017; Williams & Dixon, 2017).

Due to the tweets data availability and the information of events that can be obtained from them, is evident the utility of geospatial analysis over VGI information. Nowadays the participation of geospatial analysis in the VGI and UGC has become relevant (Saldana-Perez et al., 2018).

Many researches make use of tweets to analyze what is happening at big cities, since tweets represent updated citizens reports about what they see in their surroundings. Some other researches that make use of Twitter, locate the posts related to specific topics with a view to create maps that describe the urban problems interesting for citizens and authorities in their study areas. There exists maps that represent data from real time systems which monitor different factors in big cities such as energy saving, and dynamic regulations (Paredes-Valverde, 2015).

In Lansley & Longley (2016) approach, tweets posted in London are analyzed to identify the topics that concern to the citizens in the city related to 14 categories such as traffic, sports, culture, politic, work, pollution, among others. The research results let authorities and commercial agencies to know the main topics in the center of the United Kingdom capital.

It is true that the tweets veracity must be questioned but after a process of validation these can be considered as useful or not for the research purpose (Stehman, 2018).

One of the main computing areas that has particular interest in VGI, UGC and internet data is machine learning (Koswatte, 2018). Machine learning develops algorithms to obtain relevant data from heterogeneous information, examples of the machine learning usage are the data classification, the searching of patterns between data, the regression of data, analysis of humans behavior when buying, internet searching, or consuming multimedia content, among others (Mohadab, 2018).

Traffic Data Analysis From Social Media and RSS Notes

The present approach purpose is to make use of VGI and machine learning algorithms to obtain, classify, represent and analyze the traffic situation in the proposed study area, Mexico City. The main characteristic of the designed and implemented methodology is the use of machine learning to analyze an urban problem. The obtained results are used in a second research work where the classified data are considered as a data corpus for regressions computing.

The proposed methodology analyzes vehicular traffic, one of the main problems in big cities; it can be adapted to analyze some other city problems such as air pollution, public transport problems. The methodology extracts tweets related to the traffic situation in the area, applies a text mining process over the tweets, classify them using machine learning algorithms, and finally maps the classified tweets in order to analyze the relevance of each traffic problem and the areas that it affects the most.

Comparatively to other investigations that process a specific kind of data, in this work we structure a methodology to integrate and process data form different sources to make them useful for further investigations that need a corpus of pre-processed VGI data.

The methodology consists of five stages: *Database structuring*, databases to store data are designed; *Data monitoring*, data are obtained from the web sources; *Data management*, in this stage the text mining procedure is applied; *Data classification*, using a semi-supervised machine learning algorithm data are classified; and *Geocoding, t*he coordinates of the events are identified in order to map the analyzed event. In Figure 1, is show a blocks diagram of the proposed methodology.

Database Structuring

In this methodology stage the databases to store the treated data are structured. Since the data used in the approach proceeds from heterogeneous sources, the stage is divided into two sub-stages, *Geographic database structuring*, and *Events database structuring*. Both databases have been structured following the relational model (Kantardzic, 2011).

Geographic Database Structuring

The Geographic database is feed with two sources, an OpenStreetMaps shapefile from the OSM repository (Osm, 2018), and a *shapefile* provided by the INEGI (National Institute of Statistics and Geographic Information in Mexico.

Form the OSM shapefile is obtained information about the roads, streets, and public transport stations. The INEGI shapefile provides information about commercial entities located on the study area, such as the number of workers in each entity, its schedules, and the description of its activities.

Three tables compound the Geographic database: *roads, transport* and *infrastructure*.

Figure 1.

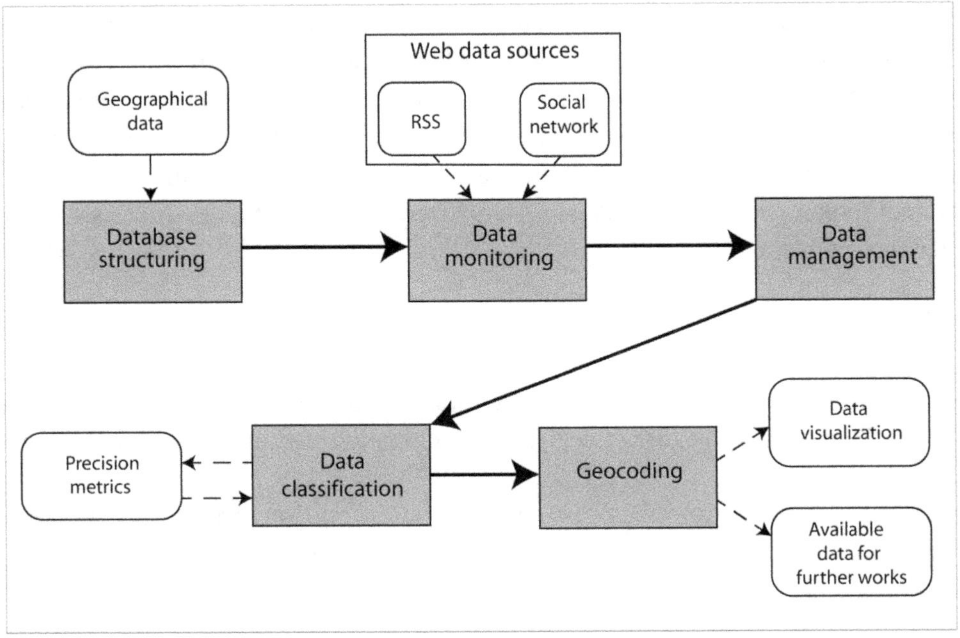

In the *table roads* is stored information about the streets and avenues in the studied area such as the name of the road, its length, the points that represent it, its beginning and ending coordinates, and the maximum number of vehicles that the road supports. The *table transport* contains information about the public transport in the area such as the coordinates and names of the bus and subway stations. The *table infrastructure* contains the information of the commercial entities in the area, its features are the same as the INEGI shapefile described above.

When working with geographic data is important to consider the following points, in order to avoid possible problems in the data processing.

- Names of the geographic elements such as streets, avenues, roads and points of interest must be normalized, this means that should not contain special characters.
- The intersection points between roads must be correct.
- Data of the geographic area must be updated when modifications are applied over their infrastructure.

Events Database Structuring

In this database are stored the posts recovered from the social media, and the RSS notes that describe traffic events in the study area.

The web sources monitoring has been limited considering the Tobler's first law of geography: "*Everything is related to everything else, but near things are more related than distant things*" (Tobler, 1970).

To treat the social media data, have been structured four tables: *usr_information* where the data of the user who published a message of interest are stored; the table *monitored_messages*, where messages recovered from the social media are stored; the table *messages_classified*, where after a text mining process the text of each message is normalized and the words of interest for the classification are identified; and finally, the table *geocoded_messages*, where the coordinates of the place where an event occurred are stored after the geocoding process.

For the RSS data, has been defined a table named *rss_data*, in this table the date, time, description, URL, latitude and longitude of the even reported in the RSS note, and its assigned classification are stored.

A fifth table named *urban_dynamics* has been structured to merge the recovered, classified, and geocoded data from the *geocoded_messages* and *rss_data* tables. The purpose of this table is to create a traffic events information data repository; this table makes possible to join the processed data from the two different monitored data sources in a normalized format. In Figure 2, are shown the fields of the *Events* database tables.

Data Monitoring

Two applications have been programmed to extract information; one of them uses the Python language program and the Twitter API to extract tweets; the other one written in PHP monitors and extract RSS notes.

Due two the differences between the consulted data sources and the applications, the data monitoring stage is divided in two sub-stages: social media monitoring, and RSS monitoring.

Figure 2.

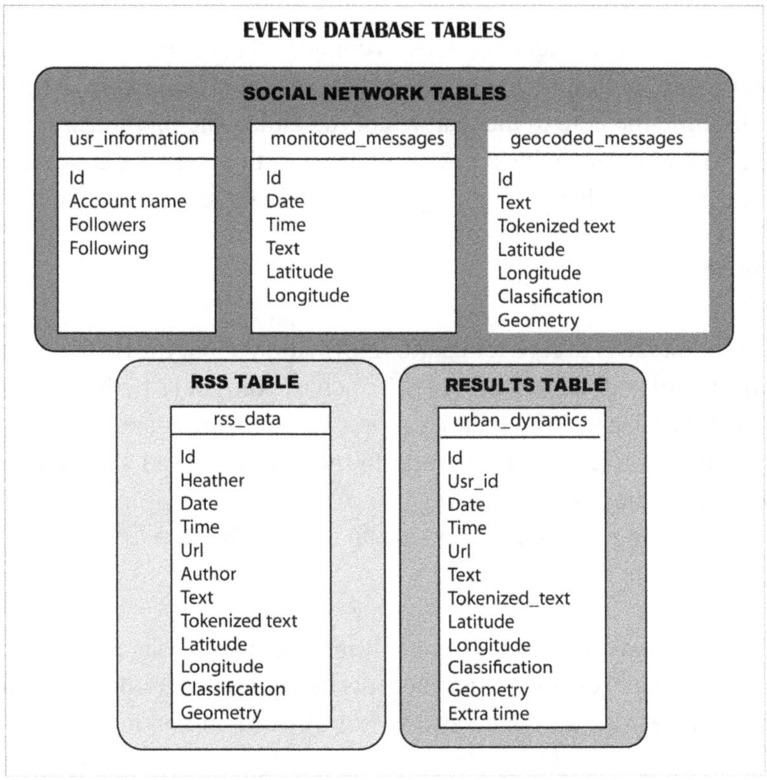

Social Media Monitoring

The messages posted in Twitter are called tweets these are 280 characters limited texts. Twitter provides a group of application interfaces for many programming languages that let them access to the tweets posted at any time in the microblogs site. The API used in this approach is the TwitterAPI (Twitterapi, 2018)

In the python program has been set a group of specific accounts to be monitored, this accounts belong to users dedicated to report the traffic conditions in the study area, their information comes from authorities and journals. Since the information they provide comes from official sources, the accounts have a large number of followers, having access to these accounts lets the application avoid repeated posts. In order to have a control about the followed users, their information is stored in the *usr_information* table.

Every time a followed account posts a tweet, this message is extracted by the programmed application and stored in the table *monitored_messages*. In the *monitored _messages* table tweets are stored preserving their original characteristics, the table represents a corpus of tweets in their original format. The monitoring of the social media is done 24 hours a day. As a future work it is considered to increase the number of monitored accounts.

RSS Monitoring

The application programmed to extract RSS notes is connected to the RSS channels of the main journals published in the study area, each time a journal posts a note the application extracts it and stores it in the table *rss_data*, dividing the note in the features of the table. The monitoring of the RSS channels is done 24 hours a day. To homogenize the extracted data to use them in the classification process, the text of the messages is treated through a text mining procedure in the data management stage.

Data Management

The main purpose of this stage is to recover useful information about traffic from the analyzed text, this is done identifying traffic related words from texts. To achieve the goal a python application programmed for the stage extracts the registries of the tables *monitored_messages* and *rss_data*, eliminates textual elements that do not provide relevant information to the approach, and store the processed text at the correspondent table in the field *tokenized_text*.

The steps of the text mining applied process are four; after the step tree, the text messages are considered processed and ready to be classified.

Step 1: Elimination of special characters. The analyzed Twitter messages are written in Spanish language, they contain special characters and accents that need to be removed in order to apply a better data analysis. Each message is treated as a vector of words if a word contains a special character (‘:,;.#$``!¿?¡/*...,) it is removed from the word, the modified word is inserted into the vector at the same the position from where it was extracted.

Step 2: Elimination of accents: In Spanish language words are written the same as they sound, so the words contain graphical accents. It is necessary to remove such accents since for many text processing tools these element may produce errors.

Step 3: Defining and removing *stop words*: One of the premises in the information recovery is to remove *stop words* from text, to make possible the extraction of relevant information. *Stop words* are all the words that usually refer to the most common ones in a language, some tools specifically avoid removing these words to support phrase search (Sidorov, 2014). Some words considered on this group are prepositions, common words, defined, and undefined articles (Silverman, 2013). On this step, we remove the *stop words* identified from the text.

Step 4: Manual classification: One of the approach purposes is to use a machine learning algorithm to classify traffic events, to make this possible a training corpus with previously classified data is needed. In this step, the third part of the treated registries is manually classified by a group of five annotators.

The annotators have been trained for the task, which consists on classify a message according to the major number of class words in its processed texts, the class words are words related to a specific traffic event.

The registry treated and its designated classification are stored at the c*lassified_messages*, in the field *classification*.

Step 5: First classification. To facilitate the classification of big amounts of registries a python script with multiple conditional sentences has been implemented, following the logic of the manual classification. In the script words related to each class of traffic event are grouped, the words of the processed text of each registry are analyzed and it is counted the number of class words that it contains; the class assigned to the message is the one with more words occurrences in the processed text. The registry treated and its designated classification are stored at the *classified_messages,* in the field *classification.*

The traffic event classes defined for the approach are seven, they have been proposed according to the most common traffic problems in Mexico city.

In Table 1, the classes have been described more detailed.

After the messages classification, data are divided into two sets, the 75% of the registries is considered as the training corpus meanwhile the other 25% is considered as the test corpus. Bot corpus are used in the semi supervised classification with a machine learning algorithm.

Data Classification

In this stage a machine learning algorithm is used to classify traffic related messages in a semi supervised way. After the classification, a geocoding process is applied over the registries to make possible their visualization in a map of the study area.

In the approach the machine learning algorithm implemented is the Naïve Bayes algorithm which, is widely used in classification and filtering of text messages and e-mails.

One of the advantages of using tweets is that their limited texts reduces the searching field of elements improving the classification precision rate; also, lets to calculate the words presence probability in a message (Lantz, 2013). The classification Naïve Bayes algorithm has been implemented considering the four steps described below.

Table 1.

Class	Definition
Traffic	The traffic is slow, even there may be a bottle neck on the road. Some words that belong to this class are: slow, not movement, traffic jam.
Favorable	The flow is appropriate, the vehicles can move through the roads with a good average speed. Some words to reference this class are: fast, good, free road.
Accident	There is a car crash on the road or someone got injured near to the road.
Closure	The movement through the road is not possible, it has been closed by a protest march or a controlled event.
Rule	The message describes something related to governmental rules or official disposals related to vehicles.
Manager	It is a message that the VGI manager published in order to give information about the application, the message does not describe a traffic event.
unidentified	The information on the message is not enough to classify it.

Step 1: Identifying data. In the Naïve Bayes process a python script extracts the training corpus been structured on the text mining process from the table *classified_messages*. The algorithm considers as relevant features the *id* of the registry, the *tokenized_text*, and its *classification*.

Step 2: Factorization. After the extraction of the training data, the classification of the registries is factorized; this means that the algorithm represent each label class as an integer number.

Step 3: Data dictionary. In an empty dictionary, each one of the words of the first registry is added as a key, then for the next registries the tokenized text words are compared with the dictionary ones, in case that a word of the tokenized text has not been included in the dictionary, such word is added as a new key.

The value of the words keys is the number of times that each word has appeared in the entire corpus; by this way the dictionary has the form *key:value*, where the key is the word found in the registries and its value is the number of times the word appeared in the corpus. At the final of this step, the dictionary would have all the words that are part of the registries texts (even if not all the words belong to the same registry), and the number of repetitions of each word.

Step 4: Vectorization. For the machine learning procedure it is needed that the tokenized text of the registries has the same length as the words dictionary, to achieve this, a words vector is created for each registry.

The words vector have the same length as the dictionary, each vector position represents a word in the dictionary. If a dictionary word appears in the registry text, such word position in the vector has as content the number of times that the specific word appeared in the text; if a dictionary word does not appears in the registry text its value in its vector position is zero.

After this process, the registry is represented as a numeric vector that has the same length as the dictionary. In Figure 3 is shown a diagram that illustrates the vectorization process.

Figure 3.

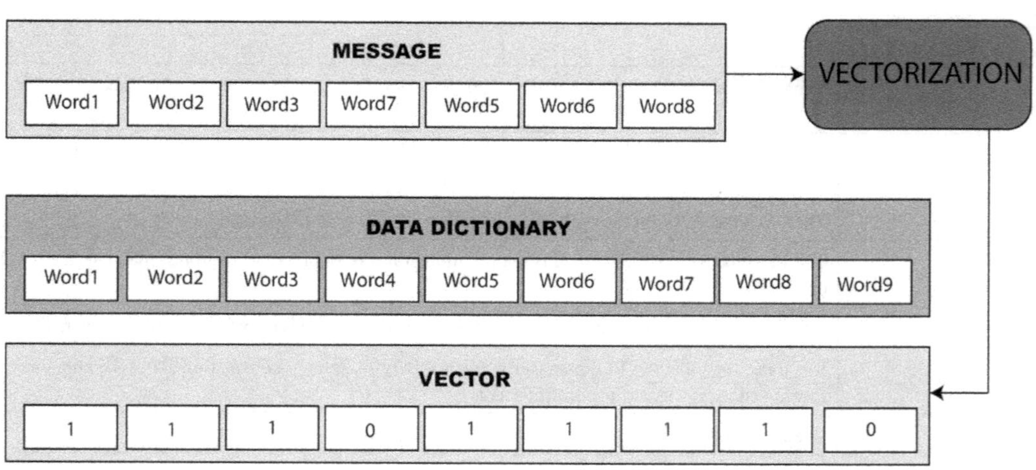

Step 5: Term-document matrix. It is possible to obtain the term-document matrix of the data corpus using the data dictionary. The words of the dictionary represent the terms, they are the fields of the matrix; meanwhile the vectors are the documents, the rows of the matrix. Inside of the matrix there are numbers that represent the number of times a word appeared in a certain registry. Most of the times the term-document matrix are sparse since the probability of occurrence of each word in all the registries is low.

Step 6: Training. In this process are considered the words from the dictionary, the belonging probability of the words to a specific class, and the probability that the registries are correctly classified according to the words they contain.

Bayes probability is based on the idea that an events probability depends on the probes at the precise moment. Naïve Bayes classification algorithm is based on two premises: the probabilities of event occurrences must be exhaustive and mutually exclusive; and that the joint probability must be considered.

Applying the Bayes theorem, it is used the Equation 1.

$$P(A_i \mid B) = \frac{P(B \mid A_i)P\left(A_i\right)}{P\left(B\right)}. \quad (1)$$

In Table 2, we describe the components of the equation 1.

Suppose there is needed to determine the probability that a registry describes an accident class event. It is known that a word of the message is '*closed*', the probability that '*closed*' was found in other messages classified as '*accident*' is the likelihood, the probability of the word '*closed*' to appear at any message is the marginal probability; the posterior probability is the probability of classify the message with the label A_i that represents the class '*accident*'.

Table 2.

Term	Meaning
B	The classification label.
A_i	The dependent event, a classification that shares features with the expected classification.
$P\left(B\right)$	Marginal probability of B, the probability that B appears at any message.
$P\left(A_i\right)$	Prior probability of event A_i
$P(B \mid A_i)$	Likelihood of B in event A_i
$P(A_i \mid B)$	Posterior probability of A_i occurrence given that B appeared.

Step 7: Learning. Considering that we are working with multiple words, there is used an extended classification Naïve Bayes. The algorithm is trained with the construction of a probabilities table, where there has been calculated the probability that each word of the dictionary belongs or not to each one of the defined classes (Bishop, 2006).

Any time a new registry is analyzed by the python script to be classified, the posterior probability must be calculated, with a view to determine the class where the message must be classified considering the probabilities of the words included or not on the message. This means that for the final probability, it is necessary to include the probability of presence or absence of each word in the message. This computation must be done for each class toward to identify the highest probability that would denote the correct classification of the event, in Equation 2 there is represented this calculation. In Equation 3 it is represented the final probability of classify a text as element of class 1, supposing that it has 4 words and the data dictionary has 7 words. In Table 3, are described the terms of Equation 2 and Equation 3.

$$Sum\,of\,probabilities\,calculated\,for\,other\,classes = P(Class1 \mid Pword1 \cap Pword2 \cap Pword3 \cap Pword4 \cap \neg Pword5 \cap \neg Pword6 \cap \neg Pword7) + \ldots + P(Class\,n \mid Pword1 \cap Pword2 \cap Pword3 \cap Pword4 \cap \neg Pword5 \cap \neg Pword6 \cap \neg Pword7) \tag{2}$$

$$P\left(text \in Class1\right) = \frac{P(Class1 \mid Pword1 \cap Pword2 \cap Pword3 \cap Pword4 \cap \neg Pword5 \cap \neg Pword6 \cap \neg Pword7)}{Sum\,of\,probabilities\,calculated\,for\,other\,classes} \tag{3}$$

The Naïve Bayes algorithm classifies the registry in the class with the highest probability. In the learning step the probabilistic model is structured as it was described above; this model is used also in the classification step.

Step 8: Models evaluation. Applying precision metrics to the obtained results is possible to determine if the algorithm performance is appropriate, or to identify in which points it should be improved.

To measure the precision of the Naïve Bayes algorithm applied in this research, the precision metrics of mean square error, the mean absolute error and the confusion matrix are computed.

In the confusion matrix, the columns represent the number of predictions for each class, and the rows the classified instances in the correct class; this is a useful tool to identify the classes where the algorithm presents more classification deficiencies (Bergstra, 2013).

Step 9: Final Classification. Once the algorithm has been trained with the training corpus and its deficiencies have been repaired, is time to apply the Naïve Bayes classification over the test data corpus. Below there are written the machine learning algorithm steps:

- The algorithm tokenizes the text of each registry from the test corpus,
- adds to the data dictionary the missing words,

Table 3.

Term	Meaning
$Sum\ of\ probabilities\ calculated\ for\ other\ classes$	Sum of the conditional probabilities calculated for each class.
$Class1$	e class that is being analyzed, there are seven different classes in this study.
$Pword1$	The probability that the word 1 appears in the registry text, this probability must be calculated for every word.
$\neg Pword1$	The probability that the word 1 does not appears in the registry text, this probability must be calculated for every word.
$P(Class1 \mid Pword1 \cap Pword2 \ldots)$	Conditional probability of the class, it considers the intersection of the words that appear and the words that do not appear in the registry.
$P\left(message \in Class1\right)$	Probability that the text of the registry that is being analyzed belongs to class 1.

- calculates the frequency for the new words added in the same way that the frequency of the training corpus words,
- analyzes the tokenized text to know the words that compound it,
- creates the text vectors and homogenizes their length with the dictionary length; obtains the term-document matrix.
- Based on the words of each vector and their calculated frequencies, the algorithm classifies the registries.
- Assigns a class label to the registry. In case that the registry came from the social media, the registry is stored in the table *geocoded _messages* with its machine learning assigned label, if the classified registry was a RSS note, the result would be saved in the *rss_data* table.

Geocoding

Geocoding of messages to identify the coordinates of the geographical place described in their texts is a common process, there exist different tools such as geographic information systems (GIS), and web services such as Google Maps that help developers in this work (Crooks, 2013).

In this last methodology stage are identified the geographical coordinates of the events described in the registries texts, this section is divided into two steps: *coordinates extraction* and *geometry.*

Coordinates Extraction

To obtain the coordinates of the registries the python script created for achieve the methodology stages, searches in a gazetteer of the study area the name of the streets, avenues, or points of reference provided in the text of the registries, the gazetteer returns the coordinates of the place. The application overlaps the coordinates assigned to the geographic elements, and after the spatial overlapping function, the python program obtains the coordinates of the place where the described event probably occurred. The coordinates of the event are stored in the registries fields' *latitude* and *longitude* of the table *geocoded_messages.*

The gazetteer and the geocoding process used in the approach are part of a geocoding methodology implemented in the research work of Salazar et al. (2016). This methodology has been used to geocoding the messages since it is not a purpose of the present approach.

After testing the geocoding results, we have noticed that in some cases there is an imprecision in the events location due to the length of some roads. To filter the events that were correctly geocoded from the ones that did not, the points of the coordinates are mapped in order to visualize those which location is not real. If an event was not correctly located, an external annotator geocodes the event manually, by searching the address of the registry text in Google Maps.

For the RSS data, the geocoding process is a little more complex. Most of the times the RSS registries description of the place where the event occurred is not clear or includes many details that generate confusion to the geocoding tools and gazetteers; to solve this problem, we have designated an external annotator to geocode such registries by using the Google Maps geocoding tool. The annotator searches in Google Maps the address of the RSS registry text, and saves the coordinates in the *latitude* and *longitude* fields of the *rss_data* table.

Geometry

The coordinates of all the registries are merged into a geometric point that represents the location of the event. This step is needed to visualize the events and to create new geographical features as polygons or heat maps from their analysis that can be useful for other investigations.

This step is implemented using the spatial extension PostGIS for the database manager Postgresql. Through a query to obtain the coordinates assigned to each registry, Postgis merges the both coordinates using the spatial projection of the analyzed geographic area, the resultant geometry data, the coordinates, and the classification of the event, are stored in the table *urban_dynamics*.

Results

The present approach surges from the needed to have a classification of the real traffic evens that every day affect to urban areas, as case of study has been chose the Mexico City, one of the most crowded cities in world.

Mexico City is one of the most important financial centers in Latin America, with a population around 21.2 million people, and an area of 1,485 square kilometers. It is the largest metropolitan area in the earth western hemisphere and the largest Spanish-speaking city in the world (Alba, 2015). Due to the increasing interest of people and local authorities in VGI and internet content, the use of internet as a data source for the present approach has been considered.

In the approach different computing technologies are merged to achieve the methodology purpose. For the database structuring the database manager Postgresql and its geospatial data tool Postgis are used. The analyzed data to obtain information about traffic events in the study area are posts from Twitter and RSS notes. The acquisition of the RSS notes from the main journals in the studied area is done using a PHP application programmed for that purpose; the extraction of tweets is done using Python language program.

After the extraction of data from internet and its storage in the database Events, the messages are treated as registries and processed through a text mining procedure where grammatical elements that do not provide information to classify the message are eliminated, and words of interest for the classification and geocoding of the message are identified.

Seven traffic eves class are proposed: *traffic, favorable, closure, accident, rule, manager, and unidentified.* The processed text of the registries is classified manually and using a python script to structure a training and a test corpus for the posterior machine learning data classification. Finally, the registries are classified by a semi supervised method using the machine learning algorithm Naïve Bayes. The classification algorithm is trained with the classified samples, and tested with text preprocessed registries.

In Figure 4, the points in the map represent the traffic related messages analyzed, these are registries from tables *rss_data* and *geocoded_messages.*

Regarding to the semi-supervised classification process, two different classification methods are compared, a script where the most common words that describe a class event are matched with the registries text words using conditional programming sentences, and a Naïve Bayes algorithm trained with preprocessed data.

From the Naïve Bayes method good classification results were obtained, the principle of the algorithm is to learn from the training corpus those words that compound the description of each class, then analyzes the words from each registry of the test corpus and according to each class words presence or absence in the registries text classifies it.

Analyzing the precision metrics of our *Data classification* procedure, the Naïve Bayes algorithm threw an error of 0.110482 when classifying social media registries and when classifying RSS registries, it presented the highest error with a value of 0.401523; however, this error rate was lower than the first classification method, and more efficient. Another advantage of the Naïve Bayes classification is that the classification is faster, even for big amount of data.

Figure 4.

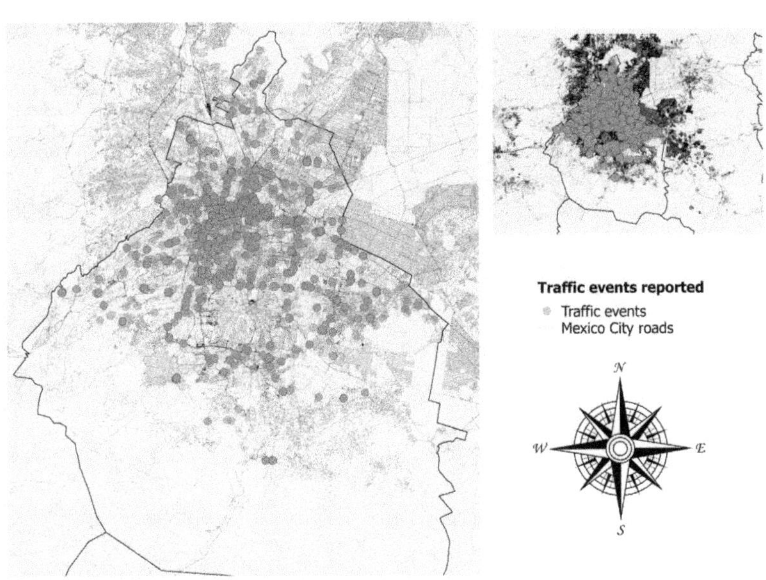

The *urban_dynamics* table, stores the geocoded traffic events of the study area. The data can be used to be classified by using other machine learning tools.

Figure 5, shows a map where the posted traffic messages density in the studied area is shown through a heat map. The color levels change according to the number of traffic events reported in the area, if few messages were posted then the color is light; if a big amount of tweets and RSS notes describes events in the same area, its color in the heat map is dark.

To have a different perspective about the traffic situation in the study area, a map that combines a heat map and points is shown in Figure 6. The heat map represents the areas where traffic problems have been reported, meanwhile the points represent reports of the favorable class. Regarding the Figure 6, is possible to see that the number of reports about good circulation on the roads is smaller than the traffic report; this has a direct relation with the peoples interest in report problems in their city, and because when the circulation on roads is speedy people prefers to save time instead of posting the traffic conditions.

CONCLUSION

The here presented method is oriented on extracting and processing social data in order to make them compatible and available for further investigations. The integrated information concerns to the cartography of the study area, bus and subway stations, and commercial business located on its roads, and the most important, the events related to traffic and vehicular problems that occurred on the area.

The present approach is guided to analyze the traffic problems that affect urban areas using VGI and machine learning algorithms. In the approach textual and geographic information is analyzed.

Figure 5.

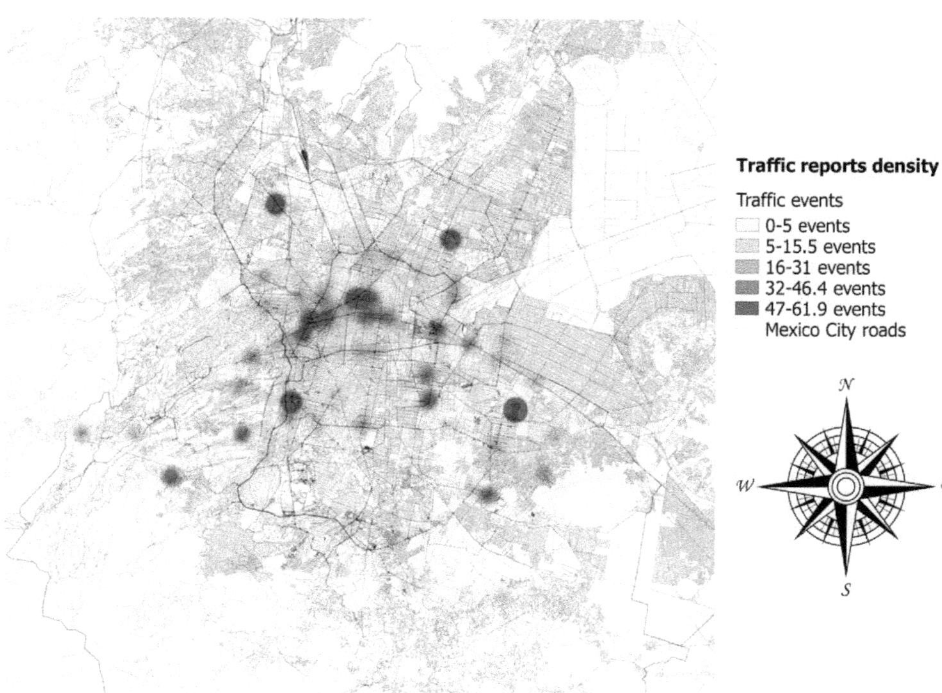

**For a more accurate representation see the electronic version.*

Figure 6.

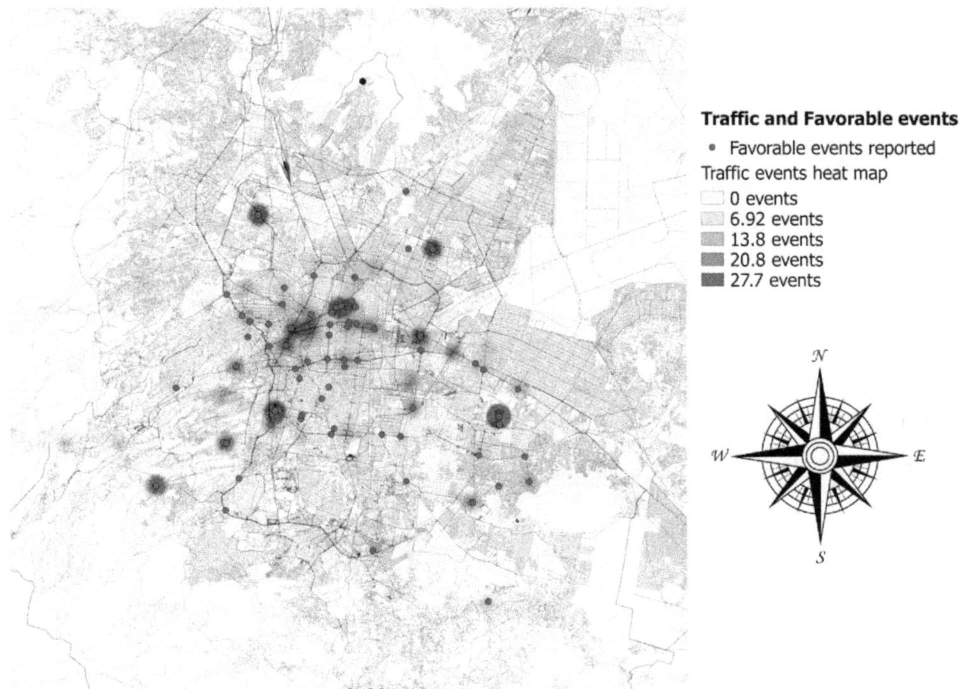

**For a more accurate representation see the electronic version.*

The traffic problems descriptions are obtained from two web sources not related between them: a social media and a RSS service. The social media data are stored in the database named *Events*, which has five tables: *usr_information, monitored_messages, rss_data*, *geocoded_messages* and *urban_dynamics.*

The geographic information is obtained from shapefiles with data about the roads in the study area; such data are stored in a geospatial database compound of three tables: *roads, infrastructure* and *transport.*

After the monitoring of the web sources, the data are preprocessed by a text mining procedure in order to extract the registries relevant information. The third part of the treated registries are classified manually, the rest of the samples are classified using a script with conditional statements in order to structuring the training corpus. The 75% of the total samples is considered as the training corpus, the rest of the samples are used in the test corpus. The training and the test corpus are used in the Naive Bayes machine learning algorithm, with a view to classify automatically the registries obtained, into one of the seven classes stablished for this investigation. The classes are: *traffic, favorable, closure, accident, rule, manager, and unidentified.*

After the classification, a geocoding process is applied to the data with the purpose of get the coordinates where each event occurred and create geometric points in order to visualize the data in a map.

When all the proposed methodology stages are complete, the registries from the two processed tables *rss_data* and *geocoded _messages,* are merged into the table *urban_dynamics*; this last table represents a repository of road and traffic events that can be useful for some other classification procedures or for further investigations related to traffic. The table urban_dynamics has information of messages that describe traffic events in the study area, such as their tokenized and cleared text, its assigned classification label, the coordinates where the event occurred, and the geometry field that allows to Geographic Information Systems to represent the registries as points.

The *geographic* and the *events* databases can be useful for GIS to represent the roads of the area for which the analysis was applied, the location of bus, subway and commercial businesses located nearby the roads; and the road events that affect the correct displacement of vehicles and people of the area.

The *events* database has been structured with the purpose of creating a data repository, about the traffic events on a city that could be useful for other works, since there are many web data sources that cannot be used together but that could provide valuable information to develop new data urban systems.

DISCUSSION

The present approach analyses VGI information and internet content. The VGI information is obtained from a social media, its main characteristic is that such information represents citizens reports of the traffic situation in the study area.

VGI lets data analysis processes to have direct contact with real world studying the users reports. People reports what they perceive and consider interesting for other users and authorities. VGI lets have updated information about topics that concern to the society where the information is produced.

Machine learning is a powerful computing area, one of its main purposes is to automatize data analysis procedures learning from the data features and achieving high precision rates.

The main purpose of the presented work is to detect the main traffic problems that affect the urban dynamics at big cities, studying tweets and RSS notes posted in Mexico City. The traffic events classes have been proposed considering the different characteristics of traffic events reported in the area through Twitter. In the methodology a machine learning classification algorithm is implemented in order to merge VGI and UGC content with machine learning.

Through the methodology process tweets and RSS notes are extracted from internet, text mining processed, classified by three different tools (manually, with conditional programming, and with a ML algorithm). Finally, data are geovisualized in maps of the studied area. The maps let identify the areas with more traffic reports, and the alternative roads that can be used to avoid traffic jams in the analyzed city.

The geovisualization of the results opens to opportunity to data interpretations related with social aspects such as the areas with more traffic reports, the relation between congested areas and air pollution, the participation of citizens reporting traffic events, the alternative routes to avoid crowded roads, and how the analyzed data and designed maps can be used by authorities and organizations to keep citizens informed about the traffic situation in their neighborhood.

As results of the implemented methodology a data corpus of tweets related to traffic in their original format, and a data corpus of text processed tweets which main topic is the traffic in Mexico city are structured. Such data corpus are used in a posterior research focused in data regressions, and represent a useful data set since there are not many compendiums of traffic related information.

As further work for the approach is proposed the text processed traffic related tweets classification using different machine learning algorithms such as K Nearest Neighbors, Support vector machine, clustering, among others, to prove their efficiency and select the one with the highest precision to classify the data. Also, a real time analysis could be develop in order to analyze tweets as soon as they are posted, and to have a more accurate traffic events visualization.

The methodology presented can be adapted to work with data from other urban area as long as there exists information about its roads, and after extracting tweets and UGC from its citizens. Once the methodology is feed with the VGI and geospatial needed information, the text mining and the classification processes can be applied.

REFERENCES

Adamko, M., Navrat, P., & Kovarova, A. (2015). Personalised Recommendation of Who to Follow Based on Fellowship of Followers. *Editorial Policy, 1*.

Alba, F. (2015). International migration: Consolidation of emerging patterns. *Demos (Mexico City, Mexico)*, (13), 10–11.

Bakerman, J., Pazdernik, K., Wilson, A., Fairchild, G., & Bahran, R. (2018). Twitter Geolocation: A Hybrid Approach. *ACM Transactions on Knowledge Discovery from Data, 12*(3), 34.

Bergstra, J., Yamins, D., & Cox, D. (2013). Making a science of model search: Hyperparameter optimization in hundreds of dimensions for vision architectures. In *Proceedings of The 30th International Conference on Machine Learning* (pp. 115-123). Academic Press.

Bishop, C. M. (2006). *Pattern recognition and machine learning*. Springer.

Cinnamon, J., & Schuurman, N. (2013). Confronting the data-divide in a time of spatial turns and volunteered geographic information. *GeoJournal, 78*(4), 657–674. doi:10.100710708-012-9458-6

Crooks, A., Croitoru, A., Stefanidis, A., & Radzikowski, J. (2013). # Earthquake: Twitter as a distributed sensor system. *Transactions in GIS, 17*(1), 124–147. doi:10.1111/j.1467-9671.2012.01359.x

Dou, W., Yu, L., Wang, X., Ma, Z., & Ribarsky, W. (2013). Hierarchicaltopics: Visually exploring large text collections using topic hierarchies. Visualization and Computer Graphics. *IEEE Transactions on, 19*(12), 2002–2011. PMID:24051766

Dredze, M., Paul, M. J., Bergsma, S., & Tran, H. (2013, June). Carmen: A twitter geolocation system with applications to public health. In AAAI workshop on expanding the boundaries of health informatics using AI (HIAI) (pp. 20-24). AAAI.

Goodchild, M. F. (2015). Twenty years of progress: GIScience in 2010. *Journal of Spatial Information Science,* (1), 3-20.

Han, Y., Hong, B., Lee, H., & Kim, K. (2017). How do we Tweet? The Comparative Analysis of Twitter Usage by Message Types, Devices, and Sources. *The Journal of Social Media in Society, 6*(1), 189–219.

Havlik, D., Egly, M., Huber, H., Kutschera, P., Falgenhauer, M., & Cizek, M. (2013). Robust and trusted crowd-sourcing and crowd-tasking in the Future Internet. In Environmental Software Systems. Fostering Information Sharing (pp. 164-176). Springer Berlin Heidelberg. doi:10.1007/978-3-642-41151-9_16

Huang, B., & Carley, K. M. (2017). *On Predicting Geolocation of Tweets using Convolutional Neural Networks*. arXiv preprint arXiv:1704.05146

Jiang, B., & Thill, J. C. (2015). Volunteered Geographic Information: Towards the establishment of a new paradigm. *Computers, Environment and Urban Systems*, *53*, 1–3. doi:10.1016/j.compenvurbsys.2015.09.011

Kantardzic, M. (2011). *Data mining: concepts, models, methods, and algorithms*. John Wiley & Sons. doi:10.1002/9781118029145

Koswatte, S., McDougall, K., & Liu, X. (2018). VGI and crowdsourced data credibility analysis using spam email detection techniques. *International Journal of Digital Earth*, *11*(5), 520–532. doi:10.1080/17538947.2017.1341558

Lansley, G., & Longley, P. A. (2016). The geography of Twitter topics in London. *Computers, Environment and Urban Systems*, *58*, 85–96. doi:10.1016/j.compenvurbsys.2016.04.002

Lantz, B. (2013). *Machine learning with R*. Packt Publishing Ltd.

Leetaru, K., Wang, S., Cao, G., Padmanabhan, A., & Shook, E. (2013). Mapping the global Twitter heartbeat: The geography of Twitter. *First Monday*, *18*(5). doi:10.5210/fm.v18i5.4366

Li, Q., Song, D., Liao, L., & Liu, L. (2015). Personalized Mention Probabilistic Ranking–Recommendation on Mention Behavior of Heterogeneous Social media. In *Web-Age Information Management* (pp. 41–52). Springer International Publishing. doi:10.1007/978-3-319-23531-8_4

Lin, W. (2013). Volunteered geographic information and networked publics? Politics of everyday mapping and spatial narratives. *GeoJournal*, *78*(6), 949–965. doi:10.100710708-013-9490-1

Mohadab, M., Bouikhalene, B., & Safi, S. (2018). *Predicting rank for scientific research papers using supervised learning*. Applied Computing and Informatics.

Osm. (n.d.). Retrieved from https://www.openstreetmap.org

Paredes-Valverde, M. A., Alor-Hernández, G., Rodríguez-González, A., Valencia-García, R., & Jiménez-Domingo, E. (2015). A systematic review of tools, languages, and methodologies for mashup development. *Software, Practice & Experience*, *45*(3), 365–397. doi:10.1002pe.2233

Pernici, B., Francalanci, C., Scalia, G., Corsi, M., Grandoni, D., & Biscardi, M. A. (2018). *Geolocating social media posts for emergency mapping*. arXiv preprint arXiv:1801.06861

Resch, B. (2013). People as sensors and collective sensing-contextual observations complementing geo-sensor network measurements. In *Progress in Location-Based Services* (pp. 391-406). Springer Berlin Heidelberg.

Resch, B., Summa, A., Sagl, G., Zeile, P., & Exner, J. P. (2015). Urban emotions—geo-semantic emotion extraction from technical sensors, human sensors and crowdsourced data. In Progress in Location-Based Services 2014 (pp. 199-212). Springer International Publishing.

Ricker, B., Daniel, S., & Hedley, N. (2014). Fuzzy Boundaries: Hybridizing Location-based Services, Volunteered Geographic Information and Geovisualization Literature. *Geography Compass*, *8*(7), 490–504. doi:10.1111/gec3.12138

Salazar, J. C., Torres, M., & Moreno, M. (2016). *Monitoreo urbano de entidades y eventos geográficos basado en censado social* (Unpublished master dissertation). Centro de Investigación en Computación, Instituto Politécnico Nacional, Mexico.

Saldana-Perez, A. M. M., & Moreno-Ibarra, M. (2016). Traffic analysis based on short texts from social media. *International Journal of Knowledge Society Research*, *7*(1), 63–79. doi:10.4018/IJKSR.2016010105

Saldana-Perez, A. M. M., Moreno-Ibarra, M. A., & Torres-Ruiz, M. J. (2018). Classification of Traffic Events Notified in Social Networks' Texts. In Encyclopedia of Information Science and Technology, Fourth Edition (pp. 6973-6984). IGI Global.

Sidorov, G., Velasquez, F., Stamatatos, E., Gelbukh, A., & Chanona-Hernández, L. (2014). Syntactic n-grams as machine learning features for natural language processing. *Expert Systems with Applications*, *41*(3), 853–860. doi:10.1016/j.eswa.2013.08.015

Silverman, K., Naik, D., Lenzo, K., & Henton, C. (2013). *U.S. Patent No. 8,583,418*. Washington, DC: U.S. Patent and Trademark Office.

Stehman, S. V., Fonte, C. C., Foody, G. M., & See, L. (2018). Using volunteered geographic information (VGI) in design-based statistical inference for area estimation and accuracy assessment of land cover. *Remote Sensing of Environment*, *212*, 47–59. doi:10.1016/j.rse.2018.04.014

Tobler, W. R. (1970). A computer movie simulating urban growth in the Detroit region. *Economic Geography*, *46*, 234–240. doi:10.2307/143141

Twitterapi. (n.d.). Retrieved from https://developer.twitter.com/en/docs/developer-utilities/twitter-libraries

Utomo, M. N. Y., Adji, T. B., & Ardiyanto, I. (2018). Geolocation Prediction in Social Media Data Using Text Analysis. *RE:view*.

Vidal-Filho, J. N., Lisboa-Filho, J., de Souza, W. D., & dos Santos, G. R. (2013). Qualitative Analysis of Volunteered Geographic Information in a Spatially Enabled Society Project. In *Computational Science and Its Applications–ICCSA 2013* (pp. 378–393). Springer Berlin Heidelberg. doi:10.1007/978-3-642-39646-5_28

Williams, E., Gray, J., & Dixon, B. (2017). Improving geolocation of social media posts. *Pervasive and Mobile Computing*, *36*, 68–79. doi:10.1016/j.pmcj.2016.09.015

Yang, L., Dong, L., & Bi, X. (2016). An improved location difference of multiple distances based nearest neighbors searching algorithm. *Optik-International Journal for Light and Electron Optics*, *127*(22), 10838–10843. doi:10.1016/j.ijleo.2016.08.091

Yılmaz, Y., & Hero, A. O. (2018). Multimodal Event Detection in Twitter Hashtag Networks. *Journal of Signal Processing Systems for Signal, Image, and Video Technology*, *90*(2), 185–200. doi:10.100711265-016-1151-4

Zhang, F., Yuan, N. J., Wilkie, D., Zheng, Y., & Xie, X. (2015). Sensing the pulse of urban refueling behavior: A perspective from taxi mobility. *ACM Transactions on Intelligent Systems and Technology*, *6*(3), 37. doi:10.1145/2644828

Zhao, Z., Cheng, Z., Hong, L., & Chi, E. H. (2015, May). Improving User Topic Interest Profiles by Behavior Factorization. In *Proceedings of the 24th International Conference on World Wide Web* (pp. 1406-1416). International World Wide Web Conferences Steering Committee. 10.1145/2736277.2741656

Chapter 9
Exploring Key E-Services Issues Toward Future Development

Abdulaziz Alshouiby
Saudi E-Government Program, Saudi Arabia

Saad Haj Bakry
King Saud University, Saudi Arabia

ABSTRACT

E-services are becoming increasingly important for government and for business services. They provide benefits to all parties concerned, including service providers, users, and the society at large. This chapter presents a framework of key issues that any e-service should enjoy. The framework is based on three main domains: the e-service strategy, its architecture, and its performance. Each of these domains is structured in three groups of issues, which are associated with three main dimensions: management, targeted features, and the people concerned. Per this structure, the framework considers "45" key e-services issues, and these issues are derived from the available literature and previous experience. Awareness of the importance of these issues among staff working in the field is investigated by means of a survey designed for this purpose. The outcome shows that there is a clear lack of awareness with regards to various key issues, and this problem would demote e-services. In response, recommendations are derived for solving the problem and toward enhanced future development.

INTRODUCTION

For introducing the work of this chapter, this section provides an overview concerned with the development of e-services. This is followed by reviewing important previous literature associated with key e-services issues. The section then identifies the work objectives in preparation for its description.

Emergence of E-Services

It was around a century between introducing of the communication technology in the mid-19th century and initiating the computer technology in the mid-20th century. In the meantime, it should be empha-

DOI: 10.4018/978-1-5225-7347-0.ch009

sized that both technologies belong to the same origin that is the electrical and electronic technology, and this facilitated their easy integration in manipulating information (Al-Sulaiym, Al-Muammar, & Bakry, 1997). Since 1970, both technologies provided us together with computer networks, and they moved toward becoming widely known as the "Information and Communication Technology (ICT)". By 1990, the Internet, the network of networks; and the cellular mobile systems appeared; followed by the world-wide-web, and the e-services, which started around the mid-1990s (Bakry & Bakry, 2001). It was apparent that interconnecting the world was on accelerated move toward increasing dependence on "e-services" in various fields both in government and in business (Bakry, 2004). Figure 1 illustrates this historical development.

At the turn of the century, from the 20th to the 21st century, e-services, which were on limited use, attracted the attention of the international community. The "World Summit on the Information Society (WSIS)" recommended wide-spreading use of ICT in activities associated with the various fields of life (WSIS, 2003); and the "United Nations Department of Economic and Social Affairs (UNDESA)" recommended the use of ICT in government services (UNDESA: United Nations Department of Economic & Social Affairs, 2001). This attention came because of the benefits of ICT in "storing, processing, and transferring information" in a "faster, cheaper, better, different, and more secure" manner leading to more efficient and more knowledgeable society (Bakry, 2004).

E-services, in both: e-government and e-business, enjoy ICT benefits in their various transactions. They provide advantages to the organizations delivering the services and to the users of the services (Bakry, Bakry, & Muhaya, 2016). These advantages lead to more efficient, more knowledgeable, more transparent, and more interactive society, contributing to its sustainable development (Bin-Abbas & Bakry, 2012).

Development of E-Services

Recognizing the importance of e-services, various independent researchers and professionals in the field have addressed the problem of e-services development both in business and in governments (Bakry & Bakry, 2001; Bakry, 2004). In addition, the international organization UNDESA has developed assessment

Figure 1. Historical development toward new era of world connectivity

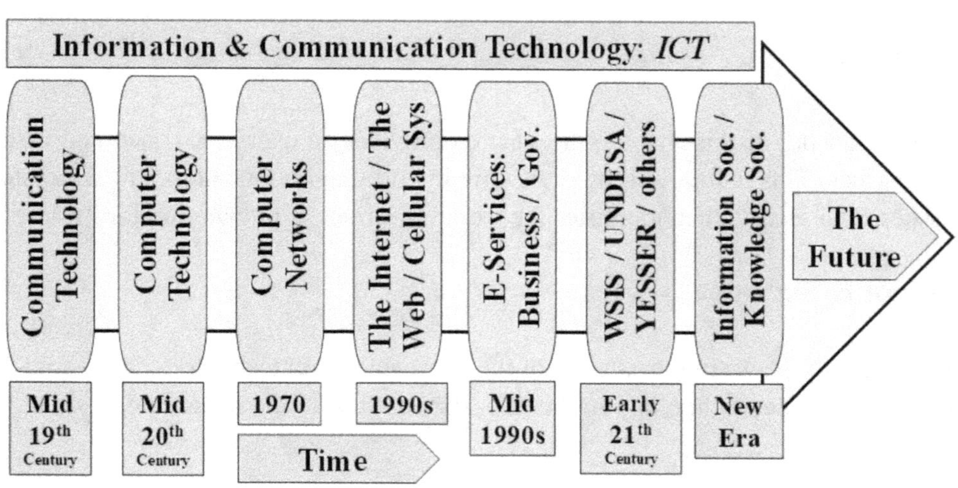

measures, for measuring e-government services, and have used these measures to assess the development of these services in most countries of the world, and to classify them per their state of development (UNDESA: United Nations Department of Economic & Social Affairs, 2016). In the meantime, national plans, in various countries, initiated government departments to provide e-government services. For example, the national ICT plan of Saudi Arabia led to the initiation of the Saudi e-government program, known as "Yesser". In turn, "Yesser" led the development of e-government services in the country; and it has also developed assessment measures to monitor their progress (Al-Almaee & Khayyat, 2010).

A relatively early published work presented a strategy for the development of e-business considering an integrated set of plans, with various development issues, and different levels of use relative to the organization and its internal and external activities and transactions (Bakry & Bakry, 2001). Another published work addressed the development of e-business at the intra-organization level, considering enterprise resource planning (Bakry & Bakry, 2005). Information security for organizations delivering e-services has also been addressed considering the main issues concerned (Bakry, 2003), and the international management standards that can be used to protect the e-services (Saleh, Alrabiah, & Bakry, 2007).

A highly cited work, still attracting attention, have considered the development e-government services per the domains of "strategy, technology, organization, people, and the environment", emphasizing competitiveness and customer satisfaction (Bakry, 2004). The work considered various development cyclic processes and underlined the use of the "six-sigma" process. The benefits of the development of e-government services have been addressed by a recently published work, from the viewpoints of the service providers, users, and the society at large, considering their development state and level of utilization (Bakry et al., 2016). A more recent paper by Alshouiby and Bakry (2017) elaborated on key e-services issues that need to be kept in mind, and these are emphasized by this chapter.

The e-government measures developed by the international organization UNDESA provided an index, known as the "e-government readiness index", which consists of three sub-indices measuring the state of communication, the state of e-government services, and the state of the human capital that is the users of these services in the assessed countries. They also included another index, known as the "e-participation index", which assesses the government interaction with its citizens (UNDESA: United Nations Department of Economic & Social Affairs, 2001; 2016).

The national e-government program, "Yesser" of Saudi Arabia, has led the effort of promoting the state of the country in UNDESA e-government readiness index, from the international position of "80" in 2005 to "44" in 2016. It has also contributed to the promotion of the e-participation index from "102" in 2010 to "39" in 2016 (UNDESA: United Nations Department of Economic & Social Affairs, 2016; 2005). In doing so, the organization drawn development plans considering different issues, and it provided measures for assessing the e-government services of the country on regular basis (Yesser, 2016). In addition, it considered accumulating and reporting experience through various workshops, and surveys (Bakry et al., 2016). Other ambitious countries would have done similar efforts.

This chapter aims at building a framework of key generic issues that are essential for e-services. The domains of the targeted framework are: the e-service "strategy", which will be concerned with the e-service "initiation requirements"; the e-service "architecture", which will be associated with the current development state of the e-service components and operation; and the e-service "performance", which will be concerned with the current state of the e-service outcome. The issues associated with each domain are concerned with three dimensions: management requirements; e-service targeted features; and the people involved. The key issues of the framework are derived from the available literature and from experience. Awareness of the importance of these issues among staff working in the field is investigated by means of a survey designed for this purpose. The outcome shows that there is a clear lack of awareness in various key issues: a problem for which solutions are recommended.

THE FRAMEWORK OF KEY ISSUES

The development of the targeted framework is described here in four main steps. The first step is concerned with addressing the principles upon which the development of the framework is based. The subsequent three steps are associated with: the e-services strategy issues; architecture issues; and performance issues.

Framework Development Principles

The principles upon which the key e-services issues framework is developed are given in the following.

- The structure of the e-services framework should be of "hierarchical nature" that permits dividing the issues into levels, while allowing their integration at the same time.
- The main e-services "domains" of the framework hierarchical structure are: "strategy", which is concerned with the essential development requirements; "architecture", which is associated with the characteristics of the current state; and "performance", which is related to the question of how well the e-services are delivered.
- Each domain will have three main "dimensions": "management dimension; e-services targeted features dimension; and people involved dimension".
- The key issues would be associated with each dimension of every domain. They would be of "generic nature", applicable to the various types and fields of e-services. They would be selected from work presented by researchers in the field, and from accumulated experience, as highlighted in the literature review given above.
- The key issues considered can be used for various development purposes including assessing the awareness of IT staff, working in the field, of their importance, as this would reflect their current practices, and consequently would help improving these practices in the future.
- The key issues considered would be of "flexible nature"; researchers and professionals in the field can update and upgrade them in the future using the main domains and their related dimensions.

Following the above principles, Figure (2) gives the general structure of the framework. The issues of the framework are addressed in the following per the main domains and per the dimensions within each of them.

The Strategy Domain

The e-services development issues, associated with the "strategy" domain, are associated with the *"main development requirements"*. They are concerned with the following.

- "Organization management requirements", including: organization main strategy; regulations; work plan; security and business continuity; and responsibility. These are explained in Table (1-a).

Figure 2. The general structure of the proposed e-services framework

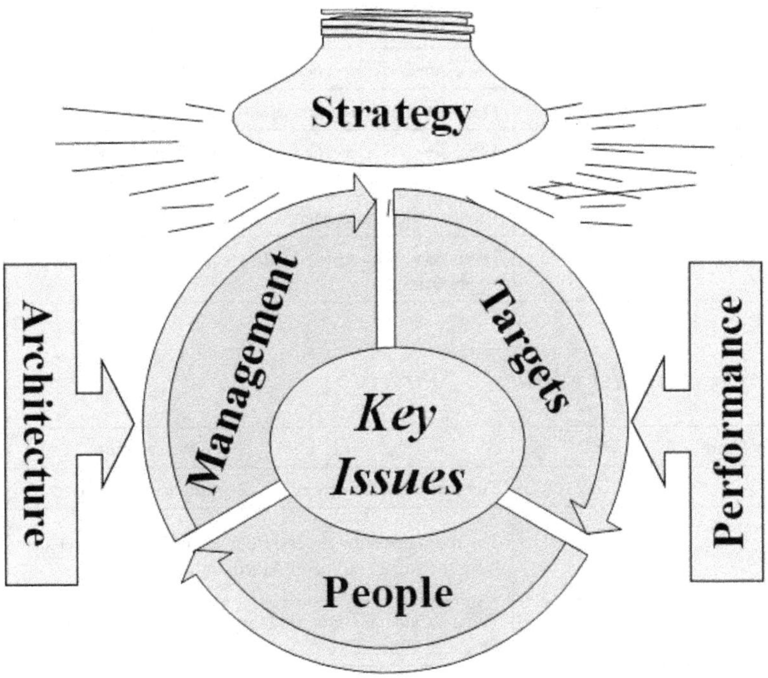

- "E-services targeted features", including: the identification of the target users; the use of re-engineering in the transformation of the manual services to e-services; the targeted access points; the maturity level of the e-service concerned; and the targeted open data facility. These are explained in Table (1-b).
- "People involved requirements", including: qualified e-services staff; users' awareness of new services; inclusion of less fortunate users; e-participation; and access control.

The Architecture Domain

The e-service development issues, concerned with the "architecture" domain, are associated with the *"current architecture state"*. They involve the following.

- "Architecture management state", including the state of: management structure; management standards; security standards; emergency and business continuity practice; in addition to responsibility and support. These are explained in Table (2-a). "E-service targeted features state", including the state of: language use; access support; trust in service provisioning; service transparency; and open data practices. These are explained in Table (2-b).
- "People involved state", including the state of: assignment of responsibilities to staff; staff training; user support; e-participation; and access control.

Table 1a. E-services key "Strategy Issues": "Management"

Key Issue		Description
(1)	Organization Strategy	The e-service concerned is associated with the strategy of the organization
(2)	Regulations	The e-service concerned is associated with the regulations of the organization
(3)	Organization Work Plan	The e-service concerned is considered by the work plan of the organization, with management, security & technical considerations.
(4)	Business Continuity Plan	The e-service concerned is considered in the organization business continuity plan
(5)	Responsibility	The e-service concerned is the responsibility of a specific management unit in the organization

Table 1b. E-services key "Strategy Issues": "Targets"

Key Issue		Description
(1)	Targeted Users	The e-service concerned has targeted users specified.
(2)	Transformation to E-Form Using Re-Engineering	The re-engineering procedure is applied to the services in their transformation from their original manual form into the e-form.
(3)	Targeted Connections to Access Points	The e-service has specified targeted connections to access points (e.g. PC, Mobile Phone & Others).
(4)	Targeted Service Maturity Level	The e-service targeted maturity level is specified (Emerging, Enhanced, Transactional, & Connected)
(5)	Targeted Open Data Feature	The e-service concerned has specified "open data" facility.

Table 2a. E-services key "Architecture Issues": "Management"

Key Issue		Description
(1)	Management Structure	Appropriate organization structure is available for the e-service concerned.
(2)	Management Standards	Appropriate management standards (such as ISO 2000) are used for the e-service concerned.
(3)	Security Standards	Appropriate security standards (such as ISO 27001) are used for the e-service concerned.
(4)	Emergency & Business Continuity	Appropriate emergency and business continuity practices are used for the e-service concerned.
(5)	Responsibility Management	Responsibilities are coordinated and enhanced through regular meetings.

The Performance Domain

The e-service development issues associated with the "performance domain" are concerned with the *"current performance state"*. They involve the following.

- "Management performance state", including the state of: providing periodic reports on current state; implementation of management requirements; implementation of technical requirements;

Table 2b. E-services key "Architecture Issues": "Targets"

Key Issue		Description
(1)	Use of Language	The e-service concerned is described clearly in a simple and correct language.
(2)	Trusted Information	The e-service concerned provides correct and trusted information.
(3)	Access Control	Access to the service concerned via the targeted access points is enabled.
(4)	Service Transparency	The e-service concerned is delivered per its maturity level with transparency.
(5)	Enabling Open Data	Targeted open data feature is enabled.

implementation of the basic security requirements; and implementation of the emergency requirements. These are explained in Table (3-a).

- E-services performance state", including the state of: monitoring the e-service; benchmarking the performance of the e-service; assessing the benefits to the e-service provider; assessing the benefits to the e-service users; and assessing the benefits of open data. These are explained in Table (3-b).

People performance issues", including the state of: staff satisfaction; users' satisfaction; satisfaction of the less fortunate users; response to users' requests; and effectiveness of access control. These are explained in Table (3-c).

USING THE FRAMEWORK

This section is concerned with using the framework to assess the awareness of ICT professionals, working in the field of e-services, of the importance of its "45 key issues", as this would reflect their current practices in e-services, with regards to these issues. For this purpose, a questionnaire based on the framework has been developed; and was used to conduct a survey of "over 100 professionals. In this main section of the chapter, the importance scale used in the questionnaire is presented; and this is followed by describing the outcome of the survey with regards to the strategy issues, architecture issues, and performance issues. Discussions and recommendations are then given in the last main section.

Table 3a. E-services key "Performance Issues": "Management"

Key Issue		Description
(1)	Periodic Reports on Current State	Perioding reports on the state of the e-service concerned are delivered (e.g. Using SWOT).
(2)	Implementation of Management Req.	Periodic revision of the management requirements is implemented.
(3)	Implementation of Technical Req.	Periodic revision of the technical requirements is implemented.
(4)	Implementation of Security Req.	Periodic revision of the basic security requirements is implemented.
(5)	Implementation of Emergency Req.	Periodic revision of the emergency requirements is implemented.

Table 3b. E-services key "Performance Issues": "Targets"

Key Issue		Description
(1)	Monitoring the State of the E-Service	The state of the e-service concerned is under continuous monitoring and assessment.
(2)	Benchmarking the State of the E-Service	The state of the e-service concerned is monitored and assessed per benchmarks.
(3)	Benefits to E-Service Provider	The benefits of the e-service concerned to the service provider are assessed.
(4)	Benefits to E-Service Users	The benefits of the e-service concerned to the service users are assessed.
(5)	Benefits of Open Data	The benefits of the open data feature, associated e-service concerned, is assessed.

Assessment Scale

The assessment scale seeks the views of the ICT staff, working in the field of e-services, with regards to the importance of every issue of the "45" issues of the framework. This is important, as it reflects the current practices of those professionals in e-services. The scale has five levels of "importance" for every issue; and it also has a space for "not important". Table (4) illustrates this importance scale.

Strategy Issues

The outcomes of the survey on the importance of the issues associated with the strategy domain, are described in the following, considering the three strategy dimensions: management; target requirements; and people.

- Figure (3-a) shows the distribution of the views of the ICT staff, for the five issues of "strategy management", per the levels of the importance scale; and Figure (3-c) gives the average importance score for those issues. These results illustrate that "67% to 80%" of the ICT staff have assigned the individual issues involved the higher importance levels of "4 & 5", with the rest assigning the levels below. The results also show that the least average score "3.7" was related to the "business continuity plan".
- Figure (4-a) shows the distribution of the views of the ICT staff, for the issues of "strategy targets", per the levels of the importance scale; and Figure (4-b) gives the average importance score for those issues. These results illustrate that "37% to 91%" of the ICT staff have assigned the individual issues involved the higher importance levels of "4 & 5", with the rest considering the levels below. The results also show that the least average score "3" was related to the need for "open data".

Table 4. The levels of the importance scale

0	1	2	3	4	5
Not Important	Very Low Importance	Low Importance	Average Importance	Important	Very Important

Figure 3. The outcome of the survey on the importance of the "strategy management issues" to the ICT staff working in the field, (a) Distribution of the views to the levels of the importance scale, (b) Distribution of the views to the levels of the importance scale, (c) Average importance

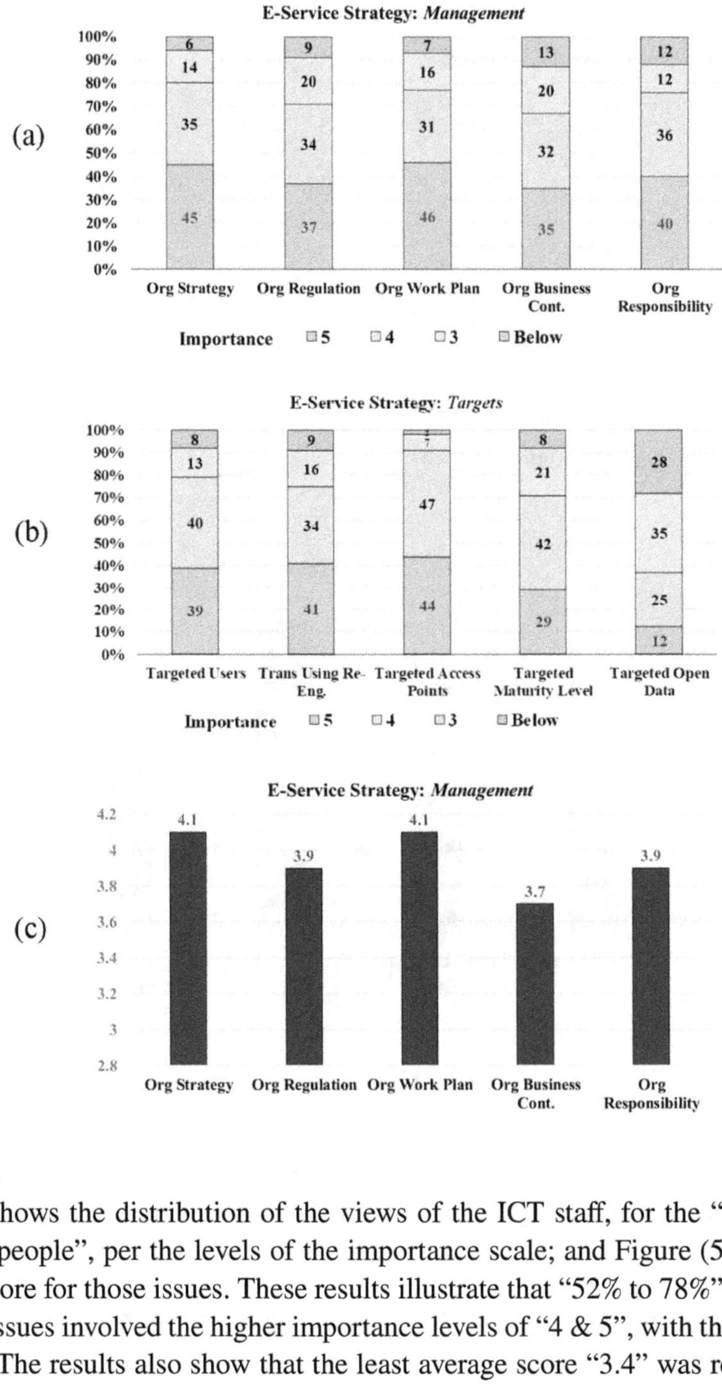

- Figure (5-a) shows the distribution of the views of the ICT staff, for the "strategy issues" concerned with "people", per the levels of the importance scale; and Figure (5-b) gives the average importance score for those issues. These results illustrate that "52% to 78%" of the ICT staff have assigned the issues involved the higher importance levels of "4 & 5", with the rest considering the levels below. The results also show that the least average score "3.4" was related to the need for "e-participation".

Figure 4. The outcome of the survey on the importance of the "strategy target requirements issues" to the ICT staff working in the field, (a) Distribution of the views to the levels of the importance scale, (b) Average importance.

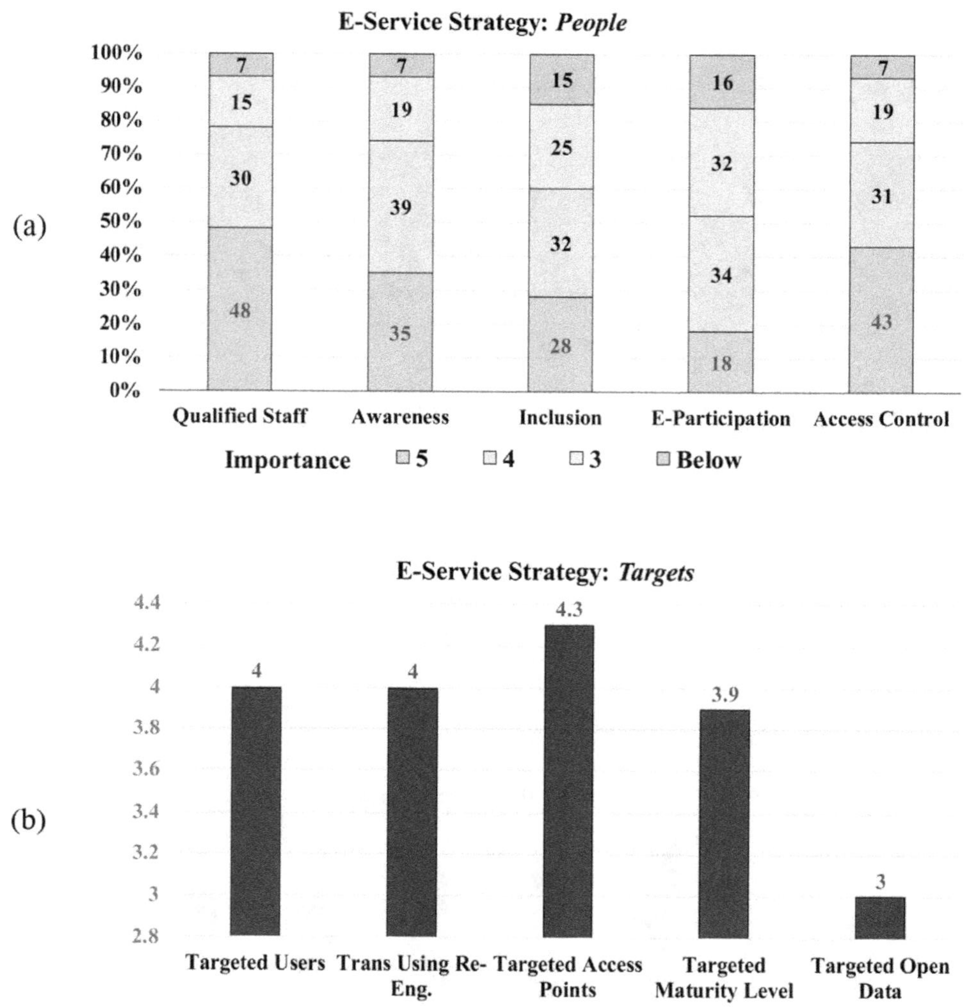

Architecture Issues

The outcomes of the survey on the importance of the issues concerned with the architecture domain, are described in the following, considering the three architecture dimensions: management; target specifications; and people.

- Figure (6-a) shows the distribution of the views of the ICT staff, for the five issues of "architecture management", per the levels of the importance scale; and Figure (6-b) gives the average importance score for those issues. These results illustrate that "52% to 77%" of the ICT staff have assigned the individual issues involved the higher importance levels of "4 & 5", with the rest considering the levels below. The results also show that the least average score "3.3" was related to using "management standards".

Figure 5. The outcome of the survey on the importance of the "strategy issues" concerned with "people" to the ICT staff working in the field, (a) Distribution of the views to the levels of the importance scale, (b) Average importance.

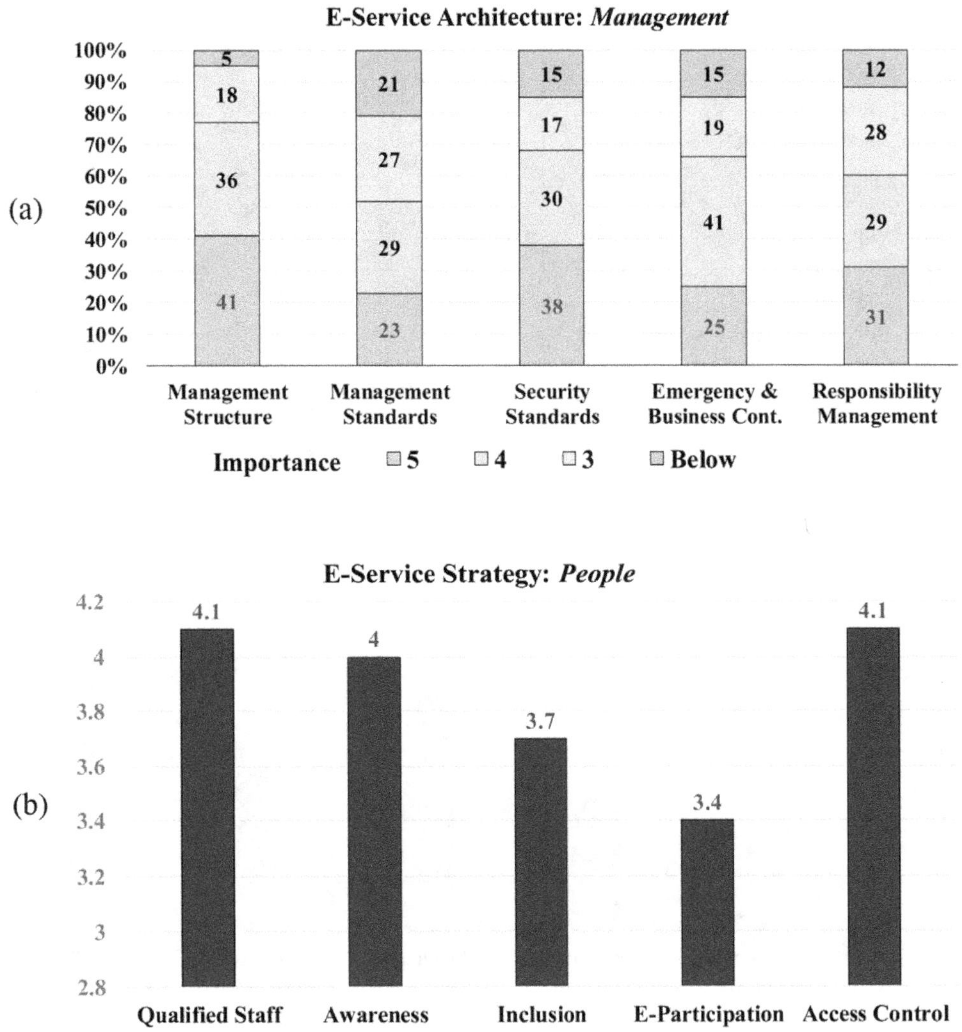

- Figure (7-a) shows the distribution of the views of the ICT staff, for the issues of "architecture target specifications", per the levels of the importance scale; and Figure (7-b) gives the average importance score for those issues. These results illustrate that "46% to 91%" of the ICT staff have assigned the issues involved the higher importance levels of "4 & 5", with the rest considering the levels below. The results also show that the least average score "3.2" was related to enabling "open data".
- Figure (8-a) shows the distribution of the views of the ICT staff, for the "architecture issues" concerned with "people", per the levels of the importance scale; and Figure (8-b) gives the average importance score for those issues. These results illustrate that "68% to 76%" of the ICT staff have assigned the individual issues involved the higher importance levels of "4 & 5", with the rest considering the levels below. The results also show that the least average score "3.8" was related to two issues: "e-participation, and access practices".

Figure 6. The outcome of the survey on the importance of the "architecture management issues" to the ICT staff working in the field, (a) Distribution of the views to the levels of the importance scale, (b) Average importance.

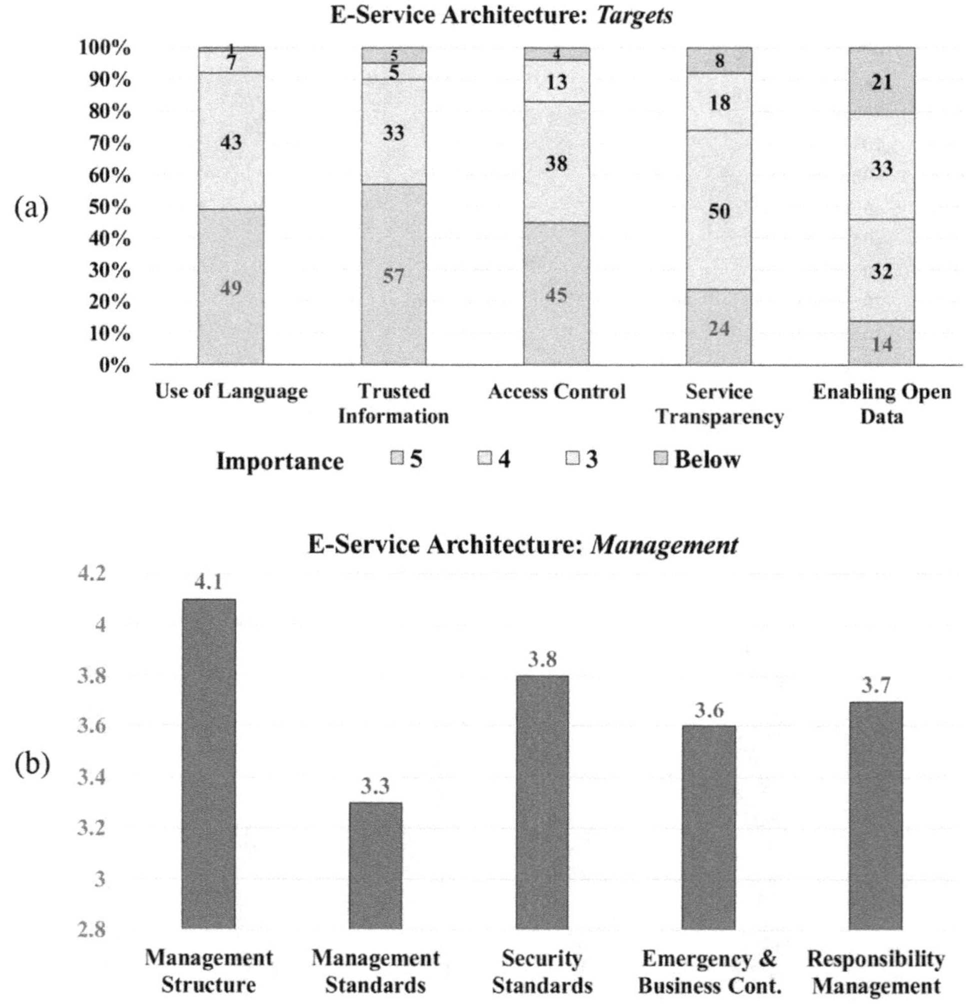

Performance Issues

The outcomes of the survey on the importance of the issues associated with the performance domain, are described in the following, considering the three performance dimensions: management; targets; and people.

- Figure (9-a) shows the distribution of the views of the ICT staff, for the five issues of "performance management", per the levels of the importance scale; and Figure (9-b) gives the average importance score for those issues. These results illustrate that "60% to 73%" of the ICT staff have assigned the individual issues involved the higher importance levels of "4 & 5", with the rest

Figure 7. The outcome of the survey on the importance of the "architecture specifications issues" to the ICT staff working in the field, (a) Distribution of the views to the levels of the importance scale, (b) Average importance.

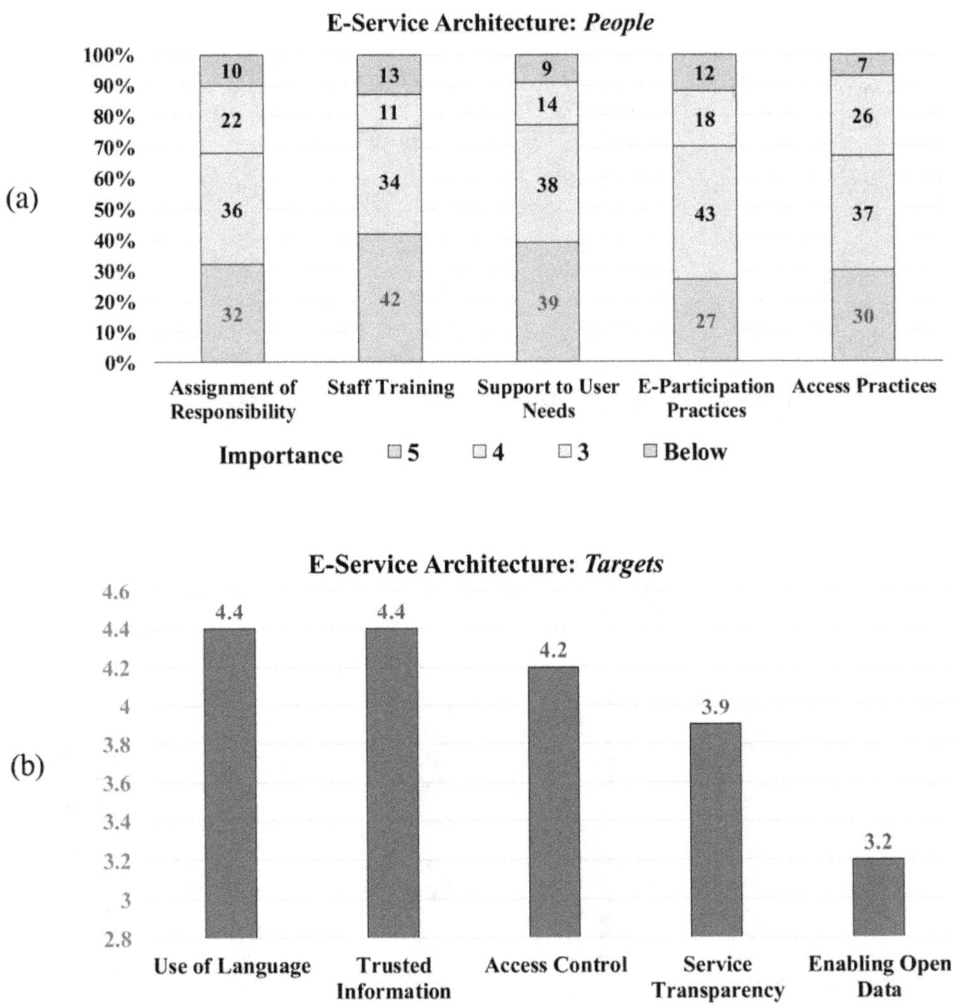

considering the levels below. The results also show that the least average score "3.5" was related to "periodic reports".

- Figure (10-a) shows the distribution of the views of the ICT staff, for the issues of "performance targets", per the levels of the importance scale; and Figure (10-b) gives the average importance score for those issues. These results illustrate that "43% to 73%" of the ICT staff have assigned the individual issues involved the higher importance levels of "4 & 5", with the rest considering the levels below. The results also show that the least average score "3" was related to the "open data" benefits.
- Figure (11-a) shows the distribution of the views of the ICT staff, for the "performance issues" concerned with "people", per the levels of the importance scale; and Figure (11-b) gives the average importance score for those issues. These results illustrate that "58% to 69%" of the ICT staff have assigned the individual issues involved the higher importance levels of "4 & 5". The results

Figure 8. The outcome of the survey on the importance of the "architecture issues" concerned with "people" to the ICT staff in the field, (a) Distribution of the views to the levels of the importance scale, (b) Average importance.

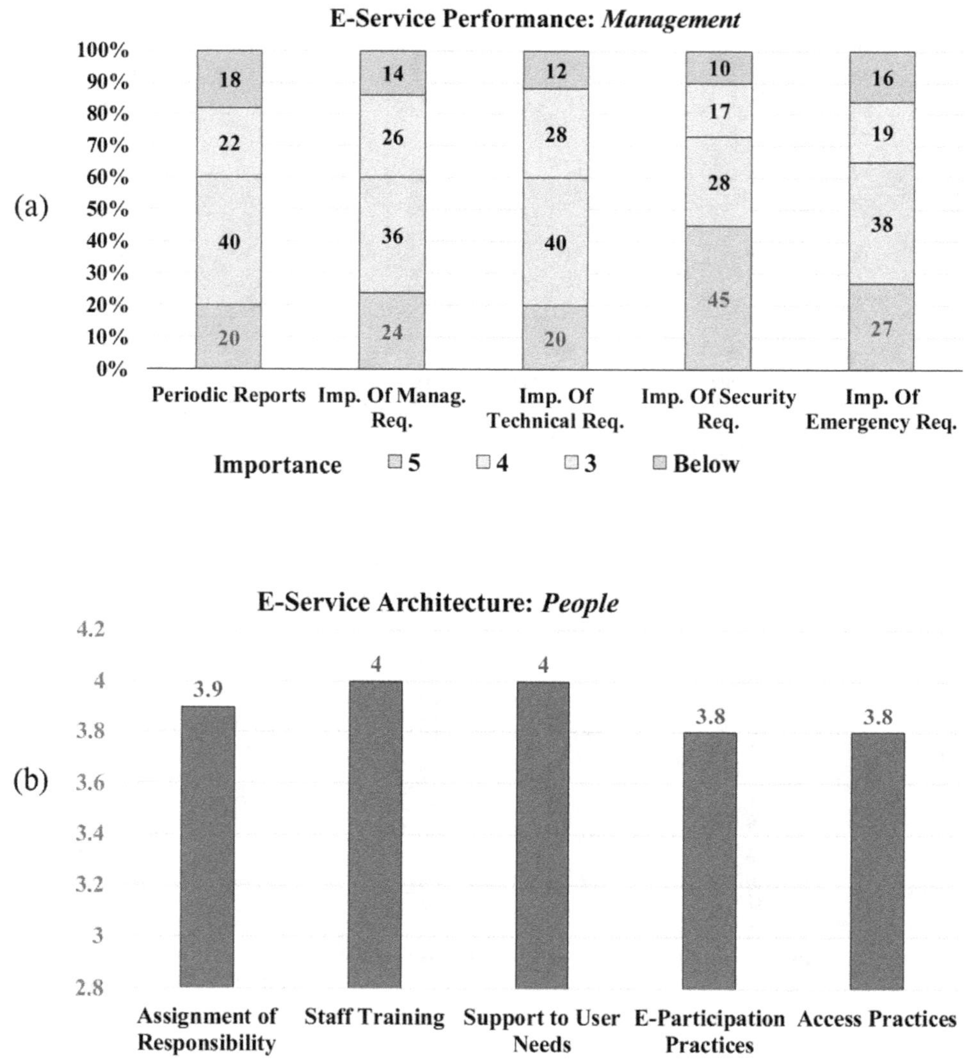

also show that the least average score "3.5" was related to the satisfaction of the "less fortunate users".

REMARKS AND RECOMMENDATIONS

The resulted outcome shows that there is a clear lack of awareness among e-services professionals with regards to some key issues addressed in this chapter. This section is concerned with deriving remarks from the resulted outcome on the hand; and providing recommendations toward future development on the other.

Figure 9. The outcome of the survey on the importance of the "strategy management issues" to the ICT staff working in the field, (a) Distribution of the views to the levels of the importance scale, (b) average importance.

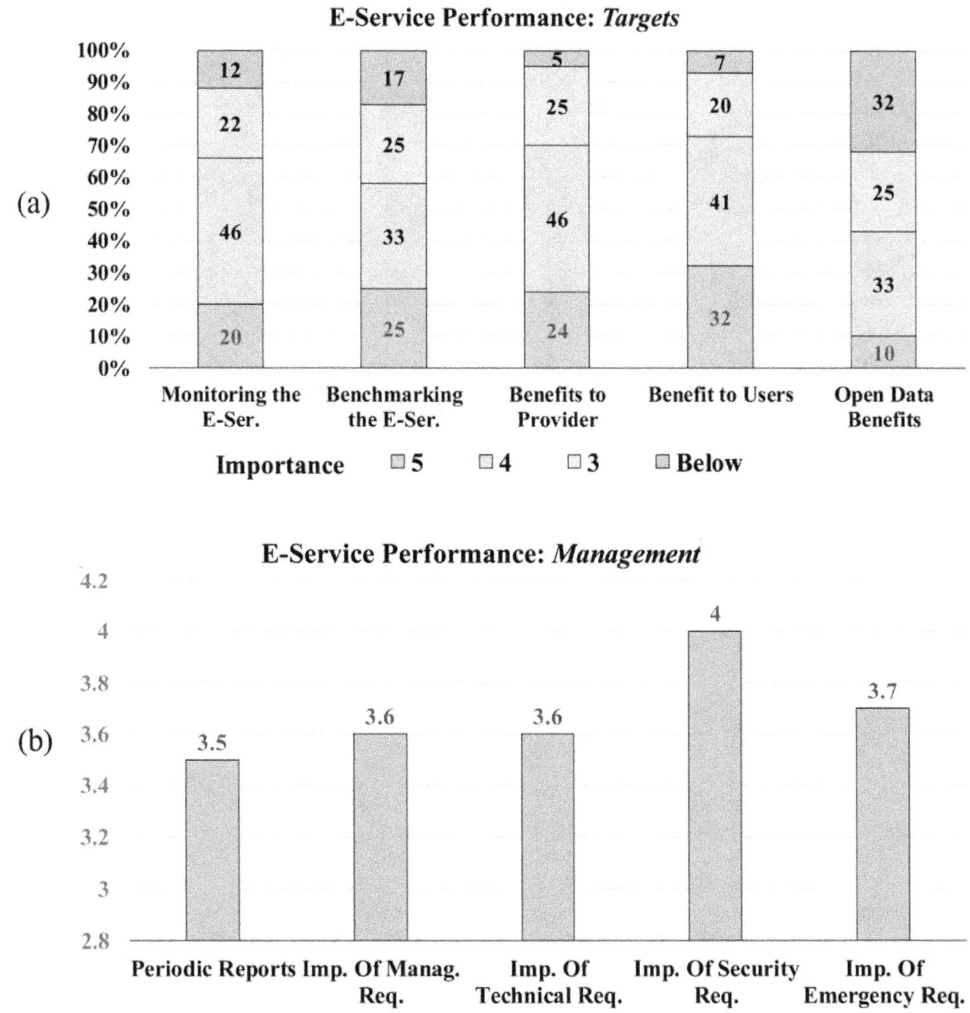

Remarks

Table (5) provides a summary of the collective assessment of the scores, and their averages, for the three main domains, and for their three dimensions. Figure (12-a) illustrates these scores per domain; while Figure (12-b) illustrates them per dimensions. The overall average importance score for each of the three dimensions of every domain is below "4", except for one which is exactly "4". In the mean time, non-of these is blow "3.6".

Among the detailed issues, various scores where even below "3.5" that is below "70%". In the following highlights are given on the issues with: an average score below "3.5"; issues with an average score ranging "from 3.5 to below 4"; and issues with an average score of "4 and above".

Figure 10. The outcome of the survey on the importance of the "strategy target requirements issues" to the ICT staff working in the field, (a) Distribution of the views to the levels of the importance scale, (b) average importance.

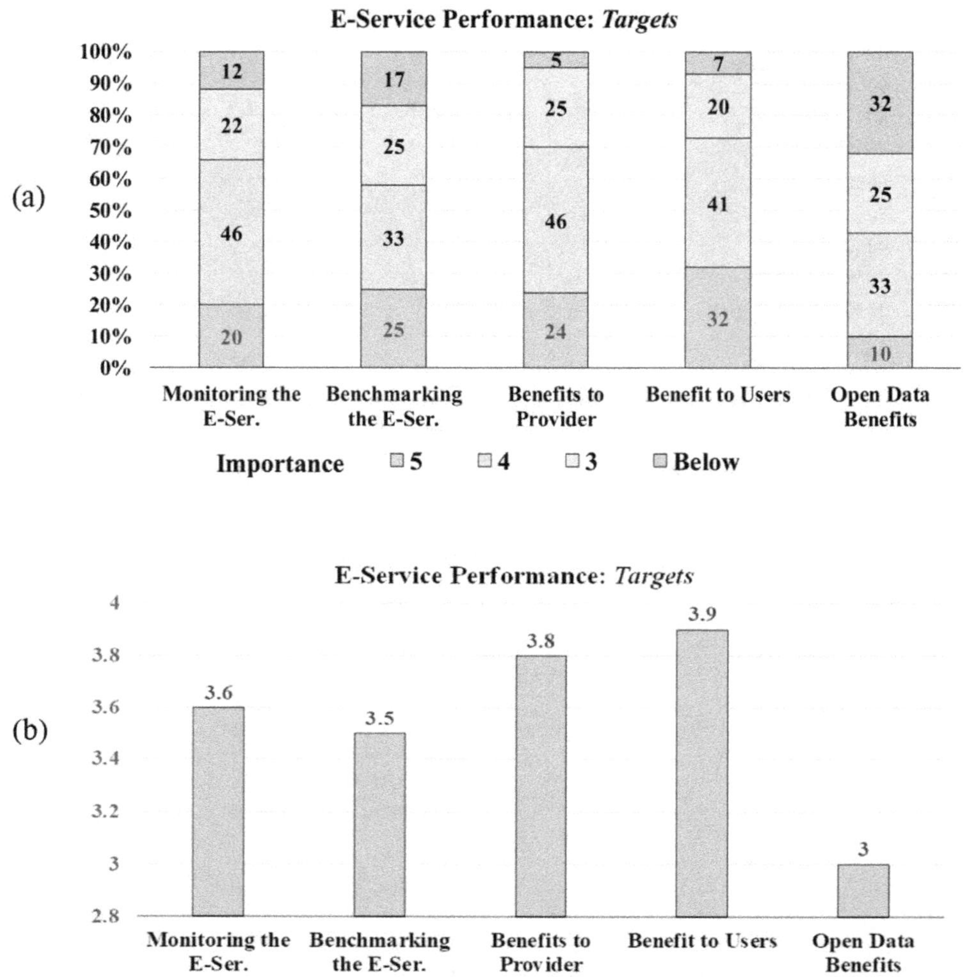

- The issues with an average importance score below "3.5" are concerned with the subjects of: "open data; e-participation; and management standards". It should be observed here that "open data and e-participation" are considered by UNDESA 2016 report as important current development trends in e-government.
- The issues with an average score below "4" but over "3.5" are associated with the subjects of: "organization regulations; business continuity; responsibility; service maturity; inclusion of less fortunate users; security standards including access considerations; transparency; reporting; technical requirements; monitoring; benchmarking; benefits; and satisfaction of the people concerned.
- The average importance scores of the rest of the key issues related to other subjects were in the range "4 to 5".

Figure 11. The outcome of the survey on the importance of the "strategy issues" concerned with "people" to the ICT staff working in the field, (b) average importance

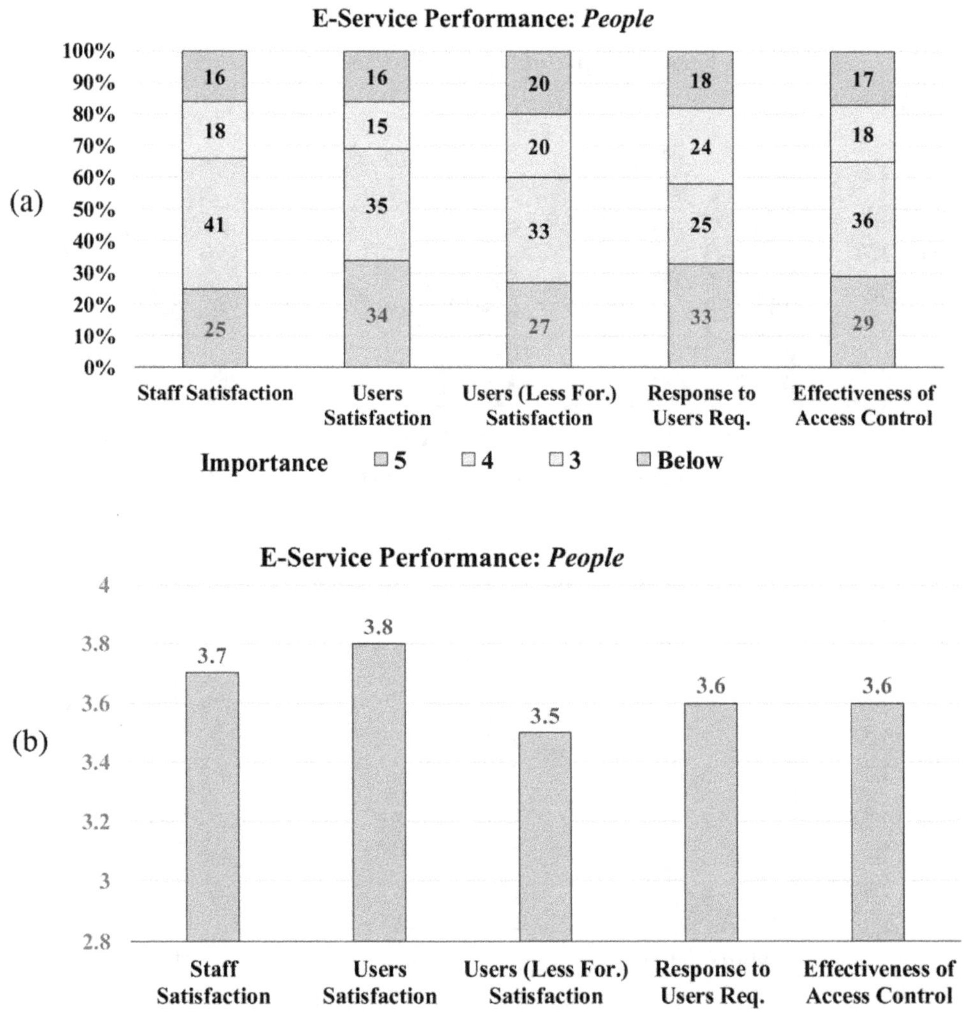

Recommendations

Lack of awareness of some key e-services issues among professionals is a problem with consequences leading to demoting e-services. Therefore, these issues need to be well-understood by those professionals. In addition, there are also other people in other positions who need to recognize the issues. Those people are: the e-services leaders that is the top management of organizations providing e- services on the one hand, and the users who use the services and benefit from them on the other. For having all people concerned knowledgeable, in what they need, with regards to e-services, the following types of e-services courses are recommended.

Figure 12. (a) The average importance score summaries for the three main domains and their three dimensions, (b) The average importance score summaries for the three main dimensions and their three domains.

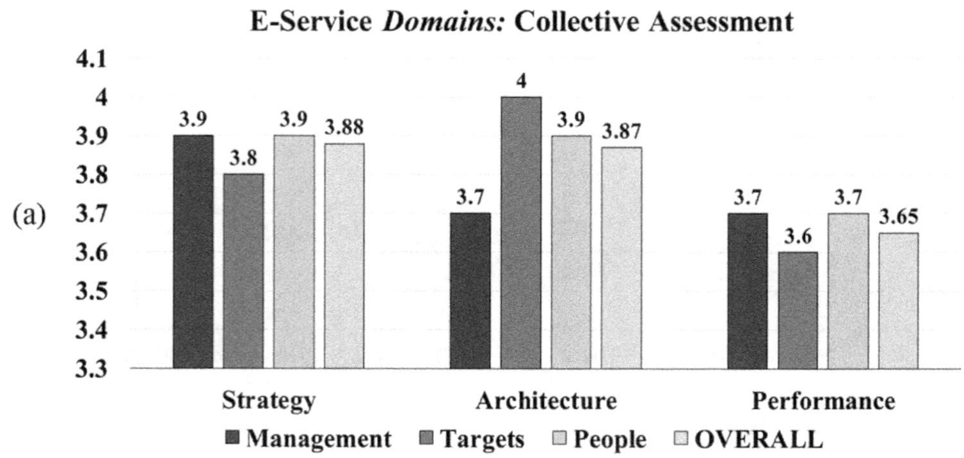

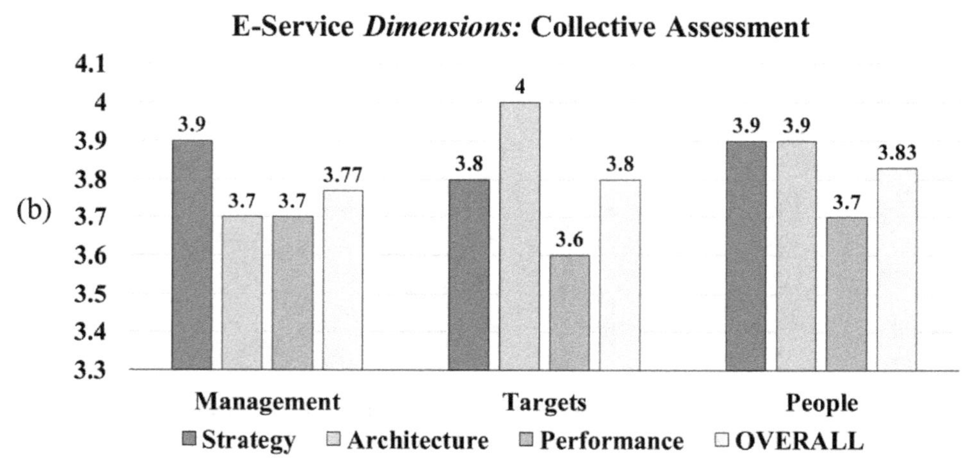

Table 5. Summary of the assessment

Domains / Dimensions	Strategy	Architecture	Performance	Results by Dimension
Management	3.9	3.7	3.7	3.77
Target Features	3.8	4	3.6	3.80
People	3.9	3.9	3.7	3.83
Results by Domain	3.88	3.87	3.65	Overall Result 3.8

- Awareness courses on e-services that address all key issues and provide the needed depth for understanding the services and using them effectively and efficiently. Such courses would be useful to leaders of organizations providing e-services and to the users of these services.
- Training courses for e-services professionals that enable them not only to understand all issues concerned, but also to be able to develop newer e-services and to update the older ones.
- Follow-up courses that would involve both: awareness and training with regards to new ICT technology development and the resulting applications. Such courses would be directed to professionals, leaders, and users depending on the needed level of knowledge.

The recommended courses above would be concerned with the whole scope of e-services issues for the people concerned. Other more focused recommendations that can support e-services are given in the following.

- Promoting "open data" and "e-participation" among the organizations providing e-services, as this would support having more informative and pro-active people and enhance digital transformation of the society. This may involve the development of special open rules by the government for all to follow.
- Developing detailed e-services international "assessment" tools, beyond that published by UNDESA, for accommodating all issues involved and ensuring effectiveness, with excellence "benchmarks" that set achievement targets.
- Developing a "rewarding system" for e-services that encourages competition among professionals to develop better e-services for the benefits of the users and the society at large.

In closing, it should be noticed that innovation and development in ICT and its applications is continuous, if not continuously accelerating. This would have a continuous impact on the development of e-services that should be considered. Therefore, all the above would need to be continuously revised as time progresses.

ACKNOWLEDGMENT

The authors wish to extend their gratitude to all ICT professionals who participated in the survey. Special thanks are extended to all members of "Yesser E-Government Assessment Team", particularly those in the team who helped in distributing the survey and managing its outcome.

REFERENCES

Al-Almaee, S. M., & Khayyat, W. R. (2010). Strategic planning and supporting initiatives: Saudi e-government program. In *Proceedings of the10th European Conference on e-Government, National Centre for Taxation Studies*. University of Limerick.

Al-Sulaiym, H., Al-Muammar, A., & Bakry, S. (1997). An Overview of Public, Research and Private Data Networks. *International Journal of Network Management*, *7*(2), 103–113. doi:10.1002/(SICI)1099-1190(199703/04)7:2<103::AID-NEM230>3.0.CO;2-0

Alshouiby, A., & Bakry, S. (2017). A Framework of Key E-Services Issues. *International Journal of Knowledge Society Research*, *8*(4), 1–22. doi:10.4018/IJKSR.2017100101

Bakry, A., & Bakry, S. (2005). Enterprise resource planning: A review and a STOPE view. *International Journal of Network Management*, *15*(5), 363–370. doi:10.1002/nem.584

Bakry, S. (2003). Development of security policies for private networks. *International Journal of Network Management*, *13*(3), 203–210. doi:10.1002/nem.472

Bakry, S. (2004). Development of e-government: A STOPE view. *International Journal of Network Management*, *14*(5), 339–350. doi:10.1002/nem.529

Bakry, S., & Bakry, F. (2001). A strategic view for the development of e-business. *International Journal of Network Management*, *11*(2), 103–112. doi:10.1002/nem.395

Bakry, S., Bakry, Z., & Muhaya, F. (2016). An Approach for the Assessment of E-Government Benefits to Governments, citizens and Businesses. *Journal of Socialomics*, *05*(02). doi:10.4172/2167-0358.1000155

Bin-Abbas, H., & Bakry, S. (2012). Knowledge Management. *International Journal of Knowledge Society Research*, *3*(3), 58–67. doi:10.4018/jksr.2012070105

Muhaya, F., Bakry, S., & AlAlmaee, S. (2015). E-Government Success Factors: Views of Saudi Professionals. *J Entrepren Organiz Manag*, *4*, 128. doi:10.4172/2169-026X.1000128

Saleh, M., Alrabiah, A., & Bakry, S. (2007). A STOPE model for the investigation of compliance with ISO 17799-2005. *Information Management & Computer Security*, *15*(4), 283–294. doi:10.1108/09685220710817806

UNDESA (United Nations Department of Economic & Social Affairs). (2001). *Benchmarking E-Government 2001: Assessing Progress of UN Member States*. Author.

UNDESA (United Nations Department of Economic & Social Affairs). (2005). *UN Global E-Government Readiness Report 2005: From E-Government to E-Inclusion*. Author.

UNDESA (United Nations Department of Economic & Social Affairs). (2016). *E-Government Survey 2016: E-Government in Support of Sustainable Development*. Author.

WSIS. (2003). Declaration of Principles: Building the Information Society: a global challenge in the new Millennium. *World Summit on the Information Society (WSIS), Geneva 2003-Tunis 2005*.

Yesser. (n.d.). *E-Government Program*. Saudi Ministry of Communication and Information Technology. Retrieved October 2016 from https://www.yesser.gov.sa/en/Pages/default.aspx

Chapter 10
Improved Indoor Geo-Localization System:
Bluetooth Low Energy Technology of iBeacons and Wi-Fi Network for Better IOT Systems Accuracy

Abdaoui Noura
Mannouba university, Tunisia

Ismahène Hadj Khalifa
Institut Supérieur des Arts de Multimedia, Tunisia

Sami Faiz
Institut Supérieur des Arts de Multimedia, Tunisia

ABSTRACT

In the concept of internet of things (IOT), physical position of smart object is very useful for relevant function over sensor networks. However, the invalid information of indoor geo-localization systems relative to these wireless sensor compromises the intelligence of IOT network. Therefore, this chapter produces the recent progress in the indoor geo-localization systems and the IOTs area. It defines the best indoor geo-localization technologies that meet their needs while respecting the constraints related to sensor networks. This framework combines between simplicity of Bluetooth low energy (BLE), popular wi-fi infrastructure, and the k-nearest neighbor (KNN) algorithm (in order to filter the initial fingerprint dataset). This new conception increases real-time detection accuracy and guarantees the low energy consumption.

DOI: 10.4018/978-1-5225-7347-0.ch010

INTRODUCTION

Nowadays, the user of smartphone becomes an interactive actor who reacts with his environment which gives birth to a new platform composed by the Customer-Business relationship. This phenomenon is known as the domain of ubiquitous marketing which defined the emergence of a new research theme called the Internet of Things. Smaller, more reliable and reconfigurable depending on the applications, this new theme has developed a new generation of networks: intelligent wireless sensor network. Our research is based on technologies existing in the literature to increase the rates of customer's loyalty in their own platforms. Related to the rapid growth of sale networks and communication channels, organizations need to readapt new customers' profiles that are always looking for new promotions based on the information of their profiles and their interactions histories with the environment. This strategy is shown by (Leite Lucas de Azevedo and Borges, 2015). In fact, the studies of personalized targeting in ubiquitous marketing are limited in the literature because of the new context of the interaction between the mobile customer and smart objects. Thereby when everything is connected, the daily user journey becomes equipped by the huge use of the smartphones. Indeed, researchers reviewed the studies related to indoor geo-localization systems in order to provide an intelligent one in the IOT area. This chapter is an enhanced work since the first publication of (Abdaoui et al., 2017). Over there, it has developed a prototype to send a personalized advertisement to loyal customers and potential customers in ubiquitous environment. According the degree of loyalty obtained by the behavioral analysis, each target received a personalized ad. In this study, they add the new part of detecting the user indoor the building in order to increase the rates of customer's loyalty in their own platforms in real time.

In section 2, researchers post a view of the customer's behavioral techniques existed to define and find the target. Researchers describe the evolution of advertising in marketing with an overview of Customer-Business relationship techniques. After, they describe the techniques used in the literature to analyze customer's behavior and create his profile. After, researchers outline the methods used in various types of marketing to look for the target. And the added part manifests in supplying a review of the main technologies discussed in the literature to solve the indoor geo-localization issue. In section 3, researchers display the whole process of personalized advertisements. Researchers validate the prototype of personalized advertisements in section 4, using the data set in question and discussing the relevant results.

BACKGROUND

Advertising has a strong connection with the development of trade. In the first subsection, researchers focus on the evolution of advertising in different forms of marketing: Direct marketing, e-marketing, mobile marketing and the ubiquitous marketing. This evolution is treated after knowing the needs of customer. These needs describe the profile of each customer. In literature, there are several techniques used to analyze the customer's behavior. In the second subsection, researchers try to supply a review of the main technologies discussed in the literature to solve the indoor geo-localization issue. This could hand over a better dealing of the state-of-the-art of the IOT and push new research efforts in this hopeful field. Then, concentrating on one of the major challenges in the indoor geo-localization field, they explain the final choice that have been revised and compared by choosing the best solution.

Customer Behavioral Analysis

In the marketing process, the datamining techniques used to analyze the consumer's behavior represent a basic step. These techniques exist in two criteria: Descriptive datamining and predictive datamining according to the nature of the data considered, as well as the type of the extracted knowledge (Heo et al., 2010). The descriptive datamining focuses primarily on the discover of models or explicit relationships that a decision maker can directly interpret and translate into the company (Piwowarski et al., 2014). The most used are the clustering and the association rules. The predictive datamining focuses on the prediction of the model's behavior and determines the future values of variables generated from the databases (Crémilleux et al., 2014). The most used are the classification, scheduling, and the association rules. In this chapter, authors are interested to discuss the datamining techniques existing in marketing literature.

Clustering

Clustering is an exploratory and critical task in the knowledge discovery process. Besides, several studies are conducted to discuss the evolution of the types of clustering algorithms from the simplest techniques such as k-means algorithm (Izquierdo-Yusta et al., 2015) to the more advanced approaches, such as kernel methods and spectral approaches (López-Catalán et al., 2015). As well, the types of constraints used in the techniques represent another topic. According to the researchers (Ho et al., 2007), customers are classified by k-means algorithm according to two criteria: The trend value of the star product's past purchases and the static attributes of each customer. But this study is performed on a short interval of time and depends on the information stored in the web log file, which does not detect a large number of customers because of its short lifetime.

Clustering technique with (Deshpand et al., 2009) is performed according to the demographic information stored by a source of loyalty to the company. This classification is limited to static information such as demographic information like name, age, etc. and does not include the evolution of the purchasing behavior. A new dimension is added by (Rezaei, 2015), which is to analyze the behavior of online visitors in their purchases. The main problem with Rezaei (2015) proposal is related with the frequency and quantity of users' purchasing. This introduces a new issue in the cases of huge quantities. Afterwards, (Seret et al., 2014) integrated the constraints of knowledge related to the company, in the classification process and assigned each one by a weight value. Depending on each weight, the prioritization approach is performed. Prioritization results are combined with the preferences of each consumer.

Association Rules

This technique focuses on the extraction of relations between reccurent values for each attribute in a transaction. According to researchers (Goswami et al., 2015), the association rule is defined by two measures: The support and the trust of the redundant itemsets. The researchers (Goswami et al., 2105) proposed that it is better to extract the non-redundant rules for predicting the frequent product. After the progression of mobile technology, data has been found in different forms. Following this heterogeneous data in the real world: Binary, digital and quantitative data, researchers solved this problem with the use of fuzzy set (Ho et al., 2007) also a new FFP-USTREAM technique (Fuzzy Frequent Pattern Ubiquitous Streams) is developed. This technique incorporates fuzzy concepts in the flow of ubiquitous data using the method of the sliding window to manage the fuzzy association rules. In electronic marketing, they used this technique to construct the customer's profile according to his dynamic evolution during his lifetime in the e-shop.

Search of Target

After analyzing customer behavior, every company wants to raise its rate of loyalty and reduce the quantity of stock for each product. There are a reciprocal and a strong relationship between the store, the customer and the product. This relationship is ensured by the targeting techniques existing in marketing.

Target Research Techniques in M-Marketing: Wap Push Technique

Personalized targeting is used in m-marketing area. The wap push technique used two targeting technique: Segmentation by scoring and segmentation by RFM (Recency Frequency Monetary) method to send messages for each mobile user (Ye et al., 2014). The researchers (Wong et al., 2015) used Bayesian network to detect loyal customers in a restaurant with personalized messages, according their favorite dishes storing on phone.

Nevertheless, the researchers (Tang et al., 2013) integrated mobile interaction of a person with other mobile devices. After each interaction, a Snapshot technique captured at every moment new contextual information. An n-gram algorithm compares the sequences for targeting viewers based on their history. In the study of (Izquierdo-Yusta et al., 2015), they determined the most loyal targets related to the consumer reactions towards certain mobile advertising (warning, buying behavior, etc.).

Target Research Techniques in Ubiquitous Marketing

However, the term of personalization in the new form of the ubiquitous marketing is relative because of the newness of this commerce type and the high costs of some devices. Two techniques are used to search the target: X-UIDH technique and QR code technique.

- **X-UIDH Technique:** The technique of X-UIDH Tracker is used in this marketing with (Jacob, 2015). It establishes a complete and a detailed profile of each browsing habits relative to the mobile user. Following this profile, personalized advertisements are sent after each detection of an Ibeacon's tag.
- **QR Code Technique:** The technique of the alphanumeric code QR is used on digifieds. After each detection between the code and the Wifi network, all the data displayed on the screen of digifieds are transferred to the mobile phone (Schmidt et al., 2011).

Thanks to the technological improvement, authors have new levers to create the customer behavior. Today, the use of ubiquitous devices like Smartphones, changes the traditional relationship between the customer and the company like it is indicated in the study of (Carlos Filipe et al., 2015). Indeed, there are different interactions between the Smartphones and the connected sensors making the exchange in the ubiquitous marketing more extensive. In fact, the personalized advertisements in ubiquitous marketing are still ambiguous. Following these conditions, authors add a new part of detecting the user indoor the building which will be studied in the following subsection.

GEO-LOCALIZATION OF THE USER INSIDE THE HYPERMARKE

General Aspect

The Internet of Things

Is an emerging technology that is studied to update the classical devices of environment and to supply autonomous, ubiquitous and pervasive computing within Wireless Sensor Networks (WSNs). Originally, the Internet of Things relates to a set of objects that are not necessarily able to communicate with each other (Sotenga et al., 2017). This new paradigm is gaining ground day by day in the world of telecommunications and computer networks. But, according to the International Telecommunication Union (ITU) in 2005, the definition and domain of IOT are evolving to expand beyond RFID technology. In addition to RFID technology, wireless sensor technology, nanotechnology and intelligent terminal technology which will be more widely used. Indeed, the appearance of IOT is due to the variety of equipment and objects used in our daily lives (computers, sensors, actuators, smartphones, connected vehicles, smart homes, etc.). IOT represents an evolution of M2M (Machine to Machine) systems. It is considered to be the extension of the Internet to things and places in the physical world. And it claims to model data exchanges from real-world devices to the Internet. In the near future, this new concept will play an indispensable role in our daily lives, and will encompass almost all areas of IT. Yet, IOT is generally attached with the acquisition of wireless sensor node information and managing of "things".

Wireless Sensors

They are based on a set of technologies required for their good functioning. They are divided into several categories, such as:

- Identification that makes it possible to recognize the object.
- Sensors that collect information about the environment.
- Connection manages communication between different systems.
- Integration allows multiple applications to communicate with each other.
- Processing uses the data to generate possible content.
- Networks are the vectors of data circulation (mainly the Internet).

The connectivity of this wireless objects passes through several technologies. Besides, the absence of geo-localization information of these wireless sensor compromises the intelligence of the IOT network. In next subsection we will summarize the different technologies of indoor localization of sensor nodes viewed in literature. Each of which has advantages and disadvantages. It is therefore useful to choose the best that will be integrated in the real world scenario of IOTs.

WI-FI Positioning System

Indoor geo-localization over Wi-Fi relies on existing public or private Wi-Fi networks. It consists of making Wi-Fi power measurements in different places of a given site, by declaring positions on a plane, and then injecting this mapping of the Wi-Fi signal constituted into the smartphones of the users. Wi-Fi is usually characterized by a large debit to stream a flow data as videos (Zhu et al., 2015).

Wi-Fi ah is the most recent standard (December 2016), was thought for IOT. Wi-Fi ah operates at a frequency of 900 MHz, which allows it to have excellent range even indoors, focusing on low consumption. Its bit rates are limited to 8 Mbit /s, which is low but not an inconvenient in the context of connected objects. Because it is new, the norm is not yet democratized.

Ultra-Wideband (UWB)

UWB characteristics detect the multipath components (García et al., 2013). Repeatability is a strong advantage for the ultra-wideband approach. This means that the positioning result stays consistent over a time period. Although high time resolution of UWB signals makes it easier to distinguish between original and multipath signals, the non-line-of-sight condition is still a challenge.

ZigBee and Z-Wave

These two positioning technologies are meant to be used by connected objects in very small spaces. They are similar in many ways. The frequency ranges are around 900 MHz, which has advantages over the range (30 meters for Z-Wave and 10 centimeters for ZigBee). But the flow is low and the energy consumption is very low.

SigFox, LoRa

They are based on a large number of antennas which can share signals over a frequency range close to 900 MHz, with very low bit rates. Connected objects log on directly to LoRa antennas (operation is similar in SigFox). These act are bridges between the servers in the network. It sends the information about user position to the application servers over the Internet. These networks have the advantage of not needing a connection to a router or a telephone so that the object under consideration can operate data sharing via the Internet (Huircan et al., 2010).

RFID

Positioning systems of Radio Frequency Identification (RFID) technology use radiofrequency radiation to identify tagged objects in complex indoor environments such as offices, hospitals, hypermarket, etc.. RFID as a wireless technology, allows flexible and inexpensive identification of physical person or device. There are two types of RFID systems, passive and active RFID systems (Soonuk et al., 2017). In passive systems, tags are small inexpensive and cover a small area. In passive systems, the tag followed is a receiver. In active systems, active tags are transmitter-receiver that constantly transmit their identification and other information. Thus, the cost of these tags is higher and the coverage area of smart tags is larger compared to passive tags.

The WhereNet system designed by Zebra Technology Company is a real-time indoor and outdoor location system using RFID technology. The period of issue of tags varies between 5 seconds and one hour. However, WhereNet offers a location accuracy of about 2 m to 3 m, which is not satisfying enough for indoor applications. The system is complex with an infrastructure composed of several elements stated at different places.

WLAN Positioning Systems

All WLAN positioning systems are based on RSS, SNR or proximity sensing measurements. RADAR (Paramvir and Venkata, 2000) developed by Microsoft Research is a 2-D WLAN positioning system with an accuracy of about 4 m. The Ekahau system is another WLAN positioning system. Due to the complexity of indoor environments, the accuracy performance of these WLAN positioning systems is not good. The use of the fingerprinting technique that requires a larger database in location estimates becomes complex and expensive as the number of users of the positioning system increases.

LIFI

It uses VLC (visible light communication) technology. LED lamps are access points to connect to the internet as well as a receiver. The radio waves here are replaced by light spectra. The data is protected because the light does not cross the walls and their transmission at very high speed (faster 100 times more than Wi-Fi). But its detection does not exceed 10 cm and the absence of LED light cancels the indoor location. In addition, the upstream flow: the technologies presented here allow to receive the data but what happens in the other direction when one wishes to send data?

(Chizari et al., 2017) designed a dimming compatible visible light communication (VLC) system with asynchronous indoor positioning method in a standard office room. It is in combination with asynchronous and indoor positioning method according to illumination standards under channel constraints.

Bluetooth

Bluetooth is another communication standard widely used on smartphones and laptops. In Bluetooth based positioning systems, the devices are arranged in various groups and the position of a Bluetooth mobile device is determined in relation to the contributions of other mobile terminals of the same group (Gowrishankar et al., 2015, Peng et al., 2016). An example of an indoor Bluetooth positioning system is the Topaz, which combines the infra-red signal with the radio signal to locate the target. There are Classic mode and Low Energy mode. Classic mode allows high debit and high power consumption. However, the Bluetooth Low Energy (BLE) mode has the opposite objective. It is available since the version 4 of Bluetooth. It is the version that will particularly interest us in the context of connected objects. Finally, BLE works on a principle of profiles which defines the mode of object operation (the transmission rate and transmissible commands). With these profiles it is unnecessary to install additional software for the object run. This facilitates the work that will not have additional cost for the design of this software. Ibeacon is the name for Apple's technology standard, which is an indoor positioning system that used BLE technology. iBeacon technology allows Mobile Apps to understand their positionon a micro-local scale, and deliver hyper-contextual content to users based on location.

The authors of (Peng et al., 2016) proposed AltBeacon as an automate IOT application that defines the format of the advertisement message that BLE proximity beacons broadcast. It is based on the proximity triggered. The other search of (Peng et al., 2016) undertake a very dense IOT environment with

BLE compatible devices and propose a repeated Weighted KNN (IW-KNN) indoor localization method based on RSS of the BLE, which has a low power consumption and a long life expectancy.

MAIN FOCUS OF THE CHAPTER

The studies of personalized targeting in ubiquitous environment are limited in the literature because of the new context of the interaction between the mobile customer and smart objects. Thus, when smart objects are connected, the quotidian user journey becomes fit by the great employ of the smartphones. Then, IOT aims to spread ultra-miniaturized sensor networks with the lowest consumption possible to guarantee the longest possible lifetimes. It has a multiplicity of applications that affect all areas: monitoring industrial environments, improving the environment, geolocating inter building, advertising push in shopping malls, guidance in smart cities, etc.

Thereby, the main focus of this chapter is to study the indoor geo-localization based services.

When the customer is detected in the hypermarket, he should receive in his Smartphone a personalized advertisement according his loyalty in real time. Nevertheless in the last 10 years, indoor geo-localization phenomenon has received an important attention in both research and industrial areas. In the face of the wide attempts made for indoor geo-localization based services using smartphone wireless sensors, its performance is yet insufficient and inaccurate in big open zones. Such as supermarkets due to the gathered errors appeared from the uncertainty of users' mobility and variation of magnetic field. This chapter is used to develop this interaction according to context of indoor localization and made his needs in real time.

In the next section authors will show the process of personalized ads in the ubiquitous environment with the enhanced part of the chosen indoor geo-localization system.

PROPOSED APPROACH: PERSONALIZATION OF THE UBIQUITOUS ADVERTISING

The researchers choose the personalized approach that uses two types of data (static data and behavioral data) to ensure a personalized targeting for each individual. Static data are extracted from the customer's loyalty card and behavioral data which vary according to the client's lifetime within the company. These data are used to build the profile of each customer. After, they are used to make the individual targeting. Researchers choose to send a personalized advertisement in the new form of the ubiquitous marketing. Figure 1 outlines the steps of our approach in the ubiquitous marketing. Experts in the field are interested to fix the star product on advertising. According to the product, the target's criteria are established. These criteria define the organization of interested customers in the store.

The next phase concerns building the profiles of loyal customers and potential clients depending on their degrees of loyalty toward the star product. First, researchers calculate the purchases of each customer for the star product to extract those who are faithful. Later, researchers analyze the demographic information registered by the loyalty card using the clustering technique to build loyal customer profiles. Thereby for the rest of the customers, analysis of association rules is performed on their purchases' transactions in order to distinguish the loyal customers from the non-loyal ones. This analysis is measured by the computing probability to purchase the star product in the near future. At the end of this phase,

Figure 1. Steps of the proposed approach

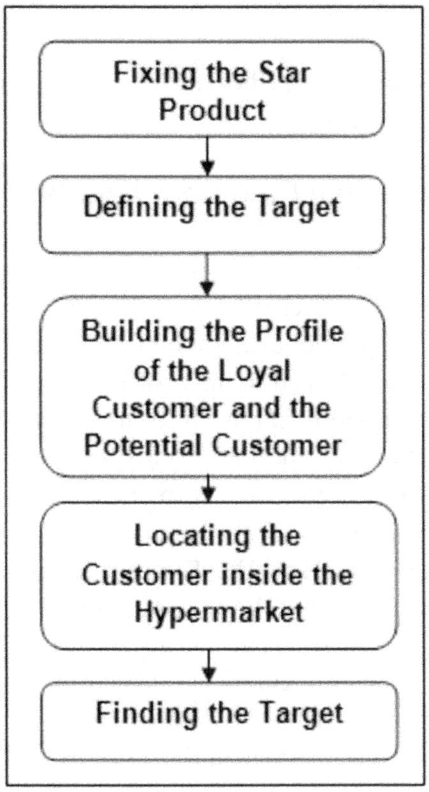

researchers obtain two types of customers: The loyal customers and the potential ones towards the star product. The last phase is devoted to send the ubiquitous ads after detecting the presence of a potential customer by a technology of Indoor location and checking the advertising constraints. Our approach uses the technique of push marketing to find the target in a better performing way.

Fixing the Star Product

The star product is the product sent in the advertising. It is chosen according to predefined criteria. Experts in the field are responsible to choose the star Product. The choice of product depends on several criteria: the product expiration date, the purchase rates, promotion, etc. This step is basic in our approach because it emphasizes on the one hand the desired product and on the other hand the company needs. Once the star product is selected according to specific selection criteria, identifying the target customer becomes inevitable.

Defining the Target

After fixing the star product, experts can find out the targets interested in this product by defining the criteria of ads. In our approach, researchers resort to the customer's behavioral information in order to define the targeted customers and send to them the advertisement of the star product. There are more than one type of customers, which necessitate an ordered sending of the personalized advertisement. To target the right customer, the database is organized according to the degree of loyalty of each client towards the star product.

The definition of both: The star product and the target is followed by another indiscernible step, which is the classification of the consumer's data

Building the Profiles of the Loyal Customer and the Potential Customer

This stage is responsible of building the profile of booth loyal customer and potential one. Our proposal is based on the principle of researchers (Ho et al., 2007) who used two types of customers. The database used the degree of fidelity toward the promotional product for classifying its customers. But the problem in the study of (Ho et al., 2007) is manifested in the choice of classification criteria. First, researchers in this study used demographic information in the clustering technique to build the profile of each customer. Such information which not permit a definite classification, could lead to a misclassification due to the uncertainty of customers' profiles. Second, researchers used association rules to extract the most frequent e-purchases in the same transaction of the star product using Apriori Algorithm. This algorithm used the last purchases stored in the web log file. The problem with this technique is the lifetime of the web log file. Once the connection is lost, all purchases are lost and the technique becomes invalid. Since, it is limited to the transactions stored in the web log file. To solve this problem, researchers make modifications in our proposal as shown in figure 2.

The step of building the customer's profile is based on the measure of similarity between the personal information of both the loyal customers and potential ones, provided by the loyalty card. According to the purchases baskets of each customer, researchers conduct an analysis of sales receipt stored in the database. Later, researchers extract those including the star product. The database is divided into two groups of customers: Buyers and non-buyers. The first group of buyers is divided into two groups according to their degree of fidelity (High Fidelity clients, low fidelity clients). Loyal customers are those whose their past purchasing tendencies (PPT) for the star product are very high. Similarly, there are two sub-groups for the non-buyers. The first sub group represents potential customers whose personal information are very similar to the profiles of the loyal customers. The second contains customers whose personal information are totally different from those of loyal customers. The Clustering technique analyzes the past purchases tendencies of loyal customers' analytical data (age, salary, number of son...).

Figure 2. Steps of building loyal and potential customers profiles

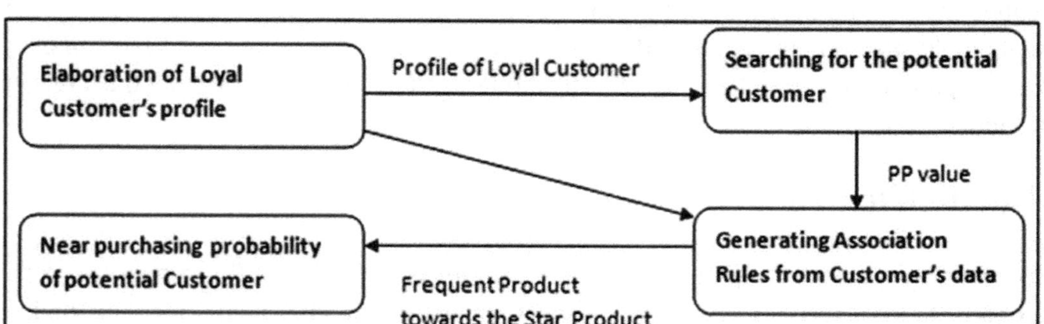

The goal here is to establish loyal customers' profiles based on personal information and past purchases tendencies of the star product.

In order to identify potential customers, researchers need to compare their personal information with loyal customers' profiles. To make this process at work, researchers use the Purchase Probabilities (PP).

Furthermore, our proposal generates the Association rules related to loyal customers to identify the recurrent products from the star product in each transaction. To predict the probability of purchasing for the potential customers in the near future, researchers detect the interest relating to each potential customer after analyzing his till receipt containing products similar to the recurrent products from the star product. Finally, researchers compare these interests with the profiles of loyal customers to increase the probability of purchases for some potential customers (calculation NPP). The figure 2 outlines the basic steps of building the customer's profile.

Elaboration of Loyal Customer's Profile

This step is responsible of constructing the loyal customer's profile. It is based on the personal information. For each loyal customer, researchers are interested to analyze the till receipts containing the star product.

Measure of Past Purchasing Tendency (PPT)

The classification of each client depends on two factors in our study: The static and the dynamic data of each customer. Static data represent the demographic information stored in the database after each registration. In our approach, researchers have chosen the loyalty card as the main source for static data. However, dynamic data is linked to the purchasing behavior of each client. This behavior changes over time due to the operations done by each consumer. To find the degree loyalty of the star product, researchers define the Past Purchasing Tendency (PPT) for each client. This measure is used to identify each client's loyalty to the star product as formula 1:

$$PPT_K = \frac{\left(\frac{Q}{Q_0}\right) * \left(F + \frac{T}{t}\right) - \min}{\max - \min} \tag{1}$$

PPT_K indicates the fidelity degree of the customer *(k)*'s past purchases to the star product. When the customer *(k)* is faithful to the star product, the value of his PPT is higher. PPT is composed of three factors: Q/Q0, F and T/t .

- Q is the total quantity of star product purchased by the customer *(k)*.
- Q_0 is the threshold quantity of the star product. It presents the average of all Q values relative to the number of loyal customers who have purchased the star product. Customers whose total quantity of purchasing the star product is over Q0 are considered the most loyal customers.
- $\frac{Q}{Q_0}$ indicates how many times the customer *(k)* has purchased the star product according to the threshold quantity.

- F the frequency of the star product purchased by the client *(k)*. The higher the F value is the more times the customer *(k)* purchases the star product.
- min represents the minimum value of all PPT values.
- max represents the maximum value of all PPT values.
- T is the period of time between the first purchase and the last purchase.
- t is the turnover period of the star product, which is decided by the system designer and it depends on the product expiry date. Generally, the non-consumable products (clothes, dishes, etc.) are more durable. They have the larger t values. In contrast, the consumable products tend to have smaller values of t. The ratio T/t indicates the number of times the customer *(k)* has purchased the star product. If the value of T/t is high, it indicates a longer time since customer *(k)* has made his first purchase. F and T/t are important for the PPT value because clients who purchase the star product few times during a long period are considered loyal customers. Therefore, according to formula 1, authors can say that customers who buy large quantities of star product many times during a long period are the most loyal customers. To restore the PPT value between 0 and 1, the PPT value is divided by the ratio (maximum-minimum). A customer with a high PPT value is more likely to purchase the star product. In addition, the customer who PPT= 1 is the one that has the most loyal profile.

Authors consider the algorithm shown in Figure 3 to calculate the PPT values of each loyal customer.

Figure 3. Algorithm for calculating PPT value

```
Input idvedette, Q0, t ;
Q=0 ;
F=0;
T=0;
For i=1 to n
  If (PanierProduit [idproduit] == idvedette) then
    Repeat
   {  F+=1;
      Q=+quantite,
       T=datefin- datedebut;
       Until atteind datefin for each purchase of client;
   }
 PPT(i) = [(Q/Q0) × (F + T/t)] ;
End if
End for
```

Indeed, every loyal customer is identified by m personal characteristics. These characteristics are in two kinds: Analytical information (age, number of children, salary, etc.) and categorical information (sex, civic status, study level, occupation, etc.).

Clustering Analysis of Loyal Customers

This step of classification based on the personal backgrounds of the loyal clients and their PPT values (calculated previously). Each customer has m personal characteristics expressed by the formula (2):

$$ch_i = \left\{ch_1, ch_2, ...ch_m\right\}, i = 1..m \tag{2}$$

There are in total m characteristics of personal information: Analytical and categorical data. The clustering analysis is used to group customers according to their analytical data. The result by the clustering technique is expressed by the formula (3):

$$ch_{i,j} = \left\{C_{i1}, C_{i2}, ...C_{in}\right\}, j = 1..n, \forall a_i, j_{\min} \le C_{ij} \le \forall a_i, j_{\max} \tag{3}$$

For each characteristics (i), loyal customers are classified into n clusters (j=1..n), then each C_{ij} represent the class number of characteristics (i). The maximum value of data in the cluster C_{ij} is $a_{ij\max}$ and the minimum value is $a_{ij\min}$. The average PPT value in C_{ij} is the total value of PPT data divided by the number of data in C_{ij}. Therefore, there are two forms to present personal data of loyal customers expressed by the formulas (4) and (5).

Presenting the static data:

$$ch_{i,j} = \begin{Bmatrix} C_{i1}, Avg\left(PPT\right)_{C_{i1}}, & a_{i1\min} \le C_{i1} \le a_{i1\max} \\ C_{i2}, Avg\left(PPT\right)_{C_{i2}}, & a_{i2\min} \le C_{i2} \le a_{i2\max} \\ & . \\ & . \\ C_{in}, Avg\left(PPT\right)_{C_{in}}, & a_{in\min} \le C_{in} \le a_{in\max} \end{Bmatrix}, j = 1..n \tag{4}$$

Presenting the categorical data:

$$ch_{i,j} = \begin{Bmatrix} C_1, Avg\left(PPT\right)_1 \\ . \\ . \\ . \\ C_n, Avg\left(PPT\right)_n \end{Bmatrix}, j = 1..n \tag{5}$$

According to formula (4), each characteristics of *(i)* is divided into n classes. C_{i1} means the first class of the characteristics *(i)* and it is defined by an interval which is limited by the two bounds ($a_{i1_{\min}}$ and $a_{i1_{\max}}$). To define the categorical data, authors calculate the average of PPT values in each category like in the static data: C_1 is the first category and C_n is the last category. Authors use the K-Means algorithm to classify the loyal customers according to their PPT values and their static information. The K-Means algorithm is able to extract the characteristics of the input in an efficient way and on a large scale as the view of (Tsang and Lam, 2012). The quality of this algorithm depends strongly on the initial nuclei. In addition the sensitivity of the algorithm to the initialization is all the greater in that the dimensionality of the data is great.

Authors choose this algorithm because it is more effective in analyzing the customer's purchasing behavior on the datawerhouse. The procedure follows a simple and easy way to classify a given data set through a certain number of clusters (assume k clusters) fixed a priori. The main idea is to define k centroids, one for each cluster. These centroids shoud be placed in a cunning way because of different location causes different result. So, the better choice is to place them as much as possible far away from each other. The next step is to take each point belonging to a given data set and associate it to the nearest centroid. When no point is pending, the first step is completed and an early groupage is done. At this point authors need to re-calculate k new centroids as barycenters of the clusters resulting from the previous step. After they have these k new centroids, a new binding has to be done between the same data set points and the nearest new centroid. A loop has been generated. As a result of this loop they may notice that the k centroids change their location step by step until no more changes are done. In other words centroids do not move any more. The algorithm is composed of the following steps:

1. Place K points into the space represented by the objects that are being clustered. These points represent initial group centroids.
2. Assign each object to the group that has the closest centroid.
3. When all objects have been assigned, recalculate the positions of the K centroids.
4. Repeat Steps 2 and 3 until the centroids no longer move. This produces a separation of the objects into groups from which the metric to be minimized can be calculated.

In addition, the K-Means in general includes a multidimensional array which represents our data: One static and the other dynamic. After using the clustering analysis to combine between the personal backgrounds and the PPT values, authors get the profiles of each loyal customer. Afterwards, they classify the customers who have not shopped the star product. According to the profiles of loyal customers resulting previously, they make a comparison with the characteristics of each non- buyer. The next step is employed to estimate the probability of buying the star product (PP) in the future of potential customers.

Searching for Potential Customer

In this step, authors are interested in identifying customers who are most likely to buy the star product. This measure is conducted through the analysis of loyal customers' profiles and potential customers' personal information taken by loyalty cards. Formula (6) expresses this relation:

$$PP = \frac{\sum Avg\left(PPT\right)}{Number\ of\ characteristics} \tag{6}$$

PP denotes purchasing probability of each potential customer who has never purchased the star product. The algorithm that calculates the PP value of each potential client is presented in Figure 4. There are in total m characteristics of personal information. For each characteristic ch_i, authors calculate PP_{Chi} in order to predict each potential customer's purchasing probability. Every potential customer's PP value becomes the average of all PP_{Chi}. The presence of characteristics is shown in either analytical data or categorical data. If it is shown in analytical data, the result of Clustering analysis obtained in formula (4) is applied for the following calculation. If it is shown in categorical data, the result of formula (5) is used for the following calculation. For analytical data, there are three possible situations:

Situation 1: Potential customer's ch_i falls into the range of no cluster. This situation indicates that this customer is not likely to buy the star product according to the observation of past shopping tendencies.

Situation 2: The potential customer's chi is within the range of one cluster.

Situation 3: The potential customer's ch_i is in the range of more than one cluster. Although the average value of PPT in a cluster demonstrates the past shopping tendency of this cluster.

As for categorical data, authors calculate the average of PPT value found in the given category $C_{i,j}$ as the PP value of ch_i ($PP_{Ci,j}$).

In this level, authors find customers who might be likely to buy the star product. Yet, the interests of customers can change at any time. It is important to trace customers' recent interests in different products at the same time. In this section, the recent intentions of potential customers are measured by analyzing their latest purchases in the database via Association Rules. These rules are generated to measure the correspondence between the included products in the same transactions of the star product and the other products bought by potential customers.

Generating Association Rules From Loyal Customers' Data

This step extracts the rules made by the loyal customer which contain the transactions of the star product. As authors are interested in identifying the transactions made by loyal customers, only the till receipts containing the star product are included in the analysis. The condition of this rule is as *"If the product x existed then the star product is also affected"*.

According to the results of analyzing the association rules that contain products similar to the star product, authors are able to identify the most likely potential customers who will buy the star product in the near future. The loyalty degree of potential customer is measured by the value of the purchase probability (PP). To make the probability of buying the star product by the potential customer soar in the near future, they use transactions made by the loyal customers as formula 7 indicates:

Figure 4. Algorithm for calculating PP value

```
For each customer input Ch_i
PP = 0
For i = 1 to m
    PP_Chi = 0
    If Ch_i is analytical data, then
    k = 0
    Array Data_number = 0
    For j = 1 to n
        If a_ij min <= Ch_i <= a_ij max then,
        k = k + 1
        PP_Chi(k) = Avg(PPT)_Cij
        Data_number(k) = number of data in C_ij
        End if
    Next
    If k = 0, then
        PP_Chi = PP_Chi(k)
        else
            Data_number_total = 0
            For x = 1 to k
            Data_number_total = Data_number(x) + Data_number_total
            Next
            For y = 1 to k
            PP_Chi = PP_Chi(y) × (Data_number(y) ÷ Data_number_total) + PP_Chi
            Next
        End if
    End if
    else (Ch_i is categorical data)
        For j = 1 to n
            If Ch_i = Ca_ij, then
            PP_Chi = Avg(PPT)_Caij
            Exit
            End if
        Next
    End if
    PP = PP_Chi + PP
Next
PP = PP ÷ m
```

$$NPP = \frac{\left(PP + NFR\right)}{NR} \tag{7}$$

- NPP presents the Probability of buying the star product by a potential client in the near future.
- NR is the number of rules discovered by the analysis of association rules.
- NRF represents the number of rules followed by the potential customer

The main idea behind this approach is that the clients who have a high value of purchasing probability (PP) and their recent data follow the rules generated by the analysis of Association Rules; are most likely to seek information about the star product in the near future. The more the potential customer followers the rules, the more he has the chance to buy the star product in the near future. The above equation resizes the purchasing probability of the potential customers who can have many rules in their recent databases. If a potential customer follows more than one rule generated by the analysis of association rules, he should receive a higher NPP value. However, a potential customer who does not follow any rules has the same NPP value as his PP.

NPP values vary between 0 and 2. Customers who are less likely to buy the star product are those who have not personal information in common with the loyal customers and do not have any intention to purchase the star product in the near future, so they have a zero as NPP value. Potential customers who have personal information similar to those loyal customers have the highest possibility to purchase the star product with a NPP value around of 2. Authors use the potential Central value NPP to distinguish potential customers from the visitors in the store. They could find the targeted customers using the NPP value as they could give them information about star product.

In this study, authors include the dimension of context to perform the ubiquitous marketing. This added part is realized following the customer's interaction with the environment through the Indoor geo-localization system. Next, authors will indicate the choice of technologies chosen in this context.

PROPOSED APPROACH OF INDOOR GEO-LOCALIZATION SYSTEM

Our approach is to collect different geo-localization data from BLE devices located in an indoor space of IOT. Therefore, we use them to optimize the data recuperated from low cost BLE technology and popular Wi-Fi RSS fingerprint-based indoor positioning systems, in order to better positioning accuracy. Implementing this concept, a new enhanced K-Nearest Neighbour (KNN) positioning algorithm was developed (KNN), which is oriented to filter the initial Wi-Fi fingerprint dataset as well as the proximity of the RSS fingerprints to the BLE devices.

By choosing to filter the initial dataset instead of simply combining BLE and Wi-Fi fingerprints, KNN utilizes an optimized and small size subset of possible user locations for the final position estimation. The KNN algorithm uses the data transmitted from the i-Beacons concerning the evaluated distance between a BLE device and the smartphone, as well as the information assigning to the nearest i-Beacon of the BLE network. These data are used as an input to the filtering processes of the KNN in order to approximately evaluate a probable area A which includes the user's position. The area A is formulated between a minimum and a maximum radius measured from a center point, where the BLE device is located. These radius values are calculated, with a predefined tolerance parameter, which accommodates any positioning error factors. The area A is then used to project the number of candidate fingerprints, extracted from the initial wifi fingerprint dataset. The filtered fingerprint data subset S is finally used as the optimized input, to typical indoor positioning algorithms (KNN). The proposed methodology serves two purposes: firstly, attaining fast positioning estimation due to the use of a fragment of the initial fingerprint dataset and secondly attaining better positioning accuracy by constraining any possible calculation errors within a very specific area A, where the user is actually located. The latter is affected due to the inherited short range of IEEE 802.15 and its capability to identify the nearest i-beacon device to each smartphone.

- **KNN Algorithm:** The filtering algorithm KNN receives as an input the Wi-Fi radiomap and BLE locations. A procedure named FINGERPRINTS to BLE DISTANCE pre-calculates distances between BLE devices and all fingerprint locations for easier rating of the final output, which is subset S.

During the real-time localization estimation, the algorithm initially scans BLEs locations by running BLE DISCOVERY procedure. Then, it collects their parameters and sorts them accordingly based on their Received Signal Strength (RSS). Upon collection of BLE information, KNN selects the nearest BLE device and utilizes the distance information broadcast by the beacon to additionally calculate the Tolerance Tol level. In the latter case, the closer the user is to the BLE, the smaller the Tol will be, and hence a smaller optimized dataset will be selected. This methodology is a more optimized option. Finally, instead of utilizing the heavy initial radiomap, the much smaller dataset S feeds the typical KNN methods in order to estimate the location of the user.

ANALYSIS OF RESULTS

To validate the approach, researchers utilize a database containing 100 products and 30 customers. Indeed, they choose the star product "CHOCOLATE ARUON" during one year. Therefore, they set the criteria for determining the interested customers: Age and civic status for each person. The results are analyzed during the year 2015. The choice of this interval enhances the research results. It gives more chance for customers who purchased small quantities of star Product. The wider the time interval is the more chances researchers get to detect a higher number of customers. This period must be wider than the advertising duration to guarantee a high number of loyal customers.

Measurement of Loyal Customers' PPT

After choosing and fixing the star product, researchers have identified that there are 15 customers who have bought the star product. Table 1 displays the PPT values of each loyal client to build the profile of the best loyal customer. As the star product is consumable, the turnaround time should take a small value t= 30. They choose the threshold quantity Q0 as the average value of purchase quantities made by the loyal customers, as it is shown in formula (8):

$$Q_0 = \frac{\sum Q_0}{15} = 42.26 \quad (8)$$

After calculating the PPT of each loyal customer as shown in Table 1, researchers notice that the client 26 is the most faithful to the star product with PPT= 245.43. He has the highest values of the three parameters (F, T and Q). However, the customer 28 is the lowest with PPT= 1.26. The client 26 has the best profile among the other loyal customers. His rank of PPT is the highest.

As it was mentioned previously, the values must be measured according the same scale. Researchers resize the PPT values on the scale [0, 1] as shown in Table 1. The min and the max in this case are the minimum and maximum value of PPT respectively. Here, Min = 1.26 and Max=245.43. Along the study, researchers use the resized PPT values to facilitate the calculation.

Table 1. Results of loyal customers' PPT

Idclient	F	T	Q	PPT	PPT Resized
3	10	122	50	16.64	0.062
7	20	210	28	17.88	0.068
10	15	214	130	68.08	0.271
12	19	100	20	10.56	0.038
13	10	242	60	25.65	0.092
14	10	151	10	3.55	0.009
15	3	15	60	4.96	0.015
17	9	154	15	5.01	0.015
20	10	91	20	6.16	0.020
23	10	20	10	2.52	0.005
24	10	200	10	3.94	0.010
25	20	199	23	14.49	0.054
26	43	374	187	245.43	1.00
28	5	10	10	1.26	0.00
30	10	10	10	2.44	0.004

Loyal Customers' Classification: Combination Between the PPT Values and the Personal Information

After knowing the list of loyal customers and their corresponding PPT values, researchers can identify better the classification of customers. They use two personal information: The age as static information and the civic status as categorical information. Following the previous results, they seek the classes to which they belong according to their ages and their PPT values. By simple query according to the average values of PPT, Table 2 presents the classification results according to loyal customers' age. They divide the age attribute in two intervals [18...25] and [26...44]. They choose this classification to differentiate between two types of customers: The young and the adults. For each age interval, researchers seek the number of customers. After, they count for each interval the quantity of data and all the PPT values resized (by function Sum) in order to measure the average of PPT.

According to the results shown in Table 2, loyal customers are classified in two classes after analyzing their shopping behavior toward the star product *'CHOCOLATE ARUON'* and considering their age. For each interval, researchers calculate the value of average PPT.

Table 2. Classifying the loyal customer according to the age and the average PPT values

Cluster	Min	Max	Quantity of Data	Sum of PPT	Avg (PPT
1	18	25	7	2.29	0.328
2	26	44	8	0.328	0.041

They can also take other categorical classification features to better identify the loyal customer (civic status, level of education, etc.). In this case, they have three categories of classification according to the civic status: M: Married, F: Engaged and C: Single. With the same principle of calculation the Avg(PPT) of each interval, authors calculate the average of the three categories. The calculation results indicate that loyal customers are classified into three classes according the civic status, as indicated in the formula (9):

$$\begin{Bmatrix} C_1, Avg\left(PPT\right)_M = 0.278 \\ C_2, Avg\left(PPT\right)_F = 0.146 \\ C_3, Avg\left(PPT\right)_C = 0.009 \end{Bmatrix} \tag{9}$$

Similarly to the loyal customers, researchers are interested to estimate the probability of purchasing relative to the other customers who don't purchase the star product chocolate. In the next subsection, they calculate the probability of purchasing relative to the potential customers.

Potential Customers' Classification: Association Rules

The potential customers, in our study, mean the customers who don't affect any purchasing of the star product. Yet, in order to push them to buy the star product, researchers estimate their probability of purchasing by calculating the PP values as shown in Table 3. Using the same attributes (age and civic status) as in the loyal customers' classification, they find that potential customers fall into two classes according to their ages and their PP. The PP values vary between classes. The variation indicates the loyalty degree of each potential customer in order to estimate the probability of buying the star product.

Table 3. Clustering the potential customers

Idclient	Civic Status)	Age	PP	Clustering (Age
1	C1	30	0.52	2
2	C1	33	0.52	2
5	C1	34	0.52	2
6	C2	25	0.21	1
8	C3	18	0.14	1
9	C3	23	0.14	1
11	C3	24	0.14	1
16	C3	28	0.39	2
18	C2	33	0.50	2
19	C2	20	0.21	1
21	C1	44	0.13	2
22	C2	39	0.50	2
27	C3	40	0.39	2
29	C2	39	0.50	2

After the classification results, researchers are interested now in identifying products similar to those found in the transactions made by loyal customers and contain the star product. They assume that the most frequent rule is the one that has the largest number of products. The most frequent product in this transaction is DANY and the most frequent rule when the confidence value is 100 per cent is: *"IF DANY THEN ARUON*". As a result, the most frequent rule which contains the highest support is: *"DANY, MASTER, YAOURT"*. In our case, researchers indicate that the number of rules (NR) made by loyal customers towards the star product is 3. Then, it is time to test whether a potential customer has made a purchase to such product or not.

The results shown in Table 4 display the possibility of buying the star product in the near future by the potential customers (NPP). This segmentation is calculated by the formula (4) where NRF represents the number of rules followed by the potential customer and PP represents the probability of purchasing the star product. In Table 4, the best potential customer profile is the customer number 1 who has the highest transaction. Client 1 is able to purchase the star product in the near future with the probability value of 0.84. Yet, the client 21 has the lowest probability with NPP=PP.

Our database is now segmented in two types of customers. Each one is differentiated by his degree of loyalty to the star product (PPT for loyal customers and NPP for potential customers).

Sending Personalized Advertising

The purpose of this last step is to target each customer according to his profile classification in the database. This step uses the push marketing technique. The push strategy is based on behavioral information for each target. Researchers suppose that the four customers identified by 1, 3, 14 and 22 are detected

Table 4. Segmentation result of potential customer EA

Idclient	Products	NFR	PP	NPP
1	DANY, MASTER, YAOURT, DCROC	2	0.52	0.84
2	YAOURT, DCROC	1	0.52	0.50
4	DANY, MASTER, DCROC	1	0.52	0.50
16	YAOURT, DCROC	1	0.52	0.43
22	YAOURT	1	0.21	0.43
5	MASTER	1	0.50	0.40
18	DANY, YAOURT, DCROC	1	0.21	0.40
6	DANY	1	0.21	0.30
8	YAOURT, DCROC	1	0.14	0.30
9	DANY, MASTER, DCROC	1	0.14	0.30
11	DANY, MASTER, DCROC	1	0.14	0.30
27	MASTER	1	0.39	0.30
29	YAOURT	1	0.39	0.30
19	YAOURT, DCRO	1	0.13	0.28
21		0	0.13	0.13

Table 5. Personalized Push Technology

Order	idclient	Fidelity Degree
1	3	0.062
2	14	0.009
3	1	0.84
4	22	0.43

simultaneously by the geolocation system. They receive the advertising in order as listed in Table 5. The clients with id 3 and 14 are loyal customers and the two others are potential ones.

After being in the store, each one of the four customers activates the WIFI on his Smartphone. Using the authentication system of the hypermarket, an analysis is made in the server which contains the database to identify the customer's category and sends him the personalized advertising. This leads to the activation of both: the interactive service for mobility and the personalized advertising system. According to the degree of loyalty, a personalized advertisement is sent by a text message, using the Push strategy. The customers are targeted by priority of PPT values: Loyal customers are the first who receive the advertisement then the potential customers.

CONCLUSION

In this chapter researchers have presented a survey of different Indoor geo-localization based systems. Then they have proposed and implemented a novel system that improves accuracy while reducing costs. They used KNN as a way to reduce the effects of unlikely location estimations due to wrong fingerprint plateforms. Thereby, our original indoor localization algorithm (KNN) combined with the simplicity of BLE technology and the popular Wi-Fi network. It treats different type of information based on fingerprint localization platforms. KNN provides an optimized small subset of possible user locations, based on which it finally estimates the user position and improve positioning accuracy by minimizing any calculation errors.

REFERENCES

Abdaoui, N., Hadj Khalifa, I., & Faïz, S. (2017). Process of Personalizing the Ubiquitous Advertisements. *International Journal of Knowledge Society Research*, *8*(2), 13–35. doi:10.4018/IJKSR.2017040102

Carlos Filipe,, F., António,, M., Lino,, F., & João,, B., António, S., & Carlos, R. (2015). Context aware middleware in ambient intelligent environments. *International Journal on Computer Science and Engineering*, *10*(4), 347–358.

Chizari, A., Jamali, M.-V., Abdollahramezani, S., & Jawad, A. (2017). Visible light for communication, indoor positioning, and dimmable illumination: A system design based on overlapping pulse position modulation. *Optik International Journal for Light and Electron Optics*, *151*, 110–122. doi:10.1016/j.ijleo.2017.08.003

Crémilleux, B., Boizumault, P., Loudni, S., & Rojas, W.-U. (2014). *Modélisation et extraction de motifs optimaux*. Reconnaissance de Formes et Intelligence Artificielle.

García, E., Poudereux, P., Hernández, A., García, J.-J., & Ureña, J. (2013). DS- UWB indoor positioning system implementation based on FPGAs. *Sensors and Actuators. A, Physical*, *201*, 172–181. doi:10.1016/j.sna.2013.06.019

Goswami, A., Das, A.-K., & Sahoo, J. (2015). An efficient approach for mining association rules from high utility itemsets. *Expert Systems with Applications*, *42*(13), 5754–5778. doi:10.1016/j.eswa.2015.02.051

Gowrishankar, S., Madhu, N., & Basavaraju, T. G. (2015). Role of BLE in proximity based automat- -ion of IoT: A practical approach. *Proceedings of the 2015 IEEE Recent Advances in Intelligent Computational Systems (RAICS)*, 400–405. 10.1109/RAICS.2015.7488449

Heo, J., Kim, M., & Chan-Olmsted, S.-M. (2010). Perceived effectiveness and business structure among advertising agencies: A case study of mobile advertising in south Korea. *Journal of Media Business Studies*, *7*(2), 1–20. doi:10.1080/16522354.2010.11073504

Ho, C.-L., Chang, H.-J., & Hung, L.-P. (2007). An anticipation model of potential customers' purchasing behavior based on clustering analysis and association rules analysis. *Expert Systems with Applications*, *32*(3), 753–764. doi:10.1016/j.eswa.2006.01.049

Huircan, J. I., Munoz, C., Young, H., Von Dossow, L., Bustos, J., Vivallo, G., & Toneatti, M. (2010). Zigbee-based wireless sensor network localization for cattle monitoring in grazing fields. *Computers and Electronics in Agriculture*, *74*(2), 258–264. doi:10.1016/j.compag.2010.08.014

Izquierdo-Yusta, A., Olarte-Pascual, C., & Reinares-Lara, E. (2015). Attitudes toward mobile advertising among users versus non-users of the mobile internet. *Telematics and Informatics*, *32*(2), 355–366. doi:10.1016/j.tele.2014.10.001

Leite Lucas de Azevedo, V., & Borges, M. (2015). More Collaboration, More Collective Intelligence. *International Journal of Knowledge Society Research*, *6*(3).

Mitilineos, S. A., Kyriazanos, D. M., Segou, O. E., Goufas, J. N., & Thomopoulos, S. C. (2010). Indoor localisation with wireless sensor networks. *Progress in Electromagnetics Research*, *109*, 441–474. doi:10.2528/PIER10062801

Paramvir & Padmanabhan. (2000). RADAR: An in-building RF based user location and tracking system. *Proceedings of the Nineteenth Annual Joint Conference of the IEEE Computer and Communications Societies (INFOCOM 2000)*, 775–784.

Peng, Y., Fan, W., Dong, X., & Zhang, X. (2016). An Iterative Weighted KNN (IW-KNN) Based Indoor Localization Method in Bluetooth Low Energy (BLE) Environment. *Proceedings of the* 2016 *Intl IEEE Conferences on Ubiquitous Intelligence Computing, Advanced and Trusted Computing, Scalable Computing and Communications, Cloud and Big Data Computing, Internet of People, and Smart World Congress (UIC/ATC/ScalCom/CBDCom/IoP/SmartWorld)*, 794–800.

Piwowarski, B., Gauthier, L.-A., & Gallinari, P. (2014). *Filtrage collaboratif et prise en compte de la polarité des jugements*. CORIA.

Schmidt, A., Shirazi, A.-S., Zurmaar, B., Bial, D., Zaidan, F., Alt, F., . . . Lewen, T. (2011), Digifieds: Evaluating suitable interaction techniques for shared public notice areas. In *Adjunct Proceedings of the 9th International Conference on Pervasive Computing*. Springer.

Seol, S., Lee, E.-K., & Kim, W. (2017). Indoor mobile object tracking using RFID. *Future Generation Computer Systems*, *76*, 443–451. doi:10.1016/j.future.2016.08.005

Sotenga, P. Z., Djouani, K., Kurien, A. M., & Mwila, M. M. (2017). Indoor Localisation of Wireless Sensor Nodes Towards Internet of Things. *Procedia Computer Science*, *109*, 92–99. doi:10.1016/j.procs.2017.05.299

Tang, H., Liao, S.-S., & Sun, S.-X. (2013). A prediction framework based on contextual data to support mobile personalized marketing. *Decision Support Systems*, *56*, 234–246. doi:10.1016/j.dss.2013.06.004

Wong, C.-H., Tan, B.-I., Tan, G.-W.-H., & Ooi, K.-B. (2015). Mobile advertising: The changing landscape of the advertising industry. *Telematics and Informatics*, *32*(4), 720–734. doi:10.1016/j.tele.2015.03.003

Ye, C., Swilley, E., & Bacile, T.-J. (2014). From firm controlled to consumer contributed: Consumer co-production of personal media marketing communication. *Journal of Interactive Marketing*, *28*(2), 117–133. doi:10.1016/j.intmar.2013.12.001

Zhu, N., Zhao, H., Feng, W., & Wang, Z. (2015). A novel particle filter approach for indoor positioning by fusing WiFi and inertial sensors. *Chinese Journal of Aeronautics*, *28*(6), 1725–1734. doi:10.1016/j.cja.2015.09.009

Chapter 11
Managing Multi-Cloud Data Dependability Faults

Mohammed A. AlZain
Taif University, Saudi Arabia

Alice S. Li
La Trobe University, Australia

Ben Soh
La Trobe University, Australia

Mehedi Masud
Taif University, Saudi Arabia

ABSTRACT

Cloud computing is considered a great paradigm that enables access to large scale, on demand, flexible computing infrastructure. Lower infrastructure cost and better application performance are some of the benefits of utilizing a cloud computing model. With the appearance of the new generation of multi-cloud computing, clients can benefit from a diversity of services. However, data security has become an important requirement for clients when dealing with clouds that may fail due to faults in the software or hardware, or attacks from malicious insiders. Hence, building a highly dependable and reliable cloud system has become a critical research problem. To that end, this chapter investigates and presents the results in relation to that how Byzantine fault tolerance (BFT) and secure cloud data storage and sharing techniques can be deployed to manage multi-cloud data dependability faults.

INTRODUCTION

Cloud computing is considered a great paradigm that enables access to large scale, on demand, flexible computing infrastructure. Lower infrastructure cost and better application performance are some of the benefits of utilizing a cloud computing model. With the appearance of the new generation of multi-cloud computing, clients can benefit from a diversity of services. However, data security has become an important requirement for clients when dealing with clouds that may fail due to faults in the software or hardware,

DOI: 10.4018/978-1-5225-7347-0.ch011

or attacks from malicious insiders (Tiwari and Joshi 2018). . Hence, building a highly dependable and reliable cloud system has become a critical research problem (Williams and Griffin 2018). It means a cloud service provider needs to make sure that two of the dependability requirements are satisfied: (1) its clients' outsourced data are stored securely and, (2) data queries and retrievals are executed correctly (in terms of reliability) and privately (in terms of three security attributes: Confidentiality, Integrity and Availability – CIA) (AlZain, Soh et al. 2013) (AlZain, Soh et al. 2011) (AlZain, Pardede et al. 2012) (AlZain, Soh et al. 2013) (AlZain, Li et al. 2015). Here, we term any data fault as data dependability fault that leads to security beaches and unmet reliability.

On the other hand, it also means businesses are willing to outsource their data to a third party (cloud service provider) only if they trust their data are not accessible and visible to the service provider and other non-authorized parties (Hore, Mehrotra et al. 2004). However, one of the main obstacles here for ensuring multi-cloud data dependability is Byzantine faults.

Byzantine fault tolerance (BFT) has received growing attention from the academic research community but not many systems use it in practice. While a great deal of recent research has focused on comparing the standard practical Byzantine fault tolerance protocol (PBFT) (Castro and Liskov 1999) and improving its performance with the development of Zyzzyva (Kotla, Alvisi et al. 2007) and Aardvark (Clement, Wong et al. 2009), very few studies on the BFT in a multi-cloud computing environment have addressed the detection of Byzantine failure to ensure data dependability within the cloud. In this chapter, we extend the original definition of Byzantine faults (Lamport, Shostak et al. 1982) to include the CIA dimensions (data Confidentiality, data Integrity and service Availability) (AlZain, Pardede et al. 2012), and cloud data intrusion by malicious insiders. Needless to say, data dependability is a major issue whenever users rely on third-party services because of the possibility of Byzantine failure in the cloud. Therefore, building a highly dependable multi-cloud system has become a critical research problem. To that end we enhance our previous investigation on how BFT and, secure cloud data storage and sharing techniques can be deployed to manage multi-cloud data dependability faults (AlZain, Li et al. 2016).

The remainder of this chapter is organized as follows. We first present the related work followed by overviews our conceptualization of multi-cloud data dependability faults as well as fault tolerant state machines. Then we detail three crucial operations in the proposed fault-tolerant architecture, and propose fault-tolerant method in managing multi-cloud data security faults. Subsequently we give a qualitative evaluation of the proposed architecture, and conclude this chapter with our research contribution.

RELATED WORK

As mentioned previously, multi-cloud data dependability faults subsume Byzantine faults that can be caused by malicious attacks or operator errors related to cloud data. Sending inconsistent results from clouds to the clients is considered a Byzantine fault as well (Castro and Liskov 1999).

Much research has been dedicated to Byzantine Fault Tolerance (BFT) since its first introduction (Lamport, Shostak et al. 1982), (Pease, Shostak et al. 1980). Although BFT research has received a great deal of attention, it still suffers from the limitations of practical adoption (Kuznetsov and Rodrigues 2009) and remains peripheral in distributed systems (Vukolic 2010),(Lamport, Shostak et al. 1982). BFT for Web services as well as distributed systems have received more attention. Zhao (Zhao 2007) presented a framework of BFT for Web services which was based on the PBFT protocol (Castro and Liskov 1999) and ran the system with Byzantine faults in a distributed environment. In addition, Merideth et al. (Merideth,

Iyengar et al. 2005) discussed a Byzantine fault tolerant middleware for Web services. Furthermore, Li et al. (Li, He et al. 2005) employed replication schemes and the N-Modular Redundancy concept to tolerate Byzantine faults and security attacks in the web services environment. While many recent research studies have often focused on comparing the PBFT (Castro and Liskov 1999) and improving its performance with the development of Zyzzyva (Kotla, Alvisi et al. 2007) and Aardvark (Clement, Wong et al. 2009), very few studies have addressed the detection of Byzantine failure in a multi-cloud computing environment to ensure the security of stored data within the cloud.

In addition to the deployment of BFT, multi-cloud data dependability also relies on secure cloud data storage and sharing. There are three techniques that can be used to achieve secure cloud data storage and sharing: (1) cryptology (Bowers, Juels et al. 2009), (2) secret sharing of private information (Stinson 1992) (Agrawal, El Abbadi et al. 2009), and (3) data replication (Agrawal, El Abbadi et al. 2009). These techniques are commonly used for secure data outsourcing.

Cryptology

HAIL (High Availability and Integrity Layer) (Bowers, Juels et al. 2009) is an example of a protocol that controls multiple clouds using cryptology. HAIL is a distributed cryptographic system that permits a set of servers to ensure that the client's stored data is retrievable and integral. HAIL provides a software layer to address the availability and integrity of the stored data in multi-clouds (Bowers, Juels et al. 2009). Cachin et al. (Cachin, Haas et al. 2010) present a design for multi-cloud storage (ICStore) as a dependable service in multiple clouds. Cachin et al. (Cachin, Haas et al. 2010) develop theories and protocols to address the CIRC attributes (confidentiality, integrity, reliability and consistency) of the data stored in clouds.

Secret Sharing

Agrawal et al. (Agrawal, El Abbadi et al. 2009) discuss the issue of information distribution (in terms of data query and retrieval) with the aim of showing that there is an orthogonal approach which is based on information distribution instead of encryption in the area of data and computer security. The need to communicate important or private information from one party to another instigated most of the work on data security. Agrawal et al. (Agrawal, El Abbadi et al. 2009) introduced Shamir's Secret Sharing algorithm (Shamir 1979) as a solution for the privacy issue. More details on using the secret approach in our model can be found in (AlZain, Soh et al. 2012).

Data Replication

Data replication is one of the important approaches for outsourced data security. The data owner divides data and replicates them into different data storage or multi-clouds. RACS (Redundant Array of Cloud Storage) (Abu-Libdeh, Princehouse et al. 2010) for instance, utilizes RAID-like techniques that are normally used by disks and file systems, but for multiple cloud storage. Abu-Libdeh et al. (Abu-Libdeh, Princehouse et al. 2010) assume that to avoid "vender lock-in", distributing a user's data among multiple clouds is a helpful solution. This replication also decreases the cost of switching providers and offers better fault tolerance. Therefore, the storage load will be spread among several providers as a result of the RACS proxy (Abu-Libdeh, Princehouse et al. 2010).

CONCEPTUALIZATION OF MULTI-CLOUD DATA DEPENDABILITY FAULTS AND FAILURES

Dependability is a major challenge facing the cloud computing environment. Clouds can undergo common and unpredictable failures (Cachin, Haas et al. 2010). To address this kind of failures, we need to understand the concept of cloud data dependability faults. In this chapter, we regard faulty cloud data as data that has been corrupted or compromised in terms of CIA.

Classification of Multi-Cloud Data Dependability Faults

There are two kinds of multi-cloud data dependability faults: non-latent faults (non-Byzantine faults) and latent faults (Byzantine faults). The issue of non-Byzantine faults has been discussed in (AlZain, Soh et al. 2012), while the object of this chapter is the latter, Byzantine faults.

Byzantine Faults

In cloud computing, Byzantine faults usually relate to inappropriate and intrusive behaviours as well as arbitrary and crash faults (Vukolic 2010). It is clear the Byzantine faults are considered to be more insidious than non-latent faults in the cloud computing environment. System components can suffer malicious failures in which they make arbitrary outputs. Such failures are known as Byzantine failures. Byzantine failures are difficult to detect before they cause damage to the system because they are hidden and lurking. This detection difficulty can be overcome with fault-tolerant operations discussed.

Fault-Tolerant State Machines and Byzantine Failures

In this sub-section, we conceptualize fault-tolerant state machines using the Fault-Error-Failure chain (AlZain, Soh et al. 2012). Lamport et al. (Lamport, Shostak et al. 1982) consider a Byzantine failure to have occurred if a component exhibits arbitrary or malicious behaviour or if there is possible complicity with other faulty components. To build a fault-tolerant state machine (FTSM) in a distributed system, a FTSM can be realized by replicating that state machine and running processors in each replica (Schneider 1990). For realizing the FTSM, the following protocols (Lamport, Shostak et al. 1982) (Schneider 1990) are applied:

- **Replica Coordination:** All replicas receive and process the same sequence of requests;
- **Agreement:** All requests should be received by all non-faulty state machine replicas; and
- **Order:** All requests that are received by each non-faulty state machine replica will process clients' requests in the same relative order.

Schneider (Schneider 1990) argues that if each failure can affect at most one state machine replica, the output for the FTSM could be obtained by matching the output of the state machine replicas of this group. When systems experience a Byzantine problem, FTSM must have at least 2f+1 replicas (f = 1, 2, …), and the output of the replicas is produced by the majority of these replicas (Schneider 1990). All communications are time-bounded and a time-out mechanism can detect the absence of a message. Each replica state machine uses timeouts to prevent waiting more than the time-bounded to receive a response from a faulty replica state machine. These are the main features of the FTSM (Schneider 1990), (Lamport, Shostak et al. 1982).

FAULT-TOLERANT OPERATIONS IN MULTI-CLOUD DATA DEPENDABILITY MANAGEMENT

The three core operations required in multi-cloud data dependability management with a fault-tolerance approach are: data fault detection mechanism, data storage mechanism, and data retrieval mechanism.

Cloud Data Fault Detection Mechanism

Here, there are three conditions for implementing data fault detection mechanism in an FTSM: (1) Replica Coordination: All replicas receive and process the same sequence of requests synchronized by the cloud manager; (2) Agreement: all requests should be received by all non-faulty state machine replicas; and (3) Order: All requests that are received by each non-faulty state machine replica will process clients' requests in the same relative order (Lamport, Shostak et al. 1982) (Schneider 1990).

The crucial part of the detection mechanism is the Agreement protocol called Byzantine Agreement protocol that has the following conditions (Lamport, Shostak et al. 1982) (Schneider 1990):

IC1: All non-faulty processors agree on the same value.
IC2: If the transmitter is non-faulty, the value they agree on must be the same value that was sent from the original source.

In our conceptualization of data fault, we regard faulty data as data that has been corrupted or compromised in terms of CIA in cloud data management. Hence, in this chapter we rephrase the above two conditions as follows;

IC1: All non-faulty clouds agree on the same data.
IC2: If the sender cloud is non-faulty, the data they agree on must be the same one that was sent from the original source.

Cloud Data Storing Mechanism

The data storing procedure in the BFT-MCDB design involves data distribution from the data provider to multi-clouds. This is done after executing the polynomial functions on the data using the Shamir's Secret Sharing scheme (Shamir 1979).

Figure 1 shows the BFT-MCDB-Store algorithm. The key idea of the BFT-MCDB-Store algorithm is to first (line 4) receive the data from the client or data provider to be stored in the clouds. Subsequently, the receiver unit inside the cloud manager divides the data into n shares (line 5). Then, (line 7) the interpreter unit inside the cloud manager generates the secret information X (line 8) and random polynomial functions with degree at the same level (line 8) (more information regarding writing shares can be found in (Agrawal, El Abbadi et al. 2009)), one for each cloud to be stored in different clouds. After that, (line 10) the sender unit sends data to the clouds then data storage stores these data in the clouds (line 11).

The following acronyms are used in the algorithms: Data Provider (DP), Receiver Unit (RU), Interpreter Unit (IU), Sender Unit (SU), Data storage (DS), Voter Unit (VU), User (U), Data (D), Clouds (Cs), the secret value wanted to be hidden (vs), and Commit Certificate message (C.C).

Cloud Data Retrieval Mechanism

The data retrieval process in the BFT-MCDB design starts with rewriting the received query from the user in the cloud manager (n number of queries) based on Shamir's Secret Sharing scheme and then sending these queries, one for each cloud, after constructing the polynomial functions and the order for the secret values. The shares will be returned to the cloud manager in sequence to compute the polynomial function on the retrieved values. Consequently, the sender unit sends the three executed results to the voter unit. As a result, the voter unit applies the Agreement procedure on the output from the sender unit inside the cloud manager to determine whether the results are consistent and if so, sends this to the user in a secure way.

Figure 2 shows the BFT-MCDB-Retrieve algorithm. The purpose of the BFT-MCDB-Retrieve algorithm is to first (line 4) receive from the client the query to be sent to the clouds through the receiver unit in the cloud manager. Consequently, the interpreter unit inside the cloud manager rewrites three queries (one for each cloud) (line 5) to retrieve the result from the clouds. After this, (line 9) the interpreter unit inside the cloud manager computes the secret value of the output from the clouds to send them to the voter. Then, the voter unit applies the Byzantine Agreement protocol (line 12) on the computed result before it is sent to the user. If the retrieved results from 2f+1 clouds are consistent (line 13) then the result will be sent to the user (line 15,16), otherwise (line 17) the sender unit inside the cloud manager sends the commit certificate with its commit messages to all 2f+1 clouds including a list of f+1 consistent clouds with their response messages (line 19). When a cloud n receives a commit message including the commit certificate, cloud n sends a local commit message to the cloud manager as proof of a non-faulty cloud (line 20). When the cloud manager receives the local commit certificate from the non-faulty clouds, the cloud manager considers the request and sends the response to the clients (line

Figure 1. BFT-MCDB-Store algorithm

Algorithm 1: BFT-MCDB-Store

Procedure BFT-MCDBStore(*D*)
Begin
DP sends *D* // DP sends D to RU.
RU ←*D* // D received by *RU*.
RU divide *D* into (*n*) shares ($D_1 .... D_n$)
    Where $n = 2k - 1$ *AND* $k<n$
IU Generates $q_{(x)}$ for each v_s Where
$Shares(v_s, 1) = q(x_1) = ax_1^{k-1} + bx_1^{k-1} ... + v_s$
...
$Shares(v_s, n) = q(x_n) = ax_n^{k-1} + bx_n^{k-1} ... + v_s$
DS=[x_1=3, x_2=1, x_3=2, ...] // X_s stored in DS.
SU Sends ($D_1 .... D_n$) into Cs // SU sends *D* to the cloud.
Cs.DS ←($D_1 .... D_n$) // DS stores *D* in the C_s.
END

Figure 2. BFT-MCDB-Retrieve Algorithm

Algorithm 2: BFT-MCDB-Retrieve

```
1  Function BFT-MCDBRetrieve(query)
2  Begin
3  U sends query // U sends query to RU.
4  RU ←query // query received by RU.
5  IU write (n) query for each C // IU rewrite n
   queries one for each Cloud.
6  Cs ←queries // Cs receives queries.
7  Cs sends results // Cs answer queries and
   sends results to RU.
8  RU ←results // RU receives results
9  IU Re-execute q(s) for each vs // IU translates
   (re-execute) the retrieved results to
   know the stored secret value vs.
10 SU sends executed vs to VU // SU sends the
   executed results to the VU for
   Agreement.
11 VU ←vs // the secret values vs received
   by VU.
12 VU voting results // Byzantine agreement
   protocol applied on vs.
13 If (2f+1 results are consistent) // If the
   retrieved results from 2f+1 clouds are
   consistent.
14 { then
15 SU ←Vs // SU receives the voted results
   to be able to send them to the
   requester.
16 U ←Vs}// }// U receives the voted
   results.
17 Else If (f+1 results are consistent) // If the
   retrieved results from f+1 clouds are
   consistent.
18 { then
19 SU sends C.C to Cs // SU sends the commit
   certificate with its commit messages to
   all 2f+1 clouds.
20 Cs sends local-commit msg to SU // non-faulty Cs
   send a local commit message to the
   cloud manager as proof of a non-faulty
   cloud.
21 SU ←Vs // SU receives the voted results
   to be able to be sent to the requester.
22 U ←Vs} }// U receives the voted results.
23 Else // If there are inconsistent
   responses from the clouds.
24 { then
25 U← error msg} // error message will be
   sent to U.
26 END
```

22). If there are inconsistent responses from the clouds (line 23), then an error message will be sent to the user (line 25).

PROPOSED SYSTEM DESIGN FOR MANAGING MULTI-CLOUD DATA DEPENDABILITY FAULTS

In fault-tolerant system design, the redundancy concept in FTSM is the central idea, while fault-tolerant operations are essential in driving the redundant parts and in our case the driver is the cloud manager.

In this chapter, each cloud is deemed a part of the replication and for simplicity we name our proposed design BFT-MCDB (Byzantine Fault Tolerance Multi-Clouds Database).

Architecture Overview

When systems experience a Byzantine fault, fault-tolerant state machine must have at least 2f+1 replicas (f = 1, 2, …), and the output of the replicas is produced by the majority of these replicas (Schneider 1990). In our BFT-MCDB design, 2f+1 clouds communicate with a cloud manager before the client can receive the results from the clouds. The three main components of the BFT-MCDB design are: the clouds side, the BFT communication protocol, and the cloud manager (see Figure3).

The Clouds' Side

The privacy of the stored data in the clouds is guaranteed by the fact that no single cloud contains any usable data by itself because it contains Shamir's Secret Sharing data which is the changed values of the original data, due to the properties of Shamir's Secret Sharing approach. This will prevent data intrusion by making it impossible for a malicious insider to re-execute revealed data that come from Byzantine clouds.

Communication Protocol

The communication protocol of the BFT-MCDB design is based on the Byzantine Agreement protocol (Lamport, Shostak et al. 1982) (Kotla, Alvisi et al. 2007) in that there is one REQUEST message from the client to the cloud manager, one ORDER-REQ message from the cloud manager for each client REQUEST which is responsible for handling the client's requests to the clouds, one SPEC-RESPONSE message from the clouds, and a COMMIT message from the cloud manager. The cloud manager receives 2f+1 REPLY messages from the clouds. Here, clients always receive replies (a COMMIT message) to their requests from the cloud manager which has received replies from all clouds. For operations that do

Figure 3. BFT-MCDB Overview

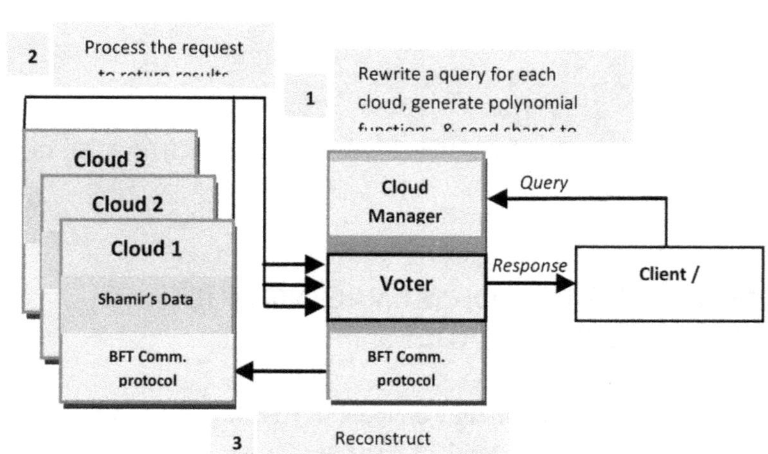

not employ the Shamir's Secret Sharing scheme, the cloud manager has to collect at least f+1 REPLY messages with the same content. Once the cloud manager has received f+1 identical results from different clouds, it can deliver the accepted result to the client.

Cloud Manager Components

The cloud manager consists of five components: receiver unit, interpreter unit, data source unit, sender unit and voter unit, as shown in Figure 4. The cloud manager is considered the most critical part in the proposed BFT-MCDB design as it is the control centre responsible for crucial fault-tolerant operations mentioned previously: cloud data storing, cloud data retrieval, cloud data fault detection.

In relation to the storing procedure in the BFT-MCDB design, the database provider (client) sends data to the cloud manager to be received by the receiver unit inside the cloud manager. The receiver unit divides the received data into n shares or clusters (in our case, three shares because we are dealing with three clouds) based on Shamir's Secret Sharing scheme. After the receiver unit has divided the data (assuming the data is a numeric value, for example, a worker's salary) into 3 shares to store them in different clouds, the receiver unit sends the shares to the interpreter unit inside the cloud manager. The interpreter unit generates random polynomial functions with degree at the same level, one for each worker's salary in the WORKER table with the actual salary as the constant part of the function. These values will then be stored in different clouds.

In relation to the data retrieval, the requester sends a query to the cloud manager. The receiver unit in the cloud manager then sends the requester's query to the interpreter unit to rewrite three queries, one for each cloud, to retrieve the converted information from the clouds. After the receiver unit has

Figure 4. Cloud Manager Components

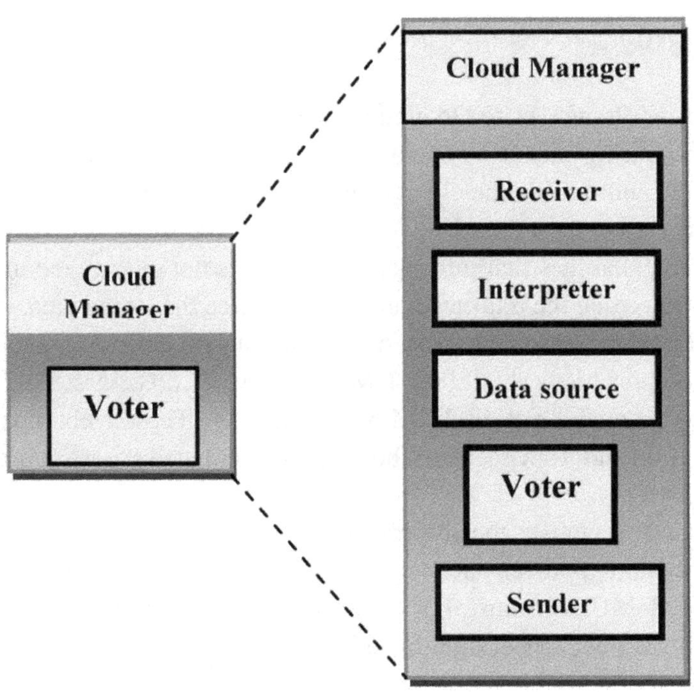

received the results from the clouds, it sends the results to the interpreter unit to re-execute the polynomial functions for each result.

After computing the secret values of the results from the clouds, the sender unit sends the three executed results to the voter unit for majority voting purposes. Consequently, the voter unit applies Byzantine Agreement protocol on the retrieved results from the sender unit inside the cloud manager. In the absence of faults, the voter unit receives matching results from all 2f+1clouds. Finally, the sender unit inside the cloud manager receives the voting results from the voter unit and then sends the results to the concerned requester. According to Yin et al. (Yin, Martin et al. 2003), to tolerate Byzantine faults, the minimum replicas required for completion is 2f+1 replicas = 3 clouds (for f = 1 with a replica of 2f+1 = 3 clouds). For instance, if each cloud contains a copy of the data, the cloud manager needs to perform a voting technique between replies and return the one that has at least f+1 majority. The final action after the majority voting has been applied on the retrieved output from the clouds is that the voter unit will send the agreed result to the concerned user or requester.

QUALITATIVE EVALUATION OF DEPENDANILITY IN MULTI-CLOUD DATA MANAGEMENT

As mentioned previously, dependability in this chapter subsumes the following attributes:

- Protect data integrity
- Protect data confidentiality
- Improve service availability
- Facilitate multi-cloud data fault detection and recovery
- Improve multi-cloud data reliability.

Protect Data Integrity

One of the most important issues related to cloud security risks is data integrity. The stored data in the cloud may be damaged during transition operations from or to the cloud storage provider. The risk of attacks from both inside and outside the cloud provider exists and should be considered. For example, data integrity was recently compromised in Amazon S3 where users' data became corrupted (Sun). Garfinkel (Garfinkel 2007) argues that information privacy is not guaranteed in Amazon S3. Data authentication, which assures that the returned data is the same as the stored data, is extremely important. Garfinkel claims that instead of following Amazon's advice that organizations encrypt data before storing them in Amazon S3, organizations should use HMAC (Krawczyk, Bellare et al. 1997) technology or a digital signature to ensure data is not modified by Amazon S3. These technologies protect users from Amazon data modification and from hackers who may have obtained access to their email or stolen their password (Garfinkel 2007).

However, as explained previously, the proposed BFT-MCDB design uses multi-clouds which is based on state machine replication approach (Schneider 1990). In addition, the use of Shamir's Secret Sharing approach makes BFT-MCDB robust. For example, the data will be distributed into three different clouds in the proposed BFT-MCDB design. In addition, the secret sharing approach will be applied on the stored data in the multiple clouds. If the Byzantine server or malicious insider wants to know the hidden information inside the cloud, they must retrieve at least two values from three different clouds with their knowledge of the secret information X to be able to know the real value which has been converted and hidden before it is stored in the multi-clouds.

This technique depends on Shamir's Secret Sharing approach with a polynomial functions technique which claims that, if there are 3 shares stored in 3 clouds (n=3, k=2), knowledge of the values of 2 shares or more with the secret information X will make the secret able to be constructed, whereas knowledge of the values of less than two shares will make the secret un-constructible even with the knowledge of the secret information X.

Another benefit of the BFT-MCDB design relating to data integrity in the cloud environment is due to the use of Byzantine Agreement protocol. As mentioned before, our design is based on the state machine replication approach (Schneider 1990). Replication of clouds in the BFT-MCDB design and the use of Byzantine Agreement protocol have a positive effect in detecting the faulty cloud which increases the level of data integrity. The results provided by the Byzantine cloud will be disregarded after being detected by the voter unit in the cloud manager. In addition, the voting technique facilitates the recovery of erroneous data. For instance, in accordance with the procedure of data retrieval in the BFT-MCDB design, the voter unit simply compares the outcomes from the clouds; if the values are different, the faulty output from the faulty cloud will be identified.

Another advantage of data replication among multi-clouds is that the voter unit inside the cloud manager in the BFT-MCDB design does not have to ask the faulty cloud to resend the results to the voter unit, because the voter can obtain the results from the non-faulty cloud. Thus, the use of Byzantine Agreement protocol in the BFT-MCDB design is helpful in addressing the issue of data integrity due to its ability to detect and identify the faulty cloud, in addition to being able to recover data in the faulty cloud. Therefore, the BFT-MCDB design is superior to a single cloud service in addressing the data integrity problem.

Protect Data Confidentiality

According to Garfinkel (Garfinkel 2007), another security risk that may occur with a single cloud provider is a hacked password or data intrusion. For example, if someone gains access to an Amazon account password, they will be able to access all the account's instances and resources. In addition, a stolen password allows the hacker to erase all the information inside the instance of the stolen account, modify it, or even disable its services. Furthermore, there is a possibility that the user's email (Amazon user name) will be hacked. (Garfinkel 2003) provides a discussion on the potential risks of E-mail). Since Amazon allows a lost password to be reset by email, the hacker may still be able to log in to the account after receiving the reset password.

However, the BFT-MCDB design, a practical model for building a system with Byzantine fault tolerance in a multi-cloud environment, is different to the single cloud service model. BFT-MCDB replicates the data among three different clouds; therefore, hackers would have to retrieve two out of the three clouds with the knowledge of the secret information X to be able to reconstruct the real value of the data in the cloud. In other words, if the hacker hacks one cloud's password even with the knowledge of the secret information X, they still need to hack the second cloud (in our case) to know the secret.

Another advantage of using the voter unit inside the cloud manager in the BFT-MCDB design is to identify the faulty cloud, depending on its response and determine in which cloud the data intrusion problem has occurred. Hence, replicating data into multi-clouds by using the state machine replication approach (Schneider 1990) may reduce the risk of data intrusion, such as in the BFT-MCDB design.

Improve Service Availability

Another major concern in cloud services is service availability. Amazon (Amazon 2006) mentions in its licensing agreement that it is possible that an Amazon user's web service may be terminated for any reason at any time if the user breaks the cloud storage policy. In addition, if damage occurs to any Amazon web service and the service fails, there will be no compensation from the Amazon Company regarding this failure. Companies strive to protect their services from system failure in several ways, such as by making backups or disconnecting any dependent cloud providers to avoid service unavailability to cloud providers (Garfinkel 2007). However, the BFT-MCDB is different from single cloud service providers in relation to service availability risk or loss of data. BFT-MCDB distributes the data into different clouds and therefore, it could be argued that the risk of data loss is reduced. If one cloud fails, users can still access their data live in other clouds.

Another benefit of the BFT-MCDB design in relation to service availability is the use of the replication technique. If one of the system's components fails, there is an alternative component instead of the faulty one and the system is still able to conduct its operations by using the alternative component. Hence, using replication techniques in the BFT-MCDB design increases the chance of the cloud services being available.

The use of Byzantine Agreement protocol in the proposed design protects the system if any components fail. The failure of a single cloud will not cause the whole system to fail because it depends on the results of two clouds out of three. In other words, the results from the voter unit to the user are generated after comparing the outputs from two clouds out of three, in our case. Hence, the use of the replication technique with Byzantine Agreement protocol in the BFT-MCDB design may reduce the risk of data loss or service unavailability.

Therefore, in light of the above discussion, our newly proposed BFT-MCDB design has been shown to have a better ability to address the three security dimensions: data integrity, data confidentiality and service availability, than a single cloud service provider. In addition, our newly proposed design offers more security in protecting the user's data from untrusted cloud service providers and from the malicious insider. The Amazon cloud service, for example, asks users to encrypt their data before storing it in their instances, whereas BFT-MCDB takes responsibility for this task.

Facilitate Multi-Cloud Data Fault Detection and Recovery

The replication of clouds in the BFT-MCDB design and the use of the voting technique have a positive impact on error detection in Byzantine clouds. The cloud manager can detect a faulty cloud if there are inconsistent replies from clouds. The faulty cloud will be identified and its result will be disregarded after it has been detected by the voter unit in the cloud manager. In addition, the voter helps to recover data which is determined to be erroneous. For example, if the voter indicates that Cloud1 is a faulty cloud, the output result of Cloud2 should not be the same as the output result of Cloud1, after the execution of the polynomial function of both retrieved shares from both clouds by the interpreter unit. This depends on the procedure of data retrieval in the BFT-MCDB design. In other words, the voter unit simply compares the outputs from the clouds with the first received output result; if the values do not match, an error has occurred at some point during the transmission. An error could occur randomly from one of the three clouds and consequently the presence of the voter unit is vitally important to detect errors in clouds.

There are many benefits of data replication among multi-clouds, since the voter does not have to ask the faulty cloud to retransmit the data, rather the voter can obtain the result from a non-faulty cloud. Thus, the use of the voter unit inside the cloud manager in the proposed BFT-MCDB design is helpful in the area of error detection and to identify the faulty cloud, in addition to data recovery in the faulty cloud.

Improve Multi-Cloud Data Reliability

Recall that there is an odd number (2f + 1) of the clouds in our proposed fault-tolerant architecture,

Acceptable results are based on the majority voting procedure, which is undertaken by the voter unit in the cloud manager of the BFT-MCDB design. For example, if f = 1, the replicated system in the clouds will not fail if two of the three clouds do not fail, or if none of the three clouds fails. In other words, the failure of a single cloud will not damage the operation of the whole system. Therefore, the result from the voter unit is more reliable because it is generated after comparing the results from at least two clouds. Hence, the majority voting procedure ensures the reliability of cloud data.

CONTRIBUTION

The contribution of this work is as follows: This work identifies the Byzantine fault tolerance problem in the multi-cloud computing environment and proposes a Byzantine fault tolerance model, named BFT-MCDB, to ensure the robustness of the multi-cloud environment. The model presented in this chapter uses a novel approach to combine the Byzantine Agreement protocols (Lamport, Shostak et al. 1982) (Kotla, Alvisi et al. 2007) (Lamport, Shostak et al. 1982) and Shamir's secret sharing approach (Johnson 1988) to detect Byzantine failure in the multi-cloud computing environment as well as to ensure the security of stored data within the cloud. BFT-MCDB is based on a state machine approach which has a replication mechanism (Johnson 1988) (Schneider 1990). Viewing our model in terms of state machines helps in understanding how our model replicates the data in the multi-clouds and centrally controls these clouds by the cloud manager.

REFERENCES

Abu-Libdeh, H., & Princehouse, L. (2010). RACS: a case for cloud storage diversity. In *Proceedings of the 1st ACM symposium on Cloud computing*. ACM. 10.1145/1807128.1807165

Agrawal, D., & El Abbadi, A. (2009). Database Management as a Service: Challenges and Opportunities. In *Proceedings of The 2009 25th International Conference on Data Engineering*. IEEE. 10.1109/ICDE.2009.151

AlZain, M. A. (2014). *Data security, data management and performance evaluation in a multi-cloud computing model* (PhD thesis). La Trobe University.

AlZain, M. A., Li, A. S., Soh, B., & Masud, M. (2016). Byzantine Fault-Tolerant Architecture in Cloud Data Management. *International Journal of Knowledge Society Research*, *7*(3), 86–98. doi:10.4018/IJKSR.2016070106

AlZain, M. A., Li, A. S., Soh, B., & Pardede, E. (2015). Multi-Cloud Data Management using Shamir's Secret Sharing and Quantum Byzantine Agreement Schemes. *International Journal of Cloud Applications and Computing*, *5*(3), 35–52. doi:10.4018/IJCAC.2015070103

AlZain, M. A., & Pardede, E. (2012). Cloud Computing Security: From Single to Multi-clouds. In *Proceedings of The 2012 45th Hawaii International Conference on System Science (HICSS)*. Maui, HI: IEEE. 10.1109/HICSS.2012.153

AlZain, M. A., & Soh, B. (2011). MCDB: Using Multi-clouds to Ensure Security in Cloud Computing. In *Proceedings of The 2011 Ninth International Conference on Dependable, Autonomic and Secure Computing (DASC)*. Sydney, Australia: IEEE. 10.1109/DASC.2011.133

AlZain, M. A., & Soh, B. (2012). A New Approach Using Redundancy Technique to Improve Security in Cloud Computing. In *Proceedings of The 2012 International Conference on Cyber Security, Cyber Warfare and Digital Forensic (CyberSec12)*. Kuala Lumpur, Malaysia: IEEE. 10.1109/CyberSec.2012.6246174

AlZain, M. A., & Soh, B. (2013). A Byzantine Fault Tolerance Model for a Multi-cloud Computing. In *Proceeding of The 2013 16th International Conference on Computational Science and Engineering CSE*. Sydney, Australia: IEEE. 10.1109/CSE.2013.30

AlZain, M. A., Soh, B., & Pardede, E. (2013). A Survey on Data Security Issues in Cloud Computing: From Single to Multi-Clouds. *Journal of Software*, *8*(5), 1068–1078. doi:10.4304/jsw.8.5.1068-1078

Amazon. (2006). *Amazon Web Services. Web services licensing agreement*. Author.

Bowers, K. D., & Juels, A. (2009). HAIL: A high-availability and integrity layer for cloud storage. In *Proceedings of the 16th ACM conference on Computer and communications security*. ACM. 10.1145/1653662.1653686

Cachin, C., & Haas, R. (2010). Dependable storage in the Intercloud. *IBM Research*, *3783*, 1–6.

Castro, M., & Liskov, B. (1999). Practical Byzantine Fault Tolerance. *Proceedings of the 1999 3rd Symposium on Operating Systems Design and Implementation.*

Clement, A., & Wong, E. (2009). *Making Byzantine fault tolerant systems tolerate Byzantine faults*. USENIX Association.

Garfinkel, S. L. (2003). Email-based identification and authentication: An alternative to PKI? *IEEE Security and Privacy*, *1*(6), 20–26. doi:10.1109/MSECP.2003.1253564

Garfinkel, S. L. (2007). *An evaluation of amazon's grid computing services: EC2, S3, and SQS*. Retrieved from http://simson.net/clips/academic/2007.Harvard.S3.pdf

Hore, B., & Mehrotra, S. (2004). A privacy-preserving index for range queries. In *Proceedings of the Thirtieth international conference on Very large data bases*. VLDB Endowment. 10.1016/B978-012088469-8.50064-4

Johnson, B. W. (1988). *Design & analysis of fault tolerant digital systems*. Addison-Wesley Longman Publishing Co., Inc.

Kotla, R., Alvisi, L., Dahlin, M., Clement, A., & Wong, E. (2007). Zyzzyva: Speculative byzantine fault tolerance. *Operating Systems Review*, *41*(6), 45–58. doi:10.1145/1323293.1294267

Krawczyk & Bellare. (1997). *HMAC: Keyed-hashing for message authentication*. Academic Press.

Kuznetsov, P., & Rodrigues, R. (2009). BFTW 3: Why? when? where? workshop on the theory and practice of byzantine fault tolerance. *ACM SIGACT News*, *40*(4), 82–86. doi:10.1145/1711475.1711494

Lamport, L., Shostak, R., & Pease, M. (1982). The Byzantine generals problem. *ACM Transactions on Programming Languages and Systems*, *4*(3), 382–401. doi:10.1145/357172.357176

Li, W., & He, J. (2005). A framework to support survivable web services. In *Proceeding of the 19th International Parallel and Distributed Processing Symposium*. IEEE.

Merideth, M. G., & Iyengar, A. (2005). Thema: Byzantine-fault-tolerant middleware for web-service applications. In *Proceeding of the 24th Symposium on Reliable Distributed Systems (SRDS)*. IEEE.

Pease, M., Shostak, R., & Lamport, L. (1980). Reaching agreement in the presence of faults. *Journal of the Association for Computing Machinery*, *27*(2), 228–234. doi:10.1145/322186.322188

Schneider, F. B. (1990). Implementing fault-tolerant services using the state machine approach: A tutorial. *ACM Computing Surveys*, *22*(4), 299–319. doi:10.1145/98163.98167

Shamir, A. (1979). How to share a secret. *Communications of the ACM*, *22*(11), 612–613. doi:10.1145/359168.359176

Stinson, D. R. (1992). An explication of secret sharing schemes. *Designs, Codes and Cryptography*, *2*(4), 357–390. doi:10.1007/BF00125203

Sun. (n.d.). Retrieved from http://blogs.sun.com/gbrunett/entry/amazon_s3_silent_data_corruption

Tiwari, P. K., & Joshi, S. (2018). A comprehensive report on security and privacy challenges in software as a service. In *Multidisciplinary Approaches to Service-Oriented Engineering* (pp. 143–167). IGI Global. doi:10.4018/978-1-5225-5951-1.ch007

Vukolic, M. (2010). The Byzantine empire in the intercloud. *ACM SIGACT News*, *41*(3), 105–111. doi:10.1145/1855118.1855137

Williams, K.Y.B., & Griffin, J.A.G. (2018). Better security and encryption within cloud computing systems. *International Journal of Public Administration in the Digital Age, 5*(2), 1-11.

Yin, J., Martin, J. P., Venkataramani, A., Alvisi, L., & Dahlin, M. (2003). Separating agreement from execution for byzantine fault tolerant services. *Operating Systems Review*, *37*(5), 253–267. doi:10.1145/1165389.945470

Zhao, W. (2007). BFT-WS: A Byzantine fault tolerance framework for web services. In *Proceeding of The Eleventh International Conference of EDOC'07*. IEEE. 10.1109/EDOCW.2007.6

Chapter 12
Retrieval of Relevant Web Pages by a New Filtering Method

Sahar Maâlej Dammak
MIRACL Laboratory, Tunisia

Anis Jedidi
MIRACL Laboratory, Tunisia

Rafik Bouaziz
MIRACL Laboratory, Tunisia

ABSTRACT

With the great mass of the pages managed through the world, and especially with the advent of the web, it has become more difficult to find the relevant pages after an interrogation. Furthermore, the manual filtering of the indexed web pages is a laborious task. A new filtering method of the annotated web pages (by a semantic annotation process) and the non-annotated web pages (retrieved from search engine Google) is then necessary to group the relevant web pages for the user. In this chapter, the authors first synthesize their previous work of the semantic annotation of web pages. Then, they define a new filtering method based on three activities. They also present their querying and filtering component of web pages; their purpose is to demonstrate the feasibility of our filtering method. Finally, the authors present an evaluation of this component, which has proved its performance for multiple domains, and they discuss the use of the extended Boolean retrieval method in the new filtering method.

INTRODUCTION

The search of the relevant pages on the Web becomes a very difficult task facing the considerable increase in the number of available Web pages, the diversity of their structures and their contents and the presence of a significant amount of useless information. It is important then to improve this difficult task in order to obtain relevant answers to users.

DOI: 10.4018/978-1-5225-7347-0.ch012

Indeed, the retrieved Web pages after the search on the Web do not often satisfy the user when making his/her interrogation. The user has to sift a long time to locate pages of interest. So, our objective in this chapter is to see how we can improve the retrieval of the relevant pages in the interrogation results, taking into account the annotations made by our semantic annotation process. In fact, with the exponential growth of the internet, it becomes more and more difficult to find the relevant pages. Therefore, a filtering process automation of the annotated and non-annotated Web pages is required.

This chapter provides then two main contributions. First, it synthesizes our previous work. In fact, we have proposed a semantic annotation approach of the Web pages (Maâlej Dammak *et al.,* 2013a, 2014b). This approach describes the Web pages by particular metadata, called "Fuzzy semantic annotations", in the semantic Web environment. The annotations are stored in RDF documents. Our annotation method (Maâlej Dammak *et al.,* 2013b, 2014b) is an enhancement of the first result of the annotation done by the Semantic Radar Plug-in (Bojars *et al.,* 2008) on the Web pages, using an enriched domain ontology, the FOAF (friend-of-a-friend) ontology (Brickley & Miller, 2015) and the SIOC (semantically-interlinked online communities) ontology (Bojars & Breslin, 2018). The additional annotation concerns the concepts of the result of the Semantic Radar. Secondly, this chapter defines a new method for filtering the indexed Web pages with the annotated ones. The new filtering method puts the retrieval Web pages in order, after the interrogation, before their display to the user. In addition, it groups and places the most relevant Web pages at the top of the extracted list. Also, the present chapter is interested in the operationalization of this method in order to automate the clustering of the relevant Web pages. So, we propose a new querying and filtering component for improving the user's search result. On the one hand, this component allows the extraction of the concepts of the domain ontology corresponding to the search area of the user. These concepts help the latter when writing the semantic query for the Web interrogation. On the other hand, the component ensures the automation of the scores calculation for the retrieved Web pages, annotated and/or non-annotated, and the hierarchical classification of these pages in order to regroup them according to their hierarchies. We set two groups in the end of the clustering; a group that contains the most relevant pages and a group that includes the less relevant pages. However, we propose presenting to the user only the group that contains the most relevant pages.

The remainder of this chapter is organized as follows. Section 2 shows the positioning of our work in relation to the literature. Section 3 presents a synthesis of our semantic annotation process of the Web pages and the RDF of the semantic and fuzzy annotation. Section 4 defines our new filtering method for the annotated and non-annotated Web pages. Section 5 describes the principal development parts of our querying and filtering component. Section 6 shows the relevance of our component, through an evaluation of these results. Section 7 discusses the use of the extended Boolean retrieval method in our new filtering method. Finally, Section 8 summarizes the contribution of the chapter and outlines future work.

RELATED WORKS

The search on the Web cannot guarantee the retrieval of the relevant Web pages to the user; the pages retrieved may not satisfy his/her request. The purpose of our filtering process is to improve the search result by the selection of the relevant pages. So, a filtering method of the indexed Web pages (including the annotated ones) that groups the relevant pages is required.

To reach this goal, we have studied various works dealing with the filtering of the resources (documents, Web pages, etc.). The vector model, introduced by Salton *et al.* (1975), was used as a content-based filtering model (Nouali & Blache, 2004). It relies on the similarity between the documents vectors and the queries vectors contained in a database for the research of the relevant documents (Manning *et al.,* 2008). Nevertheless, we cannot use such a model because we cannot create a database for each Web interrogation. The method suggested in (Jenhani El Jed, 2006) is an evaluation method to estimate the relevance of the Web pages. The steps of the filtering and the selection are ensured by an information extraction component which uses grammar and extraction patterns. As these steps may produce errors in the classification of search engine responses, they do not really lead to the correct filtering and selection of the Web pages. In (Pohl *et al.,* 2010), the authors have shown that the extended Boolean retrieval method, defined in (Fox, 1983; Salton *et al.,* 1983), is able to find a larger quantity of the relevant documents than the previous propositions in literature. Given that the previous work did not show significant improvements over the Boolean method (Karimi *et al.,* 2009), the authors in (Pohl *et al.,* 2010) have proposed to calculate the weights of the binary terms of a query, and to use the weight of the query terms in the documents, using the formulation TF.IDF (Term Frequency-Inverse Document Frequency) shown in (Salton *et al.,* 1983). In addition, these authors asserted again in 2012 (Pohl *et al.,* 2012) that the extended Boolean retrieval method produces meaningful rankings after the assignment of the scores to the documents, despite the fact that this method was proposed nearly three decades ago. Thus, we can enrich the extended Boolean retrieval method to calculate scores for the indexed Web pages annotated and non-annotated. But we must find a method that groups the relevant pages (which can be annotated and/or non-annotated), taking into account the scores of the pages.

To do this, we have studied the recent works of the resources clustering (Web page or others) such as (Rahmani *et al.,* 2009; Rossi & Villa-Vialaneix, 2011). The majority of these works are based on the "k-means" method (MacQueen, 1967) and the hierarchical classification (Lance & Williams, 1967).

The "k-means" method is a simple and effective clustering method. The disadvantage of this method is the need to provide the parameters (the "*k*" number) in the "k-means" algorithm, relying on heuristic data. This drawback prevents us from using this method for regrouping the Web pages, especially that we want to achieve an automatic regrouping of the indexed Web pages annotated and non-annotated.

The hierarchical classification is an automatic classification method. The advantage of the hierarchical classification is the clustering of the nearest individuals (by calculating the distance between the individuals according to their scores) without the need for the user's intervention and the adjustment of the parameters.

Thus, we propose following the same steps of the hierarchical classification in order to regroup the indexed Web pages according to their scores. Also, we set two (2) groups in the end of the clustering: one for the most relevant pages and another for the less relevant pages.

These studies help us to define a new filtering method for the indexed Web pages (annotated and non-annotated) and to improve the relevance of the search result for the user. In fact, the consideration of the annotated pages (with their annotation RDFs) in the filtering can satisfy the request of the user. This feature has not been considered in previous works that seek to improve the user's search results.

Our method is an enhancement of the extended Boolean retrieval method for allocating scores to the annotated and non-annotated Web pages. In addition, it uses the hierarchical classification in order to group the relevant Web pages after the interrogation.

SYNTHESIS ELEMENTS

The objective of this section is to provide a synthesis of the results of our previous work: the semantic and fuzzy annotation for the Web pages by an annotation process and the representation RDF of this annotation. In fact, these results will be used to build the specific contributions of this chapter.

An Approach to Semantic Annotation of the Web Pages

We have proposed a new approach to proceed to the semantic annotation of the Web pages (Maâlej Dammak *et al.,* 2013a). The stages of our approach are as follows (cf. Figure 1):

S1: The annotation expert has to interrogate the Web through the graphical interface of our annotation component after the specification of the field of study. This interrogation is based on the concepts of the selected domain ontology.

S2: Then, the retrieved Web pages must be transferred to Semantic Radar for an automatic and semantic analysis. This analysis retrieves an RDF file for each page containing the descriptors of FOAF and SIOC. The set of RDF files extracted by Semantic Radar represents the input of the annotation process.

S3: Subsequently, the analyzed pages have to be automatically annotated by our annotation method shown in (Maâlej Dammak *et al.,* 2013b, 2014b), which we recall hereafter to produce RDF metadata for each Web page. This method assists the expert throughout the semantic annotation phase of the Web pages. We use in this method equivalence rules and a semantic annotation model that we have defined in (Maâlej Dammak *et al.,* 2013a). The new RDF resource will be linked to the original resource on the Web and will be saved in the annotation base.

The Semantic Annotation Method

The purpose of our semantic annotation method is to seek equivalences between (i) the concepts of the analysis result made by Semantic Radar on the Web pages and (ii) the concepts of the selected enriched domain ontology. To ensure these equivalences, we have defined equivalence rules and a semantic annotation model (Maâlej Dammak *et al.,* 2013a), with a set of new semantic descriptors created for the semantic annotation process. With these rules and this model, we produce an annotation result for each page as an RDF file that represents an enhancement of the RDF result generated by Semantic Radar. Actually, the FOAF concepts of this RDF file are enhanced by the FOAF concepts of the enriched domain ontology, the domain concepts and/or the FOAF concepts of the FOAF ontology (respectively for the SIOC concepts). The first dotted part in Figure 1 shows globally the main parts of our semantic annotation method of the Web pages.

S4: This stage represents the second objective of this chapter, which will be described in Sections 4, 5 and 6. We will: (i) detail the filtering method that we propose, (ii) demonstrate the feasibility of this method in our new querying and filtering component of the Web pages, and (iii) present the different experimentations from which we evaluate the performances of our component. However, a summary description is given hereafter.

Figure 1. Synoptic schema of our approach

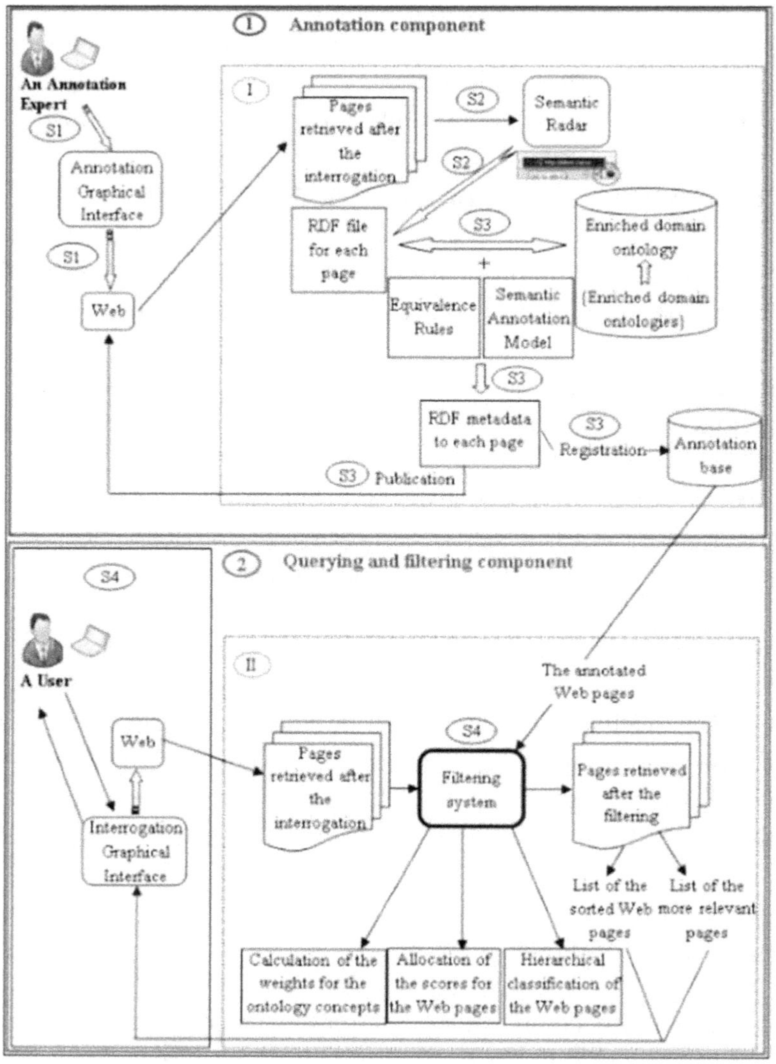

In this stage, the user (the domain expert or the non-expert) has to interrogate the Web through the graphical interface of our querying and filtering component after the specification of the field of study. This interrogation is based on the concepts of the selected domain ontology. The Web pages retrieved after the user interrogation pass to the filtering system. This system sorts these Web pages, and returns them to the user. Also, it groups the most relevant pages and places them at the top of the extracted list. The sorted pages and the relevant ones can be the annotated and/or non-annotated Web pages (the filtering system checks the annotation of the pages through the access to the annotation base). Thus, we propose using an appropriate method for filtering Web pages containing three activities. The first activity is the calculation of the weights for the ontology concepts. The second one is the allocation of the scores for the indexed Web pages (annotated and non-annotated). Finally, the third activity is the hierarchical classification of the Web pages in order to group the relevant pages. This method helps to improve the relevance of the interrogation results. The second dotted part in Figure 1 shows globally the main parts of our filtering method of the Web pages.

The RDF of the Semantic and Fuzzy Annotation

The main activity of the process model we have defined for the semantic annotation of Web pages (Maâlej Dammak *et al.,* 2014a) consists in the search of equivalence between the concepts of the result of Semantic Radar and the concepts of the selected enriched domain ontology. Following this model, we have obtained an annotation document in RDF, automatically generated for each Web page, by our annotation system. Each annotated concept, in an annotation RDF, is now linked with the semantic and fuzzy descriptors of the annotation that we have defined in the annotation model. The semantic descriptor (such as <HasChild>) frames the annotation concept while the fuzzy descriptor (such as <FuzzyValue>) frames the weight (*i.e.,* the possibility degree) of this concept in relation to the concept to be annotated. We have already defined two formulas to calculate the concept weights of the ontology (cf. Section 4.1).

The following is an extract of an annotation document (cf. Figure 2) in RDF for the page http://journal.webscience.org/532/. This page has been displayed to a user (cf. Section 5), as a relevant page, after the interrogation in the "Network of scientists" domain (Stella *et al.,* 2011).

In this example, there is the annotation of the concept "FOAF: Person" of the Semantic Radar result by the concepts of the enriched domain ontology "Network of scientists" (EmeritusFaculty, FacultyMember, NonFacultyAcademic, NonAcademic and EmeritusProfessor), using the proposed semantic annotation model. In addition, each line of the annotation receives degrees of possibilities for its concepts, with the descriptor <FuzzyValue>.

Figure 2. An extract of a generated RDF of a semantic and fuzzy annotation for a Web page

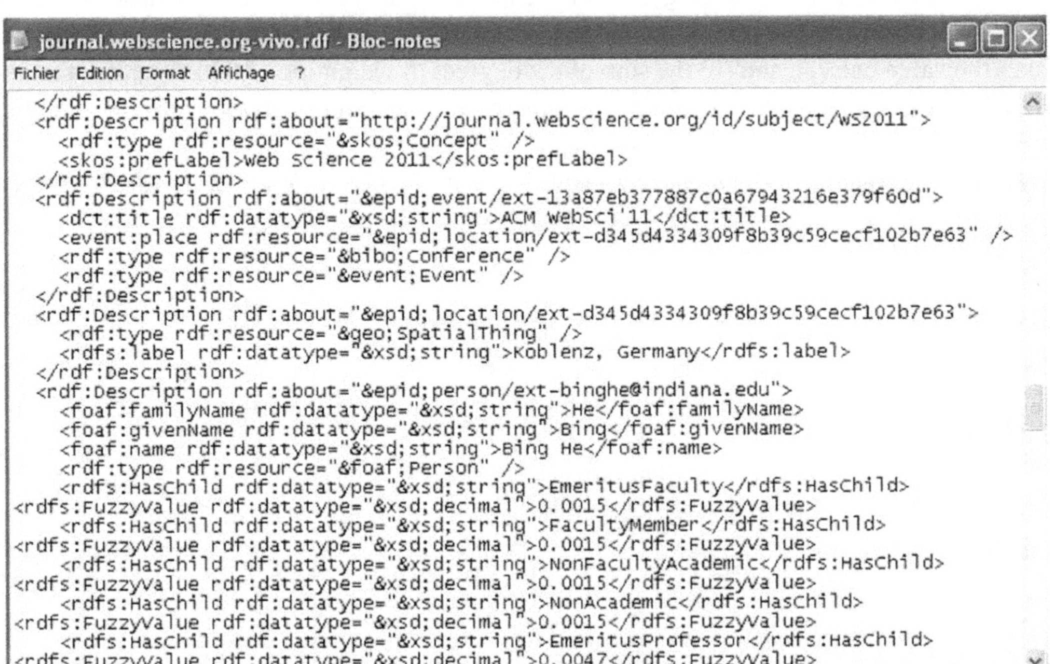

journal.webscience.org-vivo.rdf - Bloc-notes

Fichier Edition Format Affichage ?

```
  </rdf:Description>
  <rdf:Description rdf:about="http://journal.webscience.org/id/subject/WS2011">
    <rdf:type rdf:resource="&skos;Concept" />
    <skos:prefLabel>Web Science 2011</skos:prefLabel>
  </rdf:Description>
  <rdf:Description rdf:about="&epid;event/ext-13a87eb377887c0a67943216e379f60d">
    <dct:title rdf:datatype="&xsd;string">ACM WebSci'11</dct:title>
    <event:place rdf:resource="&epid;location/ext-d345d4334309f8b39c59cecf102b7e63" />
    <rdf:type rdf:resource="&bibo;Conference" />
    <rdf:type rdf:resource="&event;Event" />
  </rdf:Description>
  <rdf:Description rdf:about="&epid;location/ext-d345d4334309f8b39c59cecf102b7e63">
    <rdf:type rdf:resource="&geo;SpatialThing" />
    <rdfs:label rdf:datatype="&xsd;string">Koblenz, Germany</rdfs:label>
  </rdf:Description>
  <rdf:Description rdf:about="&epid;person/ext-binghe@indiana.edu">
    <foaf:familyName rdf:datatype="&xsd;string">He</foaf:familyName>
    <foaf:givenName rdf:datatype="&xsd;string">Bing</foaf:givenName>
    <foaf:name rdf:datatype="&xsd;string">Bing He</foaf:name>
    <rdf:type rdf:resource="&foaf;Person" />
    <rdfs:HasChild rdf:datatype="&xsd;string">EmeritusFaculty</rdfs:HasChild>
<rdfs:FuzzyValue rdf:datatype="&xsd;decimal">0.0015</rdfs:FuzzyValue>
    <rdfs:HasChild rdf:datatype="&xsd;string">FacultyMember</rdfs:HasChild>
<rdfs:FuzzyValue rdf:datatype="&xsd;decimal">0.0015</rdfs:FuzzyValue>
    <rdfs:HasChild rdf:datatype="&xsd;string">NonFacultyAcademic</rdfs:HasChild>
<rdfs:FuzzyValue rdf:datatype="&xsd;decimal">0.0015</rdfs:FuzzyValue>
    <rdfs:HasChild rdf:datatype="&xsd;string">NonAcademic</rdfs:HasChild>
<rdfs:FuzzyValue rdf:datatype="&xsd;decimal">0.0015</rdfs:FuzzyValue>
    <rdfs:HasChild rdf:datatype="&xsd;string">EmeritusProfessor</rdfs:HasChild>
<rdfs:FuzzyValue rdf:datatype="&xsd;decimal">0.0047</rdfs:FuzzyValue>
```

FILTERING METHOD FOR THE WEB PAGES

To improve the search relevance for the user, we define a new method to filter the retrieved Web pages after the interrogation (cf. Figure 1).

Our method concerns the calculation of the scores for the indexed Web pages annotated and non-annotated. It also focuses on the hierarchical classification of these pages for clustering the relevant Web pages.

The deployment of this method requires a set of tools to assist the user in his/her search on the Web. To reach our goal, we propose the following three phases.

Calculation of the Weights for the Ontology Concepts

In our annotation component, the proposed annotations, on the pages, rely on the concepts of the enriched domain ontologies (cf. Figure 1). Thus, we need to know the weights of the ontology concepts to calculate the scores of the annotated pages.

In this phase (Phase 1), we propose assigning weights to the domain ontology concepts. These weights will be used in the next phase (Phase 2).

To do this, we propose the two following steps:

- Extraction of the ontology concepts with their levels of depth.
- Assignment of the weights to the ontology concepts. Each level of depth has its own weight, and the concepts having the same level must have the same weight. The concepts are more specific in the lowest levels in the ontology, so for the lowest level we obtain the greatest weight. With an objective of standardization, we decide that (i) the weight sum of different levels of the ontology takes the value one (1), and (ii) the sum of the degrees of weighting of the concepts of an ontology also takes the value one (1). These degrees therefore take values between zero (0) and one (1).

Therefore, we define the following two equations to calculate the weights of the ontologies concepts.

$$W_i = i / \sum j \qquad (1)$$

N.B:

W_i: Weight of the depth i.
$\sum j$: The sum of the depths (1+2+...+number of the depths).

Example:

$\sum j = 1 + 2 + 3 + 4 = 10.$

$W_3 = 3/10 = 0{,}3.$

W_3: The weight of the depth 3 = 0,3.

$$Wc_i = W_i / C_n \tag{2}$$

N.B:

Wc_i: Weight of a concept in the depth i.
C_n: The number of concepts in a depth i.

Example:
In the depth 3, we have 7 concepts.

Wc_3= 0,3/ 7 = 0,042.

Calculation of the Scores of the Web Pages

In this phase (Phase 2), we propose calculating the scores for the indexed and annotated pages and the indexed and non-annotated ones retrieved after the user's interrogation. Indeed, the user interrogation is launched to the Google search engine. Our filtering system then accesses to the annotation base in order to identify the pages which are annotated by our annotation component (cf. Figure 1).

The new filtering method is a synergy between the classic information search, using the extended Boolean retrieval method (for the scores calculation of the indexed and non-annotated pages), and the semantic information search, using the ontologies (for the scores calculation of the indexed and annotated pages). Indeed, the calculation we propose on the Web pages is an enhancement of the extended Boolean retrieval method (Pohl *et al.,* 2010) by the use of the ontologies. The last method calculates the scores of the documents according to the binary concepts of a query (for example *(c1 or c2), (c1 and c2) or (not c1)*), using the TF.IDF (Term Frequency, Inverse Document Frequency) formulation. The utility of using TF.IDF is to calculate the weight of the concept in a document. This method is able to find a larger quantity of the relevant documents than the previous approaches in literature. Pohl *et al.* (2010) have proposed the calculation formulas of the scores for the documents by the extended Boolean retrieval method.

Calculation of the Scores for the Indexed and Non-Annotated Web Pages

To calculate the scores for the indexed and non-annotated Web pages (IWP), we adopt the step of scores calculation of the extended Boolean retrieval method by using the TF.IDF formulation. Indeed, the calculation for these pages is performed according to a given query (a binary query).

The following formulas are used to calculate scores for the indexed and non-annotated pages.

$Score(IWP) =$

$$Score\left(c_1,\ldots or\, c_i,\ldots or\, c_n, IWP\right) = \sqrt{\sum_{i=1}^{n} score^2\left(c_i, IWP\right)} \,/\, \sqrt{2} \tag{3}$$

or

$$Score\left(c_1,\ldots and\, c_i,\ldots and\, c_n, IWP\right) = 1 - \sqrt{\sum_{i=1}^{n} \left(1 - score\left(c_i, IWP\right)\right)^2} \,/\, \sqrt{2} \tag{4}$$

$$Score\left(not\, c1, IWP\right) = 1 - score(c1, IWP) \tag{5}$$

where:

$$Score\left(c, IWP\right) = W\left(c, IWP\right)$$

The TF.IDF formula:

$$W\left(c, IWP\right) = tf\left(c, IWP\right) * \log\left(\frac{n}{n\left(c\right)}\right) \tag{6}$$

Legend:

Score (IWP): Score of an indexed and non-annotated Web page.
Score (Binary concept(s), IWP): Score of an indexed Web page (IWP) according to the binary concepts of a query ((c_1, … or c_n), (c_1, … and c_n), (not c_1)).
W(c, IWP): Weight of a concept (c) in an indexed Web page (IWP).
tf(c, IWP): Frequency of a concept (c) in an indexed Web page (IWP).
n: The total number of the Web pages after the interrogation.
n(c): The total number of the Web pages after the interrogation containing the concept (c).

Calculation of the Scores for the Indexed and Annotated Web Pages

To calculate the scores of the indexed and annotated pages (AWP), we need two inputs: the query and the annotations RDFs for each Web page.

We also adopt the calculation formulas applied to the indexed and non-annotated Web pages to calculate the first score; the page score related to the query: *Score (Binary concept(s), AWP)*. We now propose the enrichment of these formulas by the scores sum of the RDFs of the annotations (second score: the score of the annotation RDFs of a page): $\sum_{i=1}^{n}\left(ScoreRDF\right)_i$. Indeed, each score of an annotation RDF is the weight sum of the ontology concepts in this RDF file. The concept weights are calculated according to the proposed formulas in section 4.1. Actually, we have two types of concepts in the annotation RDF. The first type represents the annotation concepts (*i.e.,* the concepts added to the annotation) (cf. Figure 2). As we have mentioned before, our annotation method improves the result of the Semantic Radar using the concepts of the enriched domain ontology and the ones of the FOAF and SIOC ontologies. The second type represents the concepts of the used ontology in the annotation RDF. Also, each score of the annotation RDF represents the weight sum of the query concepts in this RDF File. In fact, the proposed idea is to apply the TF.IDF formula for calculating the weights of the query concepts. We suggest searching the weights of the query concepts in an annotation RDF relative to the entire of the annotation RDFs of the annotated page.

We then propose the following formulas to calculate scores for the indexed and annotated pages. These new formulas uncover the value of the annotation RDFs of an annotated page.

$$\text{Score (AWP)} = \text{Score (Binary concept(s), AWP)} + \sum_{i=1}^{n}\left(ScoreRDF\right)_i \tag{7}$$

$$\left(ScoreRDF\right)_i = \sum_{j=1}^{m} W\left(CA\right) + \sum_{k=1}^{p} W\left(COnt\right) + \sum_{l=1}^{r} W\left(QC, RDF\right) \tag{8}$$

where:

$$W\left(QC, RDF\right) = tf\left(QC, RDF\right) * \log\left(\frac{n}{n\left(QC\right)}\right) \tag{9}$$

Legend:

Score (AWP): Score of an indexed and annotated Web page.

Score (Binary concept(s), AWP): Score of an annotated Web page (AWP) according to the binary concepts of a query (the same formulas of the section 4.2.1).

$\sum_{i=1}^{n}\left(ScoreRDF\right)_i$ **:** The scores sum of the RDFs of the annotations for an annotated Web page *(where n is the number of RDFs of the annotation for the concerned page).*

(ScoreRDF)$_i$: Score of annotation RDF n° i for an annotated Web page.

$\sum_{j=1}^{m} W\left(CA\right)$ **:** The weight sum of the annotation concepts in the annotation RDF *(where **m** is the number of the annotation concepts).*

$\sum_{k=1}^{p} W\left(COnt\right)$**:** The weight sum of the ontology concepts in the annotation RDF; on condition that these concepts do not exist in the added annotation (*where* ***p*** *is the number of the ontology concepts).*

$\sum_{l=1}^{r} W\left(QC, RDF\right)$**:** The weight sum of the query concepts in the annotation RDF *(where* ***r*** *is the number of the concepts in the query).*

W(QC, RDF): The weight of a query concept (QC) in an RDF of the annotation (RDF).

tf(QC, RDF): Frequency of a query concept (QC) in an annotation RDF (RDF).

n: The total number of the annotation RDFs of the concerned page.

n(QC): The total number of the annotation RDFs of the concerned page, containing the query concept (QC).

The Hierarchical Classification of the Web Pages

To improve the relevance of the interrogation results for the user, we need a method that groups the relevant pages from the retrieved pages (annotated and/or non-annotated) after an interrogation. We adopt in our case the Hierarchical Ascendant Classification (HAC) (Phase 3) as used in (Rossi & Villa-Vialaneix, 2011).

The hierarchical classification method is a method of the successive grouping of the near groups. This method groups the most similar elements in order to have groups of similar elements. In our case, we propose following the same step of this method by clustering the retrieved pages, using their scores (already calculated), according to their similarities. In our situation, we set two (2) groups in the end of the clustering: a group that contains the most relevant pages and a group that includes the less relevant pages. Both groups contain similar Web pages. However, we propose displaying for the user the group that contains the most relevant pages.

We then adopt the hierarchical classification algorithm (Rossi & Villa-Vialaneix, 2011) for clustering the relevant Web pages. The inputs of this algorithm are: "*individuals*" (in our case: list of the Web pages with their scores) and "*nbClasses*" (in our case: 2, the number of the classes (*i.e.,* groups) that we want to obtain). The output is represented by the classes "*classes*", initially empty. A class is seen as a list of individuals (in our case: list of Web pages).

A NEW COMPONENT FOR QUERYING AND FILTERING THE WEB PAGES

The querying and filtering component "Filtering Web Pages", allows the automation of the filtering of the Web pages: the clustering of the relevant Web pages.

We limit our presentation here to the principal development parts of our component for automating the scores calculation for the Web pages and the hierarchical classification of these pages through the following three activities. Figure 3 shows the interface of our querying and filtering component.

We have chosen to study the field "Network of Scientists", in the example of Figure 3, which corresponds to the enriched domain ontology "VIVO.owl". VIVO is an open source semantic Web application that, when populated with researcher interests, activities and accomplishments, enables the discovery of research across disciplines (Stella *et al.,* 2011).

The Web Interrogation

In our graphical interface, the user must specify the domain of study (left part *Domain ontologies*, cf. Figure 3). After specifying his/her domain, we automatically show to the user the hierarchy of the ontology (*Hierarchy of Ontology*, cf. Figure 3) to assist him/her in writing the semantic query. We have actually imported the "Jena API" (Apache Jena, 2018) in the java application of our querying and filtering component to read the ontologies and to extract their concepts.

Indeed, in this background, we automate the calculation of the concept weights for the selected domain ontology, in our component, following the proposed formulas in Section 4.1. These weights will be considered in the next activity. The user can take an idea on the weighting of the selected ontology concepts after the click on the *Weights of concepts* button (cf. Figure 3).

Calculation of the Scores of the Web Pages

After writing the semantic query, the user can start his/her search (cf. Figure 3). A list of the URLs of the Web pages automatically appears to respond to his/her request. We used a Web service to retrieve the results from Google in our Java application.

After the display of a URLs set, our filtering system proceeds to the scores calculation for these pages. In this case, our system acts according to two situations:

Figure 3. Example of the scores calculation of the indexed Web pages annotated and non-annotated

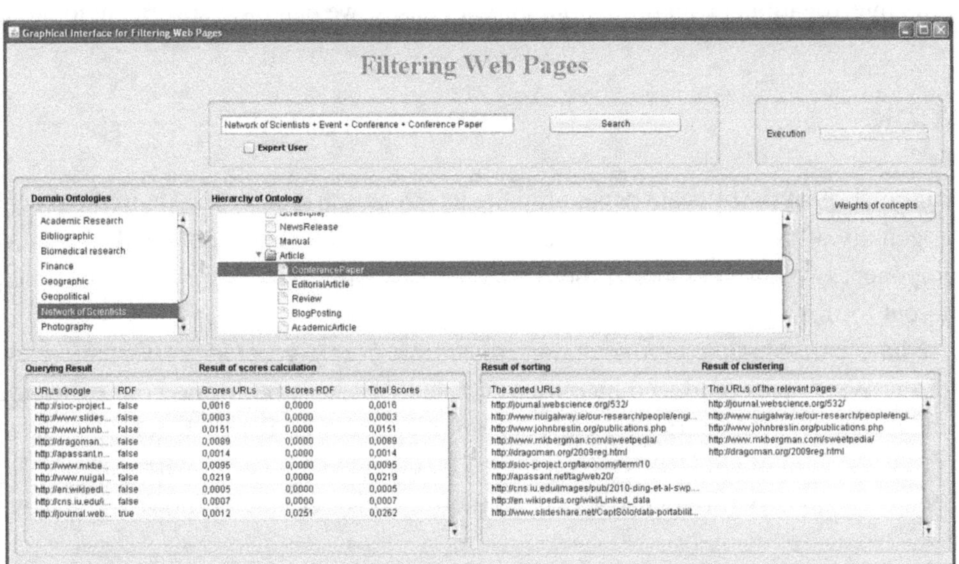

- If the page is non-annotated (the column *RDF* of the table, on the left of Figure 3, takes the value "False" because this page does not have an annotation RDF), the page score is automatically calculated, and displayed (in the column *Total Scores* of the same table, cf. Figure 3) following the calculation formulas shown in Section 4.2.1 (the formulas of the extended Boolean retrieval method depending on the Boolean query of the user interrogation). In fact, a value of *Total Scores* is equal to the sum of the corresponding value of *Scores URLs* and the corresponding value of *Scores RDF* is equal here to zero. The value of *Scores URLs* represents the page score.
- If the page is annotated (in the annotation base), the page score is automatically calculated, and displayed (in the column *Total Scores*, cf. Figure 3; *Total Scores = scores URLs + Scores RDF*) following the proposed formulas in Section 4.2.2 (that enhance the formulas of the extended Boolean retrieval method in order to take into account the scores of the annotation RDFs of annotated Web pages). In this case, the column *RDF* takes the value "True" and a value of *Scores RDF* represents the score sum of the annotations in RDF of the page.

The scores assigned for the Web pages constitute the inputs of the second functionality of the filtering system: the hierarchical classification of the Web pages for clustering the relevant pages to the user.

Hierarchical Classification of the Web Pages

Our filtering system automatically sorts the retrieved Web pages according to the assigned scores for these pages. On the one hand, it displays the result to the user (in the column *The sorted URLs* of the table on the right of Figure 3). On the other hand, this system displays the URLs of the most relevant Web pages, obtained by the hierarchical classification of the pages (in the column *The URLs of the relevant pages* of the previous table).

The hierarchical classification of the Web pages obtained by Google, after the Web interrogation by the query: "Network of Scientists + Event + Conference + Conference Paper" in the domain "Network of Scientists", gives five (5) relevant Web pages (including an annotated page), which are the most relevant pages compared to the others (in this case there is a single page that was annotated among the retrieved pages after the interrogation, cf. Figure 3, but in other cases, we can have several annotated pages).

EVALUATION

A research automation of the relevant pages after the interrogation is beneficial in a context characterised by the heterogeneity of the pages. To evaluate the relevance of the results obtained by our querying and filtering component, we have compared these results to the ones obtained by human subjects (evaluators) of different formations.

In fact, we have carried out experimentations that allow us to give an idea about the relevance of the clustering of the Web pages produced by our component. We have performed our experimentation on ten (10) domains and five (5) queries for each one (cf. Appendix 1). For each query, we have recovered five (5) lists (a list is a set of ten (10) pages) returned by Google after each interrogation. Therefore, our study has been done on fifty (50) queries and two thousand five hundred (2500) Web pages.

For each domain of study, we have recovered the (n) relevant pages obtained by our component in each list for the five queries. We have then provided the relevant pages retrieved for the ten domains to ten evaluators (an evaluator per domain). The role of each evaluator is to determine, in a first step, among the answers provided by the Google search engine for each query (the same five queries tested by our component), the number of pages per list he/she considers relevant with respect to the query. A selected relevant page by an evaluator presents a clear answer to the query of the user and contains elements that include the keywords of the query. In a second step, each evaluator must determine, among the pages identified as relevant by our component (for the five lists of the five queries), the number of pages that are irrelevant. Also, he/she should check the number of the relevant pages not retrieved by the filtering component. Depending on the outcome provided by the evaluators, we have noticed the following values (cf. Table 1).

This table shows that our component retrieves a maximum of six (6) relevant pages per list with a maximum of two (2) irrelevant pages among these six pages. Also, the search result of the relevant pages by our component can lead to a lack of a maximum of three (3) relevant pages (the pages not retrieved but relevant).

To evaluate the automation of the relevant pages research by our querying and filtering component, we have then used the following standard metrics (Van Rijsbergen, 1979): the *recall* and the *precision*.

- **The Recall (R):** It is the number of data (Web pages) correctly retrieved as relevant by the component, divided by the total number of relevant real data identified by the evaluator (*i.e.,* the sum of the number of the relevant pages retrieved by the component and the number of the relevant pages not retrieved by the component).
- **The Precision (P):** It is the number of data (Web pages) correctly retrieved as relevant by the component, divided by the total number of data retrieved by the component (*i.e.,* the total number of the pages retrieved by the component, that are deemed relevant or not relevant by the evaluator).

Depending on the outcome given by the evaluators, we have calculated the values of these metrics for the ten domains in two situations:

1. The best case (BC) obtained by our component for each domain. It is the one which maximizes the number of the Web pages correctly retrieved as relevant and minimizes the number of the retrieved Web pages but irrelevant and the number of the not retrieved Web pages but relevant.
2. The medium case (MC) obtained by our component for each domain. It is the one which gives medium values for the previous three numbers.

Table 1. Analysis of the interrogation results obtained by our component

<table>
<tr><th colspan="2">Number of the Retrieved Relevant Pages/List</th></tr>
<tr><td colspan="2">Between [1 and 6] pages</td></tr>
<tr><th>Number of the Retrieved Pages but Irrelevant</th><th>Number of the Not Retrieved Pages but Relevant</th></tr>
<tr><td>Between [0 and 2] pages</td><td>Between [0 and 3] pages</td></tr>
</table>

So, we have determined these three numbers for each list (5) for each query (5) of each domain (10). Then, we have selected the best case and the medium case for each domain. To do this selection, we have calculated the sum of the *recall* and the *precision* for each case. The best value of this sum gives the best case and the medium value of this sum gives the medium case.

To formalize this action, we have defined two algorithms: (i) algorithm of the best case and (ii) algorithm of the medium case. We present the first algorithm in Appendix 2, given that we have followed the same steps of the latter algorithm for the second algorithm but with the research of the medium cases.

Figures 4 and 5 present the values of the *recall* and the *precision* obtained in the two situations.

Figure 4. The recall and precision of our component in the best case

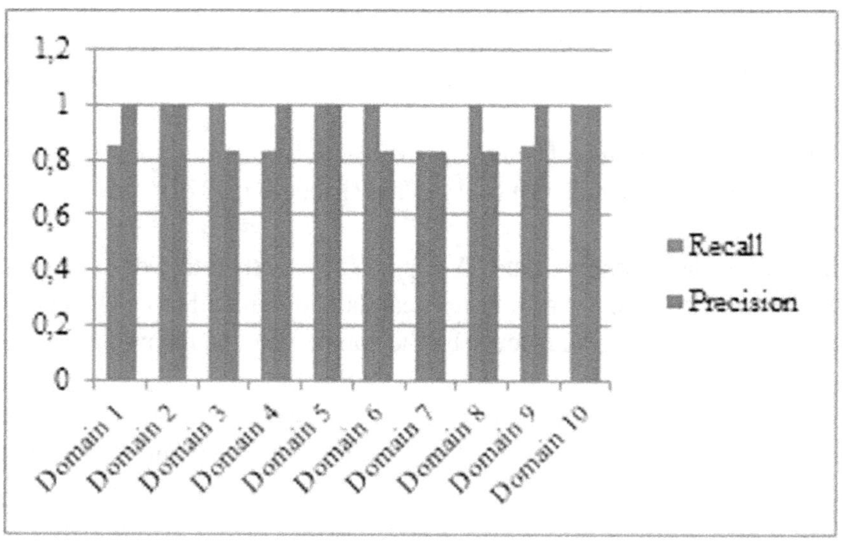

Figure 5. The recall and precision of our component in the medium cases

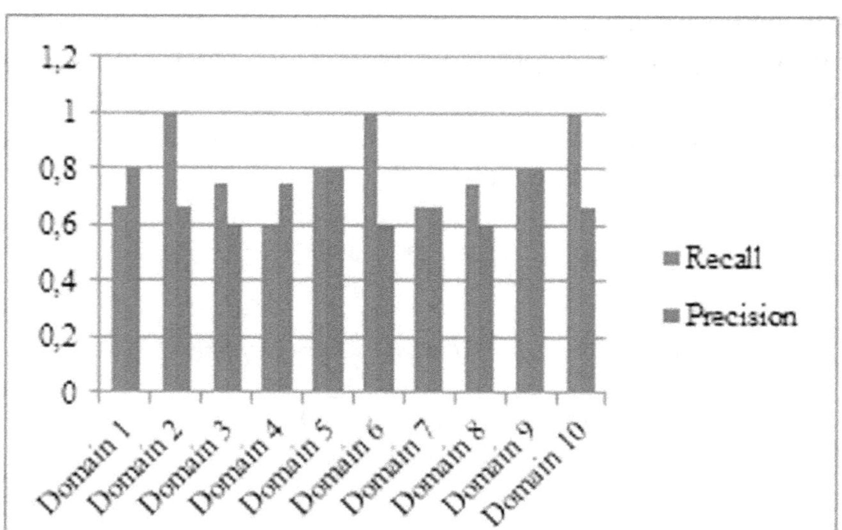

Figures 4 and 5 show for each domain (shown on the abscissa axis), as a function of all the queries (cf. Appendix 1), the *recall* values (shown in blue) and the *precision* values (shown in red), in the best case and in the medium case, respectively. The best case presents a concordance between the results of the evaluators and those retrieved by our component. In other words, the results of the relevant pages obtained by our component are very similar to the needs of the users. While the medium case presents a slight discordance between the two results.

We present in the following table (cf. Table 2) an extract of the numbers of the retrieved Web pages for the five (5) following domains: (D1) Academic research (Academic research, 2018), (D2) Photography (Camera, 2018), (D3) Network of scientists (Stella *et al.,* 2011), (D4) Requirements engineering (Lohmann & Riechert, 2010) and (D5) Biomedical research (Samwald & Adlassnig, 2008). This table shows the best cases (BC: the gray lines in the table) and also the medium cases (MC: the white lines in the table).

As we mentioned before, our component retrieves a maximum of six (6) relevant pages per list (cf. Table 1). These numbers of the Web pages retrieved by our component are used to calculate the *recall* and *precision* values, which appear in Figures 4 and 5.

Table 3 shows a summary of these values for the best and medium cases about these five domains (cf. Table 3).

To better explain the values obtained in Table 2, we show an example that details the values obtained for the domain 1 (D1) for the best case and the medium case (cf. Table 4).

From the results of these evaluations, we have noted that (i) for the best cases, the average values are as follows: 93,6% for the recall and 96,6% for the precision; the average of the two reached 95,1%, and (ii) for the medium cases, the average value of the *recall* is 76,2% and the average value of the *precision* is 72,2%; the average of the two reached 74,2%. These results prove the good performance of our querying and filtering component and show that this latter selects the very similar responses to those selected by a human being with considerable time saving. We remind that this component supports, during the clustering of the relevant pages, those that are annotated. Moreover, we plan to improve our

Table 2. Number of the Web pages based on the results provided by the evaluators in the best and medium cases

Domain		Number of the Web Pages Correctly Retrieved as Relevant (Num 1)	Number of the Retrieved Web Pages but Irrelevant (Num 2)	Number of the Not Retrieved Web Pages but Relevant (Num 3)
D1	BC	6	0	1
	MC	4	1	2
D2	BC	6	0	0
	MC	4	2	0
D3	BC	5	1	0
	MC	3	2	1
D4	BC	5	0	1
	MC	3	1	2
D5	BC	6	0	0
	MC	4	1	1

Table 3. Result of the research automation of the relevant pages by the standard metric

Domain		Recall: R	Precision: P
D1	BC	85%	100%
	MC	66%	80%
D2	BC	100%	100%
	MC	100%	66%
D3	BC	100%	83%
	MC	75%	60%
D4	BC	83%	100%
	MC	60%	75%
D5	BC	100%	100%
	MC	80%	80%

search method of the relevant pages by the calculation of the similarity between the annotated and non-annotated Web pages, in order to improve the *recall* values. This improvement step is in progress and will be published later. We are now in the phase of suggesting similarity formulas between the annotated and non-annotated pages.

In the literature, we do not find much work that evaluates the relevance of the interrogation results of the proposed system, after the classification (or the clustering). Among the related work to our evaluation approach of the relevance, we find (i) the evaluation method of the relevance proposed in (Jenhani El Jed, 2006), (ii) an evaluation work of the classification of the Web pages (using some classifiers) proposed in (Biricik & Diri, 2009) and (iii) a categorization proposals of the Web pages, proposed in (Malarvizhi & Pujeri, 2012).

The system proposed by Jenhani El Jed (2006), entitled "WebSum", classifies the responses of the Google search engine. The experimentation carried out by the author was based on fifteen (15) queries. For each query, she recovered the hundred (100) pages provided by Google and all the pages selected by "WebSum". In addition, she evaluated the obtained results based on the results given by the evaluators (ten (10) evaluators). The evaluation has then given an average value for the *recall* of 85% and for the *precision* of 80%, with some errors that can be produced in the classification (cf. Section 2).

Biricik & Diri (2009) propose a new algorithm for extracting attributes for the classification. The proposed algorithm has then given a *precision* value between 58% and 75% (66% in average).

Malarvizhi & Pujeri (2012) propose a categorization algorithm of the Web pages (but without taking into account the relevance of the pages) based on the results of the WebCrawler (WebCrawler, 2018). The experimental results have shown that the proposed approach of the categorization reaches an average value for the *recall* of 77% and for the *precision* of 81%. In 2014, Basavaraj *et al.,* affirmed, after a comparative analysis of Web pages classification techniques, that the work of Malarvizhi & Pujeri is the last one that seeks to better categorize the Web pages. However, in 2016 Maâlej Dammak proposed an interrogation system that returns the most relevant Web pages after a hierarchical classification. Additionally, in 2016 Krishnamurthy *et al.,* suggested an approach based on exploratory data analysis using the querying results of the WebCrawler. They noticed that these results return several irrelevant pages related to the interest domain. These authors had as perspective to use a classifier, which uses

Table 4. Number of the Web pages based on the results provided by the evaluators after the interrogations in the domain 1

		Number of the Web pages correctly retrieved as relevant (Num 1)	Number of the retrieved Web pages but irrelevant (Num 2)	Number of the not retrieved Web pages but relevant (Num3)
Q1	L1	3	2	1
	L2	5	1	0
	L3	3	2	0
	L4	3	1	2
	L5	3	2	0
Q2	L1	6	0	1
	L2	4	1	2
	L3	6	0	1
	L4	4	1	2
	L5	4	2	0
Q3	L1	5	1	0
	L2	3	2	1
	L3	3	1	1
	L4	3	2	0
	L5	3	2	0
Q4	L1	5	0	1
	L2	4	1	2
	L3	4	1	2
	L4	4	2	0
	L5	4	2	0
Q5	L1	6	0	1
	L2	5	1	1
	L3	4	1	2
	L4	6	0	1
	L5	5	1	1

N.B:

- **Qi:** Where i [1..5] // five (5) queries.
- **Lj:** Where j [1..5] // five (5) lists.
- The red line shows the best case for each query.
- The blue line shows the medium case for each query.
- The gray line of the domain 1 (D1) in Table 2 presents the best case for this domain, among the best cases of the five queries. This is the same for the other domains.
- The white line of the domain 1 (D1) in Table 2 presents the medium case for this domain, among the medium cases of the five queries. This is the same for the other domains.

**For a more accurate representation see the electronic version.*

the annotated pages to filter correctly the pages and reduce the number of irrelevant pages. So, they proposed in (Pham *et al.,* 2017) a categorisation system that uses historical data collected and annotated by subject matter experts to train a classifier that identifies relevant content. This system deals only with humanitarian data sources. This drawback limits the system in order to perform well on the new data sources. As for our querying and filtering component, it interrogates the search engine "Google" for any data sources and considers all Web pages types, whether annotated or not-annotated.

After the study of these evaluation works, we have noted that our experimentation gives better results at the filtering level. We have obtained better rates of the *recall* and *precision*, especially in the best case. In the medium case, the *recall* and *precision* values are slightly better compared to the values of these metrics in the studied works. Despite the fact that there is not much difference between the values of the metrics in the medium case, our component adds a feature in the clustering by taking into consideration the Web pages that are annotated. This feature has not been taken into account in the previous works. If an annotated page is retrieved after an interrogation, it is more likely to be among the relevant pages after a clustering. It presents a clear answer to the query of the user and contains elements that include the keywords of the query due to their annotation RDFs (already proposed by our annotation component). In this case, the result of our component approximates to the need of the user (the search for the relevant pages), so this result improves the values of the metrics: the *recall* and the *precision*. These experimentation results prove the good performance for our component.

DISCUSSION: WHY HAVE WE USED AND ENHANCED THE EXTENDED BOOLEAN MODEL?

First, in 2015 Rydberg performed a comparison between the extended Boolean model and the vector model while filtering the relevant results (the Web pages) in the geographical area. The applied tests have shown that the retrieved documents with the extended Boolean model are more relevant than those with the vector model. He found that the main focus of the vector model is efficiency rather than relevancy.

Second, the vector model uses a cosine score computation to rank the documents-query similarity (Manning *et al.,* 2008). And since in our case we have five queries for each domain, so it is necessary to record a pages-query matrix for each returned list (we have five lists per query). In addition, a list is a set of ten Web pages returned after interrogation. We then need five matrices per query, so twenty-five (25) matrices for the five queries. In total, we need two hundred and fifty (250) matrices for the ten (10) domains. Moreover, a registered matrix must be used to look for the similarity between the pages-query (using a cosine score between the pages vectors and the query vector) to identify the relevant pages compared to the query. To conclude, we cannot use such a model because we are not able to create matrices in a database for each Web interrogation.

As for the extended Boolean model, which uses P-norm scores (the formulas presented in section 4.2.1), it rank documents not only by matching terms, but also by taking the Boolean interconnections between the terms in the query into account (Rydberg, 2015; Kumar & Reddy, 2013). In our case, our component has been implemented on the base of the Boolean queries ("or", "and", "not" between terms, noted by "+", ".", "-" in our implementation).

The use of the extended Boolean model has then become an obligation for two reasons: (i) our querying and filtering component "Filtering Web Pages" works with Boolean queries and (ii) it aims to search the relevance between Web pages for each interrogation.

Moreover, we have enhanced the extended Boolean retrieval method to be able to calculate scores for the indexed Web pages, whether for annotated or not-annotated pages.

Table 5 summarizes and compares the characteristics of these two models.

Table 5. Comparative table between the vector model and the extended Boolean model

Models	Comparison Criteria				
	Similarity to the Query Terms	**Matching Principle and Boolean Query**	**Need for a Database**	**Efficiency**	**Relevancy**
Vector model	Yes	No	Yes	Yes	No
Extended Boolean model	No	Yes	No	No	Yes

CONCLUSION

In this chapter, we are interested in the filtering of the annotated and non-annotated Web pages in order to improve the relevance of search for the user. In fact, our first contribution here is to synthesize our annotation process of the Web pages. Our second contribution is to automate the filtering process of the annotated and non-annotated Web pages, since it is a difficult process because of the heterogeneity of these pages. We have then proposed a new filtering method of the Web pages, which contains three phases. This method helps to improve the relevance of the interrogation results. For its operationalization, we have developed a new component, which allows the automation of the filtering of the Web pages in order to group the relevant ones for the user (including the annotated pages that respond to his/her request). The results of the development of our component have shown the feasibility and the benefits of our filtering proposal, despite the fact that this automation is a delicate and difficult task. The evaluation of our component gives an average percentage of the *recall* and the *precision* reached 95,1% for the best case. In the medium case, the average percentage of the two reached 74,2%. These values are obtained after a study done in different domains. We have noticed that these results are satisfactory in comparison with the results we have found in the literature.

As future works, we will search how we can enhance the performance of our component by improving the response time of the calculation of the scores for the Web pages. Additionally, we will see how to take into consideration for this calculation of the scores the similarity calculation between the annotated and non-annotated pages.

REFERENCES

Academic research. (2018). *Protege Ontology Library*. Retrieved from http://protegewiki.stanford.edu/wiki/Protege_Ontology_Library

Apache Jena. (2018). *The Apache Jena project*. Retrieved from https://jena.apache.org/

Basavaraj, S., Anami, Ramesh, S., Wadawadagi, & Veerappa, B. (2014). Machine Learning Techniques in Web Content Mining: A Comparative Analysis. *Journal of Information & Knowledge Management, 13*(1).

Biricik, G., & Diri, B. (2009). Impact of a New Attribute Extraction Algorithm on Web Page Classification. In *The 2009 International Conference on Data Mining* (pp. 481-485). Las Vegas, NV: Academic Press.

Bojars, U., & Breslin, J. (2018). *Sioc core ontology specification*. Retrieved from http://rdfs.org/sioc/spec/

Bojars, U., Breslin, J., Peristeras, V., Tummarello, G., & Decker, S. (2008). Interlinking the Social Web with Semantic. *Journal IEEE Intelligent Systems*, *23*(3), 29–40. doi:10.1109/MIS.2008.50

Brickley, D., & Miller, L. (2015). *FOAF (2000-2015+*. Retrieved from http://www.foaf-project.org/

Camera. (2018). *Protege Ontology Library*. Retrieved from http://protegewiki.stanford.edu/wiki/Protege_Ontology_Library

Fox, E. A. (1983). *Extending the Boolean and Vector Space Models of Information Retrieval with P-Norm Queries and Multiple Concept Types* (PhD thesis). Cornell University, Ithaca, NY.

Jenhani El Jed, O. (2006). Classification des réponses d'un moteur de recherche et évaluation de leur pertinence. In *Actes du XXIVème Congrès INFORSID* (pp. 291-306). Hammamet, Tunisie: Academic Press.

Karimi, S., Zobel, J., Pohl, S., & Scholer, F. (2009). The challenge of high recall in biomedical systematic search. In *The ACM 3rd International Workshop of Data and Text Mining Methods in Bioinformatics* (pp. 89-92). Hong Kong: ACM. 10.1145/1651318.1651338

Krishnamurthy, Y., Pham, K., & Santos, A. (2016). Interactive Exploration for Domain Discovery on the Web. In *ACM KDD Workshop on Interactive Data Exploration and Analytics (IDEA)* (pp. 64-71). San Francisco, CA: ACM.

Kumar, J. N., & Reddy, M. J. (2013). Effective efficient Boolean retrieval. *International Journal of Computer Science and Mobile Applications*, *1*(5), 38–43.

Lance, G. N., & Williams, W. T. (1967). A General Theory of Classificatory Sorting Strategies: Hierarchical Systems. *The Computer Journal*, *9*(4), 373–380. doi:10.1093/comjnl/9.4.373

Lohmann, S., & Riechert, T. (2010). Adding semantics to social software engineering: (re-) using ontologies in a community-oriented requirements engineering environment. In The Social Software Engineering Workshops (pp. 485-494). Paderborn, Germany: Academic Press.

Maâlej Dammak, S. (2016). *A framework for achieving semantic annotation and interrogation for Web pages* (Unpublished doctoral dissertation). Sfax University, Tunisie.

Maâlej Dammak, S., Jedidi, A., & Bouaziz, R. (2013a). Semantic Annotation Framework for Web Resources. In *The fifth International Conference on Internet Technologies & Applications* (pp. 106-113). Wrexham, UK: Academic Press.

Maâlej Dammak, S., Jedidi, A., & Bouaziz, R. (2013b). Automation and evaluation of the semantic annotation of Web resources. In *The 8th International Conference for Internet Technology and Secured Transactions* (pp. 448-453). London, UK: Academic Press.

Maâlej Dammak, S., Jedidi, A., & Bouaziz, R. (2014a). Fuzzy semantic annotation of Web resources. In *The World Symposium on Computer Applications & Research* (pp. 253-258). Sousse, Tunisie: Academic Press.

Maâlej Dammak, S., Jedidi, A., & Bouaziz, R. (2014b). Automation of the semantic annotation of web resources. *International Journal of Internet Technology and Secured Transactions*, *5*(2), 133–148. doi:10.1504/IJITST.2014.064513

Maâlej Dammak, S., Jedidi, A., & Bouaziz, R. (2017). Filtering Method for the Annotated and Non-Annotated Web Pages. International Journal of Knowledge Society Research - IJKSR, 8 (1), 1-22.

MacQueen, J. B. (1967). Some methods for classification and analysis of multivariate observations. In *The fifth Berkeley Symposium on Mathematical Statistics and Probability* (pp. 281-297). University of California Press.

Malarvizhi, P., & Pujeri, R. V. (2012). Distributed approach to web page categorization using map-reduce programming model. *International Journal of Engineering and Technology*, *3*(6), 373-386.

Manning, C. D., Raghavan, P., & Schutze, H. (2008). *Introduction to Information Retrieval*. Cambridge University Press. doi:10.1017/CBO9780511809071

Nouali, O., & Blache, P. (2004). A semantic vector space and features-based approach for automatic information filtering. *International Journal Expert Systems with Applications*, *26*(2), 171–179. doi:10.1016/S0957-4174(03)00118-0

Pham, K. T., Sattigeri, P., Dhurandhar, A., Jacob, A. C., Vukovic, M., Chataigner, P., … Varshney, K. R. (2017). Real-time understanding of humanitarian crises via targeted information retrieval. *IBM Journal of Research and Development, 61*(6), 7.1-7.12.

Pohl, S., Moffat, A., & Zobel, J. (2012). Efficient Extended Boolean Retrieval. *IEEE Transactions on Knowledge and Data Engineering*, *24*(6), 1014–1024. doi:10.1109/TKDE.2011.63

Pohl, S., Zobel, J., & Moffat, A. (2010). Extended Boolean Retrieval for Systematic Biomedical Reviews. In *The 33rd Australasian Computer Science Conference* (pp. 117-126). Brisbane, Australia: Academic Press.

Rahmani, A. M., Hossaini, Z., & Setayeshi, S. (2009). Link Processing for Fuzzy Web Pages Clustering and Classification. *European Journal of Scientific Research*, *27*(4), 620-627.

Rossi, F., & Villa-Vialaneix, N. (2011). Représentation d'un grand réseau à partir d'une classification hiérarchique de ses sommets. *Journal de la Société Française de Statistique SFDS*, *152*(3).

Rydberg, C. (2015). *Time Efficiency of Information Retrieval with Geographic Filtering. Digitala Vetenskapliga Arkivet*. Stockholm, Sweden: Digital Scientific Archive.

Salton, G., Fox, E. A., & Wu, H. (1983). Extended Boolean Information Retrieval. *Magazine Communications of the ACM*, *26*(11), 1022–1036. doi:10.1145/182.358466

Salton, G., Wong, A., & Yang, C. S. (1975). A vector space model for automatic indexing. *Magazine Communications of the ACM*, *18*(11), 613–620. doi:10.1145/361219.361220

Samwald, M., & Adlassnig, K. (2008). The bio-zen plus ontology. *Journal Applied Ontology*, *3*(4), 213–217.

Stella, M., Shanshan, C., Mansoor, A., Brian, L., Paula, M., Nick, R., . . . Ying, D. (2011). The VIVO Ontology: Enabling Networking of Scientists. In The ACM WebSci'11 (pp. 1-2). Koblenz, Germany: Academic Press.

Van Rijsbergen, C. J. (1979). Information Retrieval (2nd ed.). Butterworth-Heinemann.

WebCrawler. (2018). Retrieved from www.webcrawler.com

ENDNOTES

1 http://protegewiki.stanford.edu/wiki/Protege_Ontology_Library

2 http://www.geonames.org/ontology/documentation.html

3 https://github.com/zazi/pmkb-id3-tag-extractor/blob/master/src/main/resources/smiy/pmkb/vocabulary/bibo.owl

4 http://www.fao.org/countryprofiles/geoinfo/geopolitical/resource

APPENDIX 1

Table 6. List of the ten (10) domains with the five (5) queries

The Domains of Study	Domain Ontologies	Queries
D1: Academic research	Ka.owl	• Academic research + Publication • Academic Research + Project + Development project • Academic Research + Research topic • Academic Research + Event + Conference • Academic Research + Organization + Department
D2: Photography	Camera.owl	• Photography + Camera • Photography + Camera + Digital • Photography + Camera + Digital + Large format • Photography + Camera + Large format • Photography + Range + Camera
D3: Network of scientists	Vivo.owl	• Network of scientists + Event + Conference + Conference paper • Network of scientists + Conference paper • Network of scientists + Conference paper + Person • Network of scientists + Conference paper + Person + Faculty member • Network of scientists + Conference paper + Person + Emeritus professor
D4: Requirements engineering	Swore.owl	• Requirements engineering + Mailing list • Requirements engineering + Subscription list • Requirements engineering + Reference point + Application state • Requirements engineering + Online Ecommerce Account • Requirements engineering + Project
D5: Biomedical research	Bio-zen-plus.owl	• Biomedical research + Physical object • Biomedical research + Agent + Endurant • Biomedical research + Non physical object • Biomedical research + User group • - Biomedical research + Community
D6: Tourism	Travel.owl[1]	• Tourism + Urban area • Tourism + Town • Tourism + Capital • Tourism + Farmland • Tourism + Hotel + Luxury Hotel
D7: Geographic	GeoNames.owl[2]	• Geographic + Map • Geographic + Names • Geographic + Names + Feature • Geographic + Feature • Geographic + Feature + Code
D8: Bibliographic	Bibo.owl[3]	• Bibliographic + Document + Thesis • Bibliographic + Document + Article • Bibliographic + Book • Bibliographic + Document • Bibliographic + Manual
D9: Geopolitical	Geopolitical.owl[4]	• Geopolitical + Group + Economic region • Geopolitical + Group + Special group • Geopolitical + Group + Geographical region • Geopolitical + Territory + Self governing • Geopolitical + Territory + Non self governing
D10: Finance	Finance.owl[1]	• Finance + Party customer of issuer • Finance + Party data provider • Finance + Client risk profile • Finance + Party fiscal agent • Finance + Capitalization type

N.B We propose preceding the keywords of the query with the name of the study domain in order to have direct and clear pages to the query. The keywords of the queries represent concepts in the selected ontology.

APPENDIX 2

Algorithm of the best case:

```
Input: the ten domains; the five queries of each domain; the five lists of
each query
Output: the best case of each domain
Begin:
For i in [1..10] // ten (10) domains
 For j in [1..5] // five (5) queries
 // seek the best case of the query (BCQ) in the five lists
 If (the number of BCQj> 1)
BCQj ← the case that has the best value of the sum of the recall and the
precision among the best cases of the five lists
 Else
 BCQj ← the found best case
 End If
 End For
// seek the best case of the domain (BCD) in the five queries
BCDi ← the case that has the best value of the sum of the recall and the
precision among the best cases of the five queries
End For
End
```

APPENDIX 3

Our research work is carried out within the MIRACL research laboratory (Multimedia Information Systems and Advanced Computing Laboratory), and more specifically within the research team TADB (Team of research in Advanced DataBases). This work is in the field of advanced information systems (IS).
In fact, in an increasingly open economic world, IS are becoming more and more a fundamental tool for development, competition and business expansion. Their objective at the beginning is to collect process and store information, which is often considered by companies as important as financial and human resources. But today their role is no longer limited to offering automated procedures to operational systems and dashboards to steering systems; they must also be in direct interaction with external partners and extend to more global decision-making systems. As a result, traditional databases (BD) have proved insufficient to meet all informational needs for many applications (e.g., e-commerce, project management, medical records management, technical parameters management).
Our research work is therefore about the study of temporal and versioning aspects, fuzzy aspects and methodological and modelling aspects of information systems and Databases. It mainly deals with:

- Managing temporal facts in DB (temporal specifications, designing methodologies and development tools);
- Data and Schema versioning of Temporal DB;
- Schema versioning of data warehouses, ontologies and workflows;
- Information systems engineering for temporal, real time, fuzzy and medical types;
- Semantic Web: semantic interrogation and annotation, information retrieval and information filtering;
- Service oriented approach and systems.

Chapter 13
Review on Keyword Search and Ranking Techniques for Semi-Structured Data

Dayananda P.
JSSATEB, India

Sowmyarani C. N.
Rashtreeya Vidyalaya College of Engineering, India

ABSTRACT

The size of semi-structured data is increasing continuously. Handling semi-structured data efficiently is a challenging task. Keyword search is an important task, and required information can be retrieved without having knowledge of data storage hierarchy. There are several challenges in handling XML data. This chapter discusses various challenges in terms of lowest common ancestor (LCA) semantics, processing of queries efficiently, retrieving top-k results for user needed data. The existing approach is defined under many classes based on how the problem and solution are tackled. Analysis of keyword search and ranking techniques for retrieving desired information are discussed in detail.

1. INTRODUCTION

Considering information society and information science development, storage of data in XML format is increasing over internet day by day. Standard way of data representation and exchanging of data over internet by making use of XML format. Effective way of extracting useful information is challenging job for researcher. There are many challenges to handle the data efficiently; many research methods are defined in the field of retrieving information on semi structured data. Basic understanding of how data is stored, semantics of handling data, ranking top k answers for query is very important task for researcher to give optimal solution for existing problem. Thus, taking into consideration of these requirements, in this chapter of the presented work, a number of literatures have been studied and explored for keyword search on semi-structured data.

DOI: 10.4018/978-1-5225-7347-0.ch013

2. TREE-BASED XML KEYWORD SEARCH

When XML documents do not contain IDREF, they can be modeled as trees. Approaches to handle such documents are called tree-based approaches because they are based on tree model. Inspired by the hierarchical structure of the tree model, most of existing tree-based approaches are based on the LCA (Lowest Common Ancestor) semantics, which returns the lowest common ancestors of matching nodes to keyword queries.

There are many subsequent semantics to filter less meaningful answers. Existing works either improve the effectiveness by proposing a new semantics or improve the efficiency by proposing a new method for certain semantics. The widely accepted LCA-based semantics include LCA itself, SLCA, VLCA, MLCA, ELCA, and etc, among which, SLCA and ELCA are the most popular semantics. We classify the existing research works into these semantics and result of our classification is shown in Figure 1. Some research work involves study of more than one semantic such as XRANK (Dayananda, 2016; Guo, 2003), Set-intersection (Bao et al., 2012), and Top-K (Dayananda, 2016).

2.1 LCA Semantics

The LCA semantics for XML keyword search was first proposed in XRANK (Guo et al., 2003). By the LCA semantics, for a set of matching nodes, each of which contains at least one query keyword and each query keyword matches at least one node in this set, the lowest common ancestor (LCA) of this set is a returned node. An answer is a subtree rooted as a returned node (i.e., an LCA) or a path from the returned node to matching nodes. XRANK is extended from Google's Pagerank algorithm for ranking. It takes into account the proximity of the keywords and the references between attributes. XRANK implements a naive approach, and three optimized approaches afterwards to improve the search.

2.2 SLCA Semantics

The SLCA (Smallest LCA) semantics was first proposed in XKSearch(Dayananda, 2016; Papakonstantinou & Xu, 2005). The SLCA semantics defines an SLCA to be an LCA that does not have any other LCAs as its descendants. There are many works on finding the set of SLCAs for a keyword query. XKSearch(Dayananda, 2016; Papakonstantinou & Xu, 2005) proposes two efficient algorithms to compute SLCAs, namely Indexed Lookup Eager and Scan Eager. To find all SCLAs, there are two tasks, namely finding all LCAs and remove all ancestors among LCAs to get the SLCAs. It is costly to find all LCAs.

XKSearch optimizes as follows. Firstly, for each matching node u of the keyword which has the least number of matching nodes XKSearch finds its left and right match. The left (right) match v of u refers the matching node of the other queries or the keyword and among all nodes in u's left (right) side, v is the nearest one (by pre-order). Only the LCA of u and v is a candidate SLCA. Thereby, it greatly reduces the number of computation of LCAs. The predominant characteristics of the approach SLCA refers that with provided two queries or keywords k1, k2 with one node u which comprises keyword k1, the one keyword seeks not to explore the complete keyword list so as to find the optimal solutions. In addition, one merely requires finding the left and right match of u in the k2 list, where the right or left match is the node with the minimal or maximum Dewey ID that is greater or smaller as compared to the Dewey ID of the node u. Multiway-SLCA (Dayananda, 2016; Chan et al., 2007) in addition attempted to enhance the performance of XKSearch scheme. The principal impetus behind this paradigm is the provision to avoid the redundancy of existing XKSearch by means of estimating the SLCAs through multiple intermediate SLCAs. Multi model SLCAs scheme estimates the individual SLCA by taking into consideration of one data node from the individual keyword list in single phases, rather dividing the SLCA into multiple intermediate SLCA phases.

2.3 ELCA Semantics

The ELCA (Exclusive LCA) semantics is also widely accepted. ELCAs are a superset of SLCAs, and it can find some relevant information that SLCA cannot find. An ELCA is an LCA with its own witnesses, i.e., matching nodes. In other words, consider a node u, if u contains matching nodes of all query keywords after removing all subtrees rooted at its descendant ELCAs, then u is an ELCA. This semantics is first introduced in XRANK (Guo et al., 2003) with the DeweyInvertedList algorithm, which reads match nodes in a preorder traversal, and uses a stack to simulate the postorder traversal. Many other algorithms are proposed to find ELCAs of a keyword query. The researcher in (Dayananda, 2016; Papakonstantinou & Xu, 2008) proposes an Index Stack algorithm to find ELCAs more efficiently. The algorithm to find all the ELCAs can be decomposed into two steps: first find all ELCA candidates, and then find ELCAs in those candidates. The first step can be leveraged the algorithm IndexedLookupEager in XKSearch(Cohen et al., 2005).(Dayananda, 2016; Li et al., 2010) Presents an efficient algorithm to find ELCAs named Hash Count. This approach can be sufficiently classified into two sub-steps: the first, as it retrieves the ELCA potential candidate; and second it performs validation of the retrieved candidates, irrespective of the false positives and thus obtains the final results. Here it must be noted that this paradigm is the same as that of the Indexed Stack algorithm which is well described in (Feng et al., 2007), but it employs some different techniques. Besides proposing algorithms for finding SLCAs, Top-k (Li & Guo et al., 2003) and Set-intersection (Dayananda, 2016; Chan et al., 2012)also presents algorithms for finding ELCAs with the similar methods with those of finding SLCAs.

Effective ELCA XML Keyword Query Processing Technique

One of the Effective XML Keyword Query Processing Technique (Chatur et al., 2017) is defined for ELCA query semantic based keyword search technique. In this technique the user input is provided by choosing XML file to which processing of query is performed. The next process is to create index label, assigned label for node is dewey label. Inverted list is prepared based on each query keyword, inverted list contains all the matching nodes and the path from root to matching nodes. Then check is performed for any keyword missing with help of IDDewey labels generated. If missing keyword is not found in query, the ELCA finding will be formed for a given query using indexed label, if it contains missing keywords, then remove the missing keyword from search, a new subquery will be performed based on it. ELCA step returns all the occurrence of keyword in XML with respect to root. If answered is returned as root, then it intimates the user and tell user to give subquery from subset of query and find ELCA based on sub query. In this way meaning full results are found with possible existence of keyword in XML R.

2.4 VLCA Semantics

The VLCA (Valuable LCA) semantics is introduced in (Bao eta al., 2007). According to the VLCA semantics, any two matching nodes in an answer must be homogeneous, that is there are no two nodes of the same elementary type (i.e., label, tag) on the paths connecting the two matching nodes, except themselves. Two algorithms, the Brute-Force algorithm and the Stack-based algorithms are proposed in (Dayananda,2016; Cohen, 2003) to finds VLCAs for a keyword query. There are two variants of VLCA semantics, namely XSEarch(Cohen,2003) and RLCA (Relevant LCA) (Cao & Nguyen, 2010).

XSEarch(Dayananda, 2016; Cohen, 2003) is a variant of VLCA semantics. The whole algorithm is based on a property, called interconnection. Consider, n and n_0 be the nodes in certain tree and T, $T \mid n, n'$ be the respective paths existing from n to n_0 in tree T. Then n and n_0 are considered to be connected with each other, in case any one of the conditions reserves the condition,: $T \mid n, n'$ and does not comprise two different nodes having same level; or the only two distinct nodes $T \mid n, n'$ are there with the same label are n and n_0. The intuition of such a property is that it differentiates the attributes that belongs to different entities. XSEarch tries to find sets of match nodes, such that each set contains all keywords and every two keywords in a set is interconnected. XSEarch returns the path of each set as the search result. However, the complexity is NP-complete. So XSEarch only requires that each node in one set should be interconnected with one node. This looser condition is called star-interconnected and makes it possible to find all the results in polynomial time.

RLCA (Cao & Nguyen, 2010) is similar to XSEarch. RLCA is different from XSearch into two aspects: (1) it accepts that two nodes with the same type can be meaningfully connected in a subtree, due to the fact that a user may be interested in finding more than one entity with the same type. For queries related to only single entity, RLCA uses node types to detect the relevancy of fragments rather than simply uses node labels. Hence, it can detect that some nodes are still homogeneous although there are some nodes of the same types on the path connecting them, such as the two attributes of the same object type.

2.5 MLCA Semantics

Meaningful LCA (MLCA) (Dayananda, 2016; Jagadish et al., 2004) introduces the concept of meaningful relationship between two nodes. According to the MLCA semantics, two nodes are meaningfully related to each other if condition (1) they have the hierarchical relationship (ancestor-descendant relationship), or condition (2) the two nodes belong to the same types, or condition (3) the LCA of matching nodes in the data tree belongs to the LCA of their node types in the schema tree. Otherwise, the two nodes are not meaningful. An MLCA is an LCA in which any two matching nodes have a meaningful relationship. Although the MLCA semantics is similar to the VLCA semantics, conditions of the MLCA semantics is looser than that of the VLCA semantics. They have two main differences. Firstly, for MLCA, two matching nodes of the same types always provide a meaningful answer, while for VLCA, the meaningful answer still depends on whether any nodes between them are of the same type. Secondly, for VLCA semantics, there must be no two nodes on the paths connecting matching nodes are of the same type, while for MLCA semantics, the nodes on the paths connecting matching nodes cannot be of the same type with matching nodes only For example, two matching nodes 'n1', 'n2' in Figure 1(a) and Figure 1(b) are meaningful related because they satisfy condition (1) and condition (3) respectively; while they are not meaningful related as shown in Figure 1(c) if N2' has the same type of n2. This is because 'n1', 'n2' in Figure 1(c) do not satisfy any of the three conditions. Particularly, they obviously do not satisfy condition (1) and condition (2). For condition (3), the LCA of n1 and n2 in the data tree is 'n', which is different from the LCA of their node types in the schema tree, which is the type of N' because N2'has the same type of n2.

Figure 1. Structural relationships among nodes

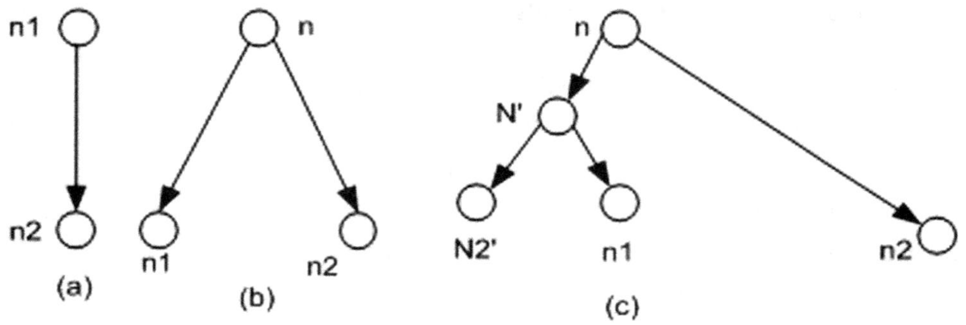

2.6 Object-Relationship-Attribute (ORA)-Semantics Based Keyword Search in XML

In order to solve the problems like meaningless answer, Missing answer, duplicate answer, Problem related to relationship and schema dependency the ORA semantics in used in XML. The object tree construction for XML data is based on ORA semantics considering only object nodes excluding all non object nodes. Tree contains very less number of nodes, where each node specifies object. For a given query, the search is done for lowest common ancestor object for a given tree. All information of an object is represented in LCOA, so we achieve the objective of solving problem like duplicated answers and common relative scheme is proposed for giving solution to schema independent keyword search. ORA semantics is defined for both XML and RDB and it capture ORA semantics for XML by constructing the Object tree and construction of Object-Relationship-Mixed (ORM) data graph for RDB. Extension the keyword query language is done by matching the metadata, i.e., the names of tags in XML and the names of relations and attributes in RDB and also aggregate functions and GROUPBY function is used for queries to retrieve from XML and RDB (Dayananda,2016; Bao & Wu, 2011; Mong et al., 2016)

2.7 Relationship and Comparison on the LCA-Based Semantics

The classification of the existing research work based on the semantics is as shown in figure 2. In addition, it is found that for the same query 'Q', the relationships among the set of answers by the LCA-based semantics are follows.

The Figure 2 shows the classification of the existing research work based on the semantics like ELCA, MLCA, SLCA, VLCA and others. The relation between different LCA based semantics is as shown in equation (1).

For the same query Q, the LCA semantics provides for most answer for the queries. However, many of them are contained by the other and are not really relevant. Therefore, the other semantics have constraints to filter out such answers. However, they may filter out meaningful answers as well.

$$LCA(Q) \supseteq ELCA(Q) \supseteq SLCA(Q) \tag{1}$$

and

Figure 2. The classification for tree-based approaches based on the semantics used

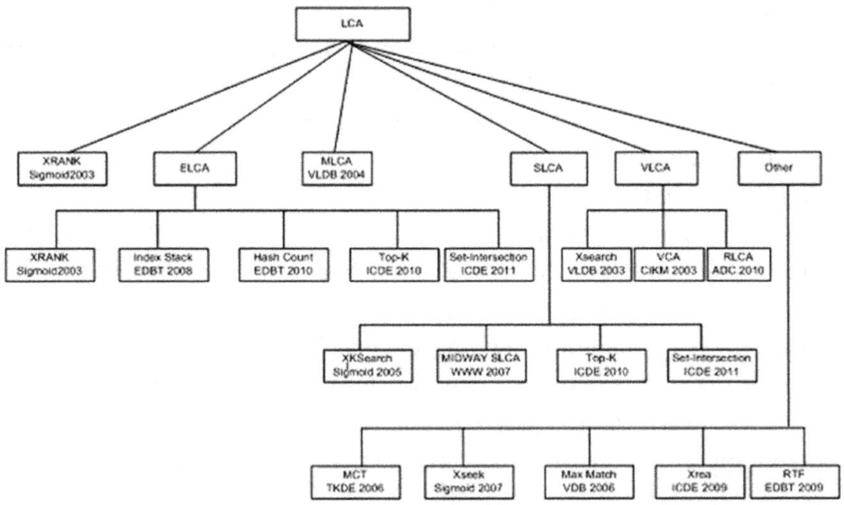

$$LCA(Q) \supseteq MLCA(Q) \supseteq VLCA(Q)$$

Example 1: Consider keyword query {Q = Clinton, Kennedy} issued against the XML data tree in Figure 3, in which we circle and label some nodes as (&o1), (&o2), (&o3), (&o4) and (&o5) for discussion. Two nodes (&o4) and (&o5) are LCAs, SLCAs, ELCAs, MLCAs, and VLCAs. LCAs of the query are nodes (&o1), (&o2), (&o3), (&o4) and (&o5). Among LCAs, only the two nodes (&o4) and (&o5) are SLCA nodes while the other do not because they are ancestors of either node (&o4) or node (&o5). Nodes (&o2) and (&o3) are not ELCA nodes either because they do not have their own witnesses. Although, node (&o1) is not an SLCA node, it is an ELCA node because after removing the two nodes (&o4) and (&o5), it still has Kennedy and Clinton as its descendants (under node (&o2) and node (&o3)). In this example, all LCA nodes are MLCA nodes because none of LCA nodes fall into the case in Figure 2(c). Among LCA nodes, node (&o1) is not a VLCA node because there exists nodes of the same types (student) on the path connecting matching nodes. The remaining nodes are VLCAs. As we can see, LCA (a) ELCA (b) S LCA (c) LCA (d) ⊇ MLCA (e) VLCA . Returned nodes of the semantics for query Q are also summarized in Table 1.

2.8 Annotation Paths for Matching XML Schemas

The semantic annotation of XML-Schemas uses technique called as annotation paths. Annotation paths capture significantly more semantics than plain model references, the declarative annotation method of the W3C standard SAWSDL. This annotation path was introduced in earlier works. New formation of their structure and define their semantics and introduce matching methods to derive simple and complex value correspondences is represented. In many application domains such as health care, e-government, e-commerce or inter-organizational business processes in general, transformations of data centric XML instance documents of some source schema to instances of some target schema are common needs to provide interoperability between heterogeneous systems. Matching can be based on logical reasoning over the annotations and the reference ontology. In this paper we present annotation paths, an annotation processed or on top of SAWSDL that is fully declarative and provides a good balance between expressiveness and ease of use for annotators. Extended annotation paths are defined and provide a strongly refined formalization of their structure. This allows us to newly introduce a formalization of their semantics in terms of description logics concepts.

The common architecture of Clio system is to produce transformation scripts that are used to transform document in to some XML source schema. Clio works in two phases matching and mapping. Matching is the process of finding value between source schema nodes and target scheme nodes so the value are either manually founded by the user are they are automatically derived by matching algorithms. revised correspondences are used as an input for mapping schema. The major disadvantage of this approach is automatic schema magic approaches. Are restricted to one to one correspondences and it also suffers from low match quality here the aim is to produce a good quality correspondences between target schema and semantically annotated source. It consists of semantic annotation here annotating data carrying elements of source and target schemas is done, schema matching is based on a extended reference ontology which contains all annotation concepts of source and target schema. Schema mapping will match the result of all simple and complex correspondence between nodes of target and source schema. Annotation path are implemented in the form of prototype, matching system for XSD Julius Köpke(2017).Implementation works in five phases, initially the schemas are passed into expanded schemas, schema matching, Further steps are complex schema matching and recommendation generation. Evaluation of annotation paths are based on higher match quality than standard SAWSSDL, higher match quality than classical matching system, support for document transformation. Match quality are accessed based on recall usual precision and f measure metrics. Precision is the total number of correspondences recall is the fraction of correctly identified correspondences plus negatives. F- measure is the harmonic mean of recall and precision. Result of APM are higher match quality than standard SAWSSDL, higher match quality than classical matching system, support for document transformation

2.9 Common Problems of the LCA-Based Semantics

Although different LCA-based semantics (e.g., LCA, SLCA, ELCA, VLCA, etc) provide different answers, they all are based on the concept of LCA. Moreover, they all ignore the Object Relationship Attribute (ORA) semantics (Bao et al., 2013), which includes the semantics of relationship, object, object and relationship attribute.

Hence, it can be found that these approaches suffer from a number of generic issues. In the following section, some of the key issues with LCA based semantics have been discussed. For reference, the XML details as illustrated in Figure 4 can be considered. The relevant XML schema is presented in Figure 5. Here it is significant to acknowledge that as illustrated in the Figure 4, the Course (11) and Course (35) has similar object even though they represent nodes with altogether different structure. The common problems of LCA are as discussed below from section (a) to (e).

Issue 1: Meaningless Answer

Let, Q1 be the attribute {Bill}, here as illustrated in Figure 4 the LCA-based schema returns Bill (6), then while, it can't be significant as it does not facilitate any significant information about the attribute Bill.

It is common in the case when certain returned node refers towards a non-object node that can be either certain value or attribute. Here the predominant reason can be that the semantics based on LCA don't possess any concept of object or certain attribute and this is the reason that it is unable to distinguish object and certain distinct variable or non-object nodes. The Table 1 illustrates the returned nodes with respect to different semantics for query.

The Table 1 shows nodes retrieved (column 4) by different methods (column 1) for keyword query {Q = Clinton, Kennedy} issued on XML data represented in form of tree structure in Figure 3. Column 2 and 3 shows the definitions and different algorithms existing under each method.

In fact returning certain object node is significant while returning certain non-object node can't be of any significance. Here, the desired query answer might be targeted to just Student (1), with respect to the variable Bill (6) since it contains additional information related to Bill such as major and student No.

Issue 2: Missing Answer

Let, an XML keyword query be represented as Q2 = {Bill, John} is given to XML data presented in the following Figure 4, where the query matches the name of the initial two students. It can be found that the LCA semantic approaches retrieve the data rather than the root as the answer for the defined query Q, which is in fact intuitively insignificant as retrieving the root means it is nothing but the complete XML tree.

Table 1. Returned nodes of the semantics for query

Semantics	Definition	Existing Algorithms	Existing Algorithms Returned Nodes
LCA	An LCA is a lowest common ancestor of a combination of matching nodes, i.e., each keyword corresponds to at least one matching node in the combination	XRANK Sigmod 2003	{ &o1, &o2, &o3, &o4, &o5 }
ELCA	*An ELCA is an LCA of a combination of matching nodes *An ELCA has its own witnesses, i.e., it does not share its matching nodes with its descendant ELCA nodes	*Index Stack EDBT 2008 *Hash Count EDBT 2010 *Top-K ICDE 2010 *Set-intersection ICDE 2011	{ &o1, &o4, &o5 }
SLCA	*An SLCA is an LCA of a combination of matching nodes *There is no LCA node as its descendant	*XKSearchSigmod 2005 *Multiway-SLCA WWW 2007 *Top-K ICDE 2010 *Set-intersection ICDE 2011	{ &o4, &o5 }
VLCA	*A VLCA is an LCA of a combination of matching nodes *For each pair of matching nodes, all nodes in the path connecting them are of different types, except themselves	*XSEarch VLDB 2003 *VLCA CIKM 2007 *RLCA ADC 2010	{ &o2, &o3, &o4, &o5 }
MLCA	*An MLCA is an LCA of a combination of matching nodes *The LCA of matching nodes in the data tree belongs to the LCA of their node types in the schema tree	*MLCA VLDB 2004	{ &o1, &o2, &o3, &o4, &o5 }

Figure 3. Example on the LCA-based semantics: LCA, SLCA, ELCA, VLCA, MLCA

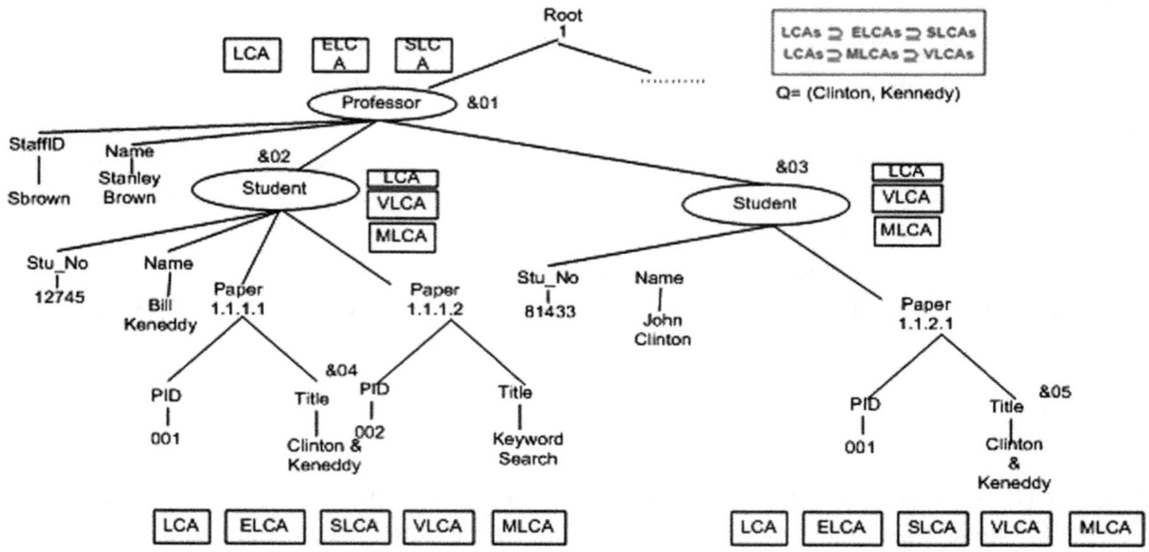

Figure 4. An XML data tree about student and course of a university

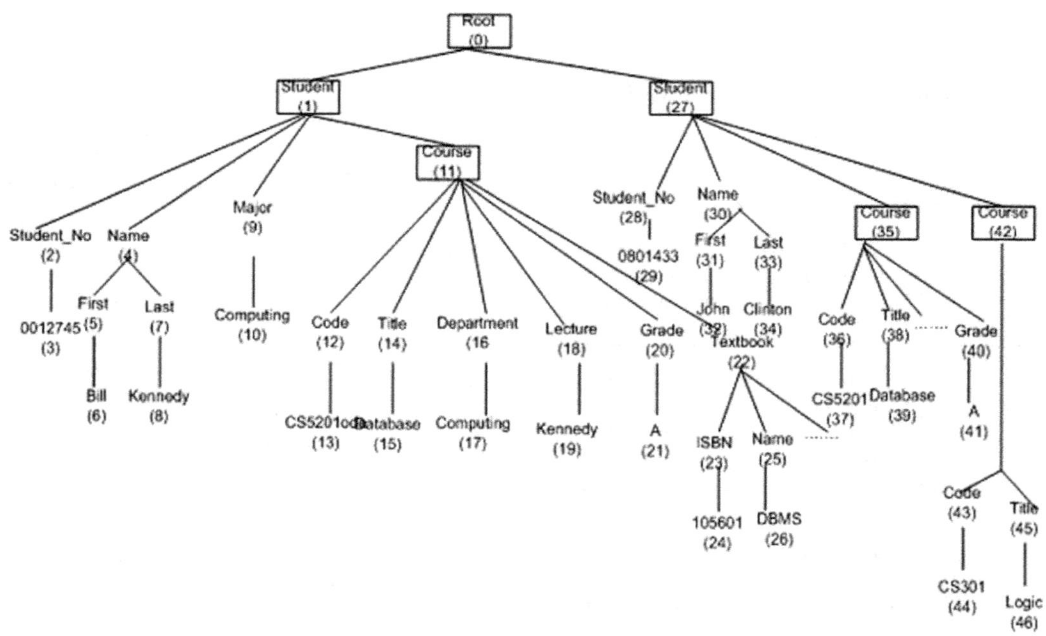

Here it must be noticed that the two objects or variables can be similar only in the case when these variables or objects belong to the similar class and doesn't have the similar object ID. In such cases, the variable Course (11) and Course (35) refer to the same object 1 because they belong to the same object classCourse and have the same object ID value CS5201.Hence, it is generic that the common course be considered by both.

Figure 5. XML data schematics

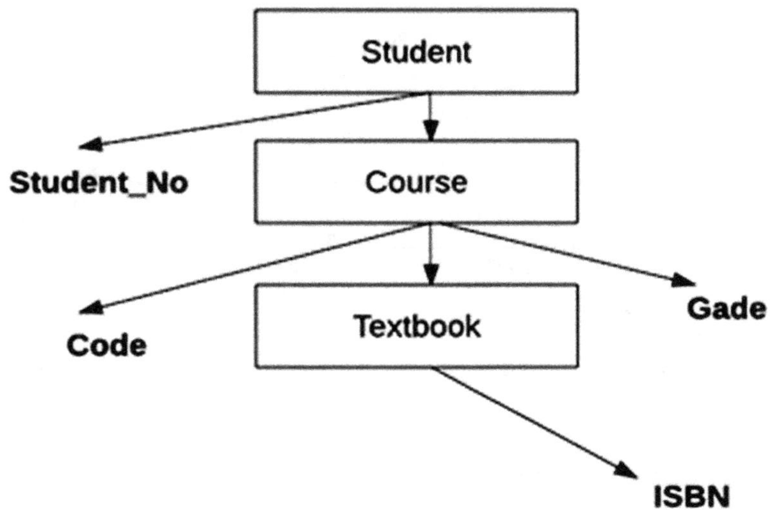

The query answer for Q2 must be students Bill and John. Then while, the LCA-based schemes don't get the answer due to the reason that these are not aware of the variable object ID. Therefore, the common courses considered by the students can't be found.

Issue 3: Duplicated Answer

Let a query be Q3 referring the variables {CS5201, Database} is given to XML data presented in the following Figure 4, where the two responses Course (11) and Course (35) are similar as the two nodes point out towards the similar object. Such issues are raised because of the unawareness of object duplication in case of multiple occurrences. In fact while making query the user expects either Course (11) or Course (35) must be returned as query result, but it should not be the both as they are varied frequencies of the same object. The matter of fact is, in case the specific course has 400 students, then such query results or answers might be duplicated 400 times, which might annoy the user.

Issue 4: Incomplete Answer

Let certain query be Q4 referring {Database A}is given to XML data presented in the following Figure 4. The LCA semantic returns Course (11) and Course (35) as the response of the query. In fact such response of query is incomplete as the grade 'A' is in fact not any course's attribute and it is in fact the grade of a student enrolled with the course. On the other hand, grade refers the inter-relationship attribute existing in between the course and the student and it can't be any attribute related to node object. It is due to the reason that the LCA-based schemes may not be able to distinguish between the relationship attribute and an object attribute.

The proper answer should be all students taking course Database and getting an 'A' grade. To do that, the answer should be moved up to contain other objects (e.g., students) participating in the relationship that 'A' grade belongs to.

Figure 6. Schema of the XML data tree in Figure 4

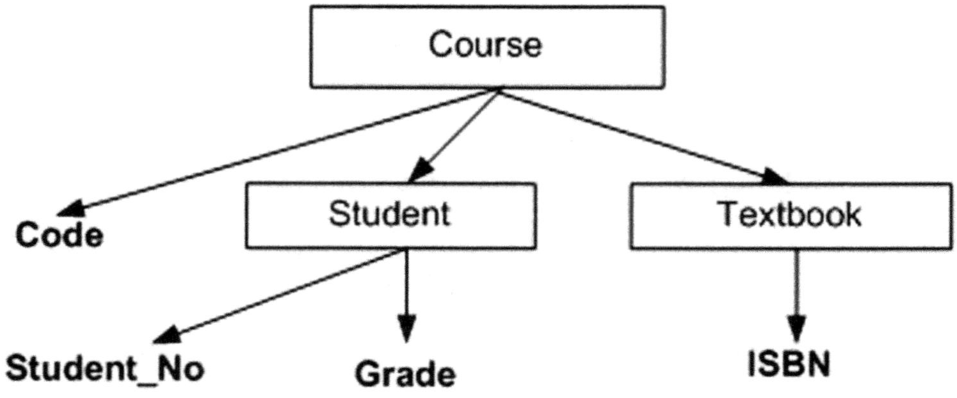

Figure 7. Another design for the university XML data in Figure 4

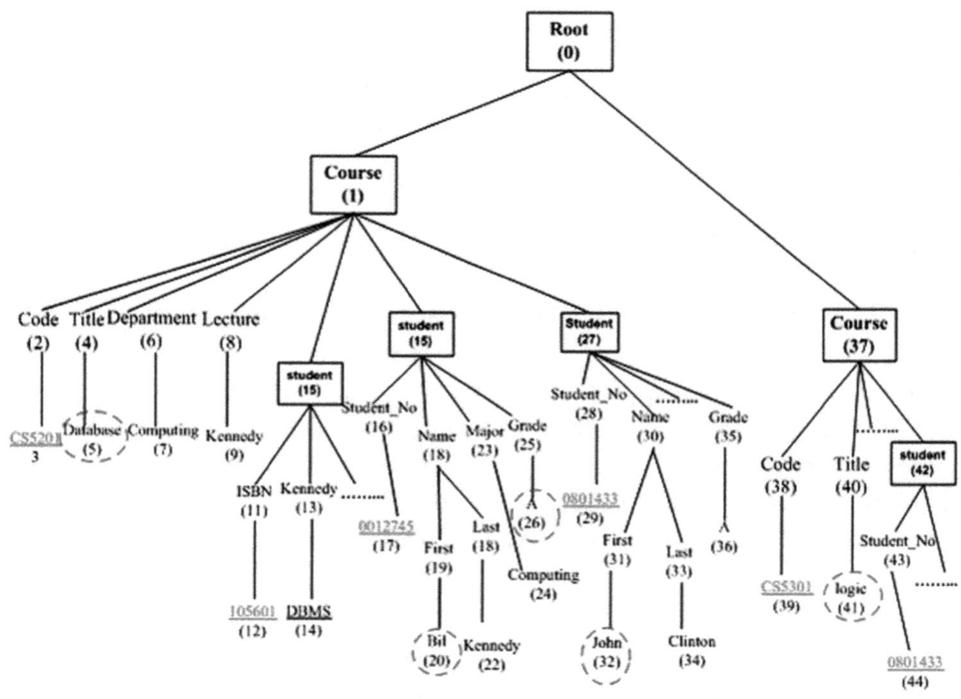

Issue 5: Schema-Dependent Answer

There could be various structural hierarchies for the same document. The XML document as illustrated in above mentioned Figure 4 could be represented as shown in Figure 7., to represent hierarchical structure for multiple object classes.

Let, Q5 be a query representing keywords {Bill, Database}. Referring to Figure 4, the LCA semantic will return Student (1). Observing another structural design as shown in Figure 7, the Course (1) is in fact obtained as query response. As illustrated in Figure 7, the query responses for various structural designs may be different, such structural designs represent precisely the same information and in fact the user is dealing with the exact same query, because the query response from the LCA semantics depends on the hierarchical structure of the considered XML document. Varied hierarchical structures might return varying query responses for the exact same query as input.

2.10 Prefix and Interval Labelling Schemes for Emerging Computation of Overflow

Dynamic XML tree is constructed based on labelling schemes. Classification for labelling is done as Multiplicative Labelling Scheme, Prefix Labelling scheme and Interval Labelling Scheme. Pre order and post order tree traversal, number pairs are assigned to node location represents labelling for Interval labelling scheme. The interval between two numbers that constitute lab identification for parent, descendant or ancestor node. Using two integer node relationship is built for labels generation. Containment labelling scheme is made used in new interval scheme labelling. The labels are allotted for suggested scheme for each node which consists of three sections: Start, End, and Position to define the node correlations. Labels are produced sequentially based on depth first tree traversal and its not supports XML document dynamically. To overcome the disadvantage for node insertion space is reserved. Space allotted will be insufficient for all nodes inserted and in few situation overflow may occur, for such situation relabeling is required. Dewey decimal coding libraries are used for Prefix Labelling Scheme (Dayananda, 2016; Samer & Siobhán, 2017) and it can be used to define the structural relationships through node label. Labels are assigned during depth first traversal of XML tree labels are assigned for prefix labelling. Each label is separated by delimiters ',' or '.'. The node's label in prefix section is the parent label, which starts from root. Dewey Encoding provides labels for dynamic update of XML documents. It requires more storage space for trees. Multiplicative Labelling Scheme uses integers as labels and uses mathematical operations such as division, modulus and multiplication. New dynamic labelling scheme is designed and it is called as vector order-centric. Prefix and interval labels are used to produce static scheme for vector order centric schemes. Node labels can't be changed for static labelling scheme even when XML tree is changed. Updation of tree can be done using two process i.e skewed and uniform insertion. Classification of two classes is done for skewed insertion, random and ordered skewed insertion. Repeatedly inserting new nodes before or after a particular node repeatedly is known as ordered insertion. Adding new nodes repeatedly between two nodes in random order is known as random skewed insertion.

3. GRAPH-BASED XML KEYWORD SEARCH

Identifier and Identifier Reference (ID/IDREF) is an XML standard and is often used in XML documents. With IDREF, XML is modeled as a graph because it is no longer a tree(Dayananda, 2016; Bhalotia et al., 2002). Existing graph techniques can be applied for XML graph-structured data such as (Dayananda, 2016; Bhalotia et al., 2002; Bao et al., 2010). Semantics applied in the existing graph-based approaches can be classified into (1) sub tree, (2) sub graph and (3) bi-directed tree and are discussed in section 3.1 to 3.3.

3.1 Subtree Based Semantics for Directed Graphs

It is natural to develop an XML datasets as a directed graph where the edges are in unambiguous context in parent-child relationships or IDREFs (reference edges). Most approaches for this kind of data model find a minimal rooted tree containing all keywords, in which the path from the root to each content node is directed. This kind of semantics includes the minimal Steiner tree semantics (Dayananda, 2016; Golenberg et al., 2008)and the distinct root semantics (Dayananda, 2016; He et al., 2007). Intuitively, these semantics are similar to the LCA semantics and they also suffer from the same problem of missing answers as the LCA semantics does. Particularly, even with IDREF, the common object appearing as the child (or the descendant in general) of two nodes cannot be found by these semantics. This is because the directed tree based semantics only search backward (i.e., follow the reversed direction of the directed edges), but never search forward to find common information which related to all matching nodes.

For example, consider query {CS1, CS2} against the directed XML graph in Figure 8, where the keywords match the two objects course 2 and course 5. Note that in this example, user matches keywords with the whole object rather than a single value node. Both pieces of information in Figure 8(a) and in Figure 8(b) are meaningful to users. Intuitively, the first one (in Figure 8(a)) means the two courses are taught by same lecturer 'Albert', and the second one (in Figure 8(b)) means there are two courses and both are taken by Student named 'Bill'. However, the directed tree based semantics only return the first one in Figure 8(a), but not able to return the student named Bill' in Figure 8(b).

3.2 Subgraph Based Semantics for Undirected Graphs

An XML document can also be modelled as an undirected graph by ignoring the direction of edges. For undirected graph, an answer is commonly either a subgraph such as the r-radius semantics (Feng et al., 2008) and the r-clique semantics (Dayananda, 2016 ; An at al., 2011) or minimum cost connected tree (Dayananda, 2016; Ding et al., 2007). These semantics can provide more answers than the directed tree based semantics do, including common descendants because they search for all directions and they not only follow the reversed direction of edges as the sub-tree based semantics do. However, they may also provide answers which can be hardly interpreted (or even meaningless) because many answers contain matching nodes which are very far in the tree or even not related at all.

Consider that an illustrative XML document is represented in Figure 8 is modelled as an undirected graph by ignoring the direction of edges. Consider keyword query = {S1,S3} where the keywords match two students. Figure 10 shows an answer under the subgraph based semantics. This answer means the two students study two courses which are both taken by another student. Intuitively, the relationship of the two students is too weak and users do not expect such answer. Although several recent work (Dayananda,2016; Feng et al., 2008) take the distance between each pair of (content) nodes into account, these work carried out still return such answer because the relationship between the two nodes may still be meaningless even if the distance between them is not far. More precisely, for this example, if we treat the weight of IDREF to be 0 and all parent-child relationships have the same size of 1, then the distance between the two content nodes is 4. Then, if the distance threshold is set to 4, the answer is returned as S2. Even if the content nodes are closer, the returned answer may be still meaningless. For example, for another keyword query {Student1, Course2}, the path student1-course1-student2-course2 is also a meaningless answer, because the relationship between student1 and course2, i.e., student1 takes a course2 which is already taken by student 2.

Figure 8. XML data graph

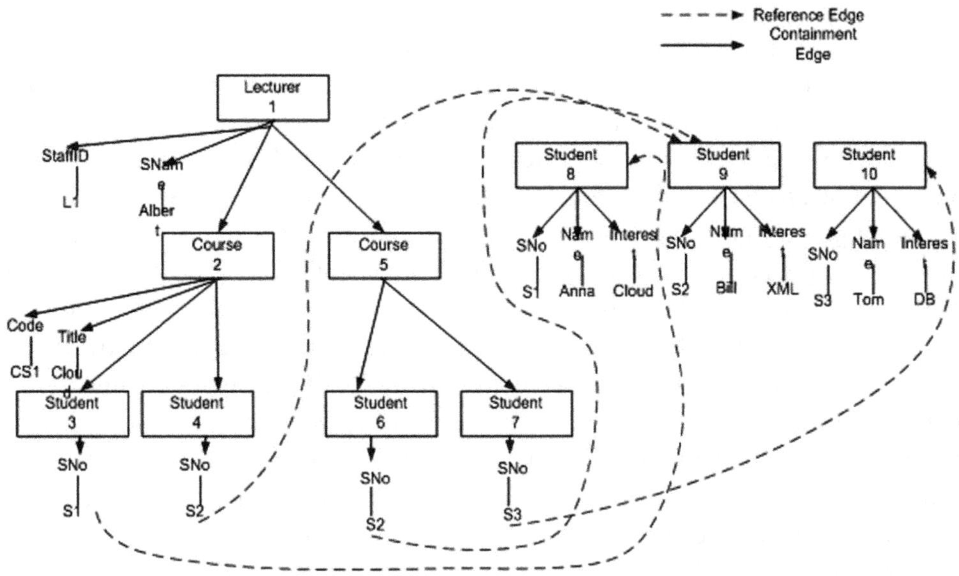

Figure 9. Illustration for query {CS1, CS2}

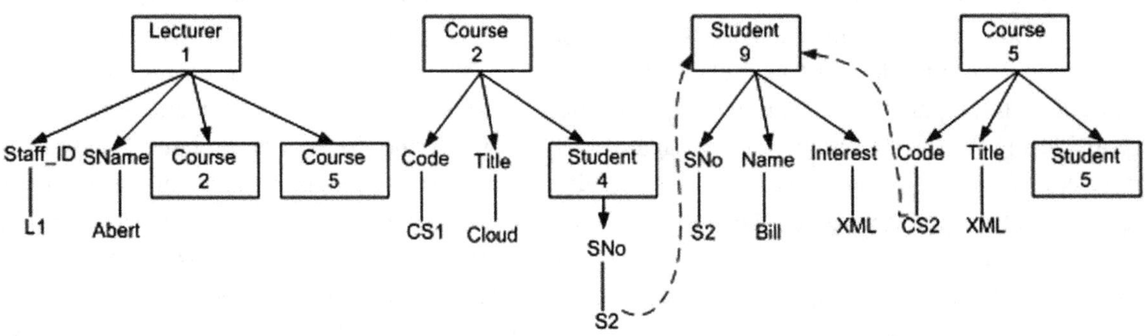

Figure 10. A meaningless answer of the subgraph based semantics

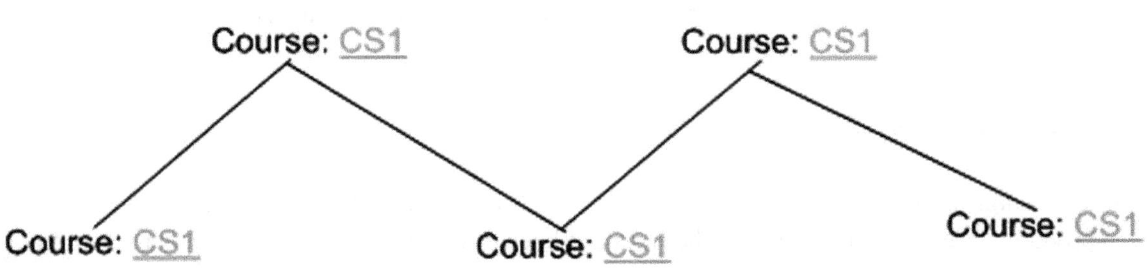

3.3 Hybrid Approach for XML References

XML emerged as the (meta) mark-up language for constituting, interchanging, and caching semi structured data. The structure of an XML document may be specified either through DTD (Document Type Definition) language or through the particular language XML Schema. While the fluency of XML Schema allows one to specify both the structure and constraints for XML documents, DTD does not allow the detailing of integrity constraints for XML documents. On the other side, DTD has a very tightly packed notation opposed to the convoluted notation and syntax of XML Schema. Thus, it becomes important to consider the issue of how to express further constraints on DTD-based XML documents, still preserving the simplicity and conciseness of DTDs. According to this framework, Xhyb focus on a (as much as possible) simple logic, named XHyb, demonstrative enough to allow the enumeration of the most common coherence and reference constraints in XML documents. In particular, we focus on constraints on ID and IDREF(S) features, which are the common way of logically connecting parts of XML documents, besides the familiar parent-child relationship of XML elements. Distinctly from other previously proposed hybrid logics, in XHyb IDREF(S) (Combi et al., 2017). features are explicitly expressible by means of suitable syntactical constructors. Moreover, we propose a rectification of the usual graph representation of XML documents in order to represent XML documents in a remote and instinctive way without flatten accessibility through IDREF(S) to the standard parent-child relationship. Model checking algorithms are then proposed, to evidence that a given XML document gratifies the considered constraints. A new hybrid logics explicitly de- fined for designating XML documents. XHyb authorises code constraints in terms of (as much as possible) elementary modal formulas. Moreover, the Kripke-style XHyb models instinctively fir the shape of XML documents, representing explicitly and definitely both the parent-child relationship and reference specification for XML document.

considering XML documents rational against a given DTD and do not comprehend DTD to allow further constraints (e.g. key constraints of ID attribute with narrower scope than the entire document). Emphasis on the specification of ID/IDREF(S) reference constraints that cannot be extracted through DTD but that sustain the original validity of XML documents. Encode interesting part of XML integrity constraints with an explicit emphasis on constraints related to ID/IDREF(S) mechanisms. This approach models checker XCheck, a simple, iteratively defined algorithm, which executing the model checking problem in theoretical setting. Model checker algorithm XCheck, as a salient step towards the automatic constraint verification. Initiating and implementing a proper extension of the DTD language for the proclamation of reference constraints, based on the provided XHyb formulas, is a further and interesting quest.

4. RESULT RANKING

From the background study presented in this chapter, it is found in the keyword search processing or query processing. Number of ranking schemes have been developed and proposed, where the emphasis has been made on query processing or keyword search over XML data.

In order to evaluate the score metrics of the query result, various ranking factors are used. Some of the significant ranking factors are discussed from section 4.1 to 4.4.

4.1 Term Frequency and Inverse Document Frequency

In the process of information retrieval, the term frequency (TF) and inverse document frequency (IDF) are employed so as to evaluate the significance of a word in the document. TF of certain term t_i in a document d_j can be retrieved using the following equation (2).

$$tf_{i,j} = \frac{n_{i,j}}{\sum_k n_{k,j}} \tag{2}$$

where $n_{i,j}$ defers the total number of incidence or the occurrences of t_i in the considered document d_j. The denominator in the above mentioned expression indicates the number of occurrences of all the terms encompassed in the document d_j. Similarly, the inverse document frequency (IDF) of term t_i can be obtained by the following equation (3).

$$idf_i = log\frac{|D|}{\left|\left\{d_j : t_i \in d_j\right\}\right|} \tag{3}$$

where $|D|$ refers to the considered document. The specific data where there is the occurrence of t_i is represented by $\left|\left\{d_j : t_i \in d_j\right\}\right|$.

As inevitable fact states that the XML query search returns the structured data, the respective two significant ranking factors represented by Term Frequency (TF) and Inverted Document Frequency (IDF) are in general estimated with an assumption that the individual node present in the retrieved result are a document. As for illustration, XSEarch (Dayananda, 2016; Cohen et al., 2003) determines the respective weight of certain leaf node for certain keyword by equation (2.4). Thus the obtained function can be represented in the same ranking factor TFIDF.

$$w\left(k,n_l\right)=tf\left(k,n_l\right)\times ilf\left(k\right) \tag{4}$$

where $tf\left(k,\ n_l\right)$ represents the term frequency of the keyword k in $k,\ n_l$.and $ilf\left(k\right)$ represents the inverse leaf frequency of keyword k represented in equation (5) and (6). More precisely,

$$tf\left(k,n_l\right) = \frac{occ\left(k,n_l\right)}{\max\{occ\left(k',n_l\right) \mid k' \in words\left(n_l\right)\}} \tag{5}$$

$$ilf\left(k\right) = log\left(\frac{\left|N\right|}{\mid \{n' \in N \mid k \in words\left(n'\right)\} \mid}\right) \tag{6}$$

where, $\mathrm{occ}\left(\mathrm{k},\ \mathrm{n}_1\right)$ represents the total number of kth occurrences in n_1, and the variable $\mathrm{words}\left(\mathrm{n}_1\right)$ represents the set of words which is present in n_1. In the above expression, N refers the set of all leaf nodes. Similarly in XBridge (Dayananda, 2016; Li et al., 2010), the respective keyword weight k was estimated with the equation,

$$\mathrm{weight}\left(\mathrm{k}\right) = \log_2\left(1 + \ \mathrm{tf}\right) \times \log_2\mathrm{ief}\left(\mathrm{n}\right),$$

where tf was considered to be unity (1); ief was considered as the total number of XML nodes over the number of nodes that encompasses the terms in the sub tree rooted at 'n' where 'n' represents the node having keyword 'k'.

4.2 Vector Space Model Ranking Factor

Vector space illustrates the algebraic framework which refers certain data by employing a defined set of features. In fact the comprising individual elements of the vector are nothing else but the word having significance in the document TFIDF. Meanwhile, a query is stated in terms of a vector, and therefore the document relevancy of certain query can be estimated with respect to the similarity existing between the query vector and the document vector. This model has been adopted this approach in XSEarch (Cohen et al., 2003). XSEarch employs these derived scores for estimating the results scores. XSEarch performs it by means of the representation of the individual node in the result and each term in the query as a vector, and further it employs the vector similarity for estimating the results relevance. Particularly, all derived vectors possess a size of $\mathrm{L} \times \mathrm{K}$ where L represents the set of all labels, and K refers the keywords.

Consider, V_n be certain node vector and V_2 represents the search term vector. Thus, the individual component of the node vector can be obtained by the following equation (7).

$$V_r\left(l,k\right) = \begin{cases} \sum_{n' \in N_{leaf}} w\left(k,n'\right) iflable\left(n\right) = l \\ 0 & otherwise \end{cases} \tag{7}$$

where $\mathrm{N}_{\mathrm{leaf}}$ represents the set of leaf offspring of n.

The individual component of the keyword search term vector is stated as following:

In case, t is having the form k and then V_t would possess 1 in every comprising components which is in relation with the k and otherwise it would result as 0. In case, t belongs to l form then V_t could possess the respective value $\mathrm{w}\left(1\right)$ in all comprising components allied with l, where $\mathrm{w}\left(1\right)$ represents TF. In case, t represents the form $1:\mathrm{k}$ then V_t possesses $\mathrm{w}\left(1\right)$ in the component in correspondent to

$\left(1,\ k\right)$. Estimating the node vectors and query terms, the query result score can be estimated as the sum of cosine similarity of nodes in the retrieved result.

4.3 Proximity Ranking Factors

Proximity ranking factors refer towards the perception that any two neighboring nodes having certain inter-relationship, can deliver the relevant response. In fact, there is an array of techniques for estimating the result propinquity. In case of Minimal Tree Semantics (MTS), the proximity can be stated as the size of individual minimal tree. In case of other semantics, where the individual result might have multiple keywords, it becomes inevitable to use varied approaches to estimate the computational proximity. For illustration, XRank estimates the root node score of the individual result and on this basis performs ranking of a result root 'v' with respect to keyword k_i. Mathematically the score vector can be obtained by the following equation (8).

$$r\left(v,k_i\right) = ElemRank\left(v_t\right) \times decay^{t-1} \quad (8)$$

where the two variables $\mathrm{ElemRank}$ and decay estimate the proximity of keyword matches and the result root that in general, exit in between 0 to 1. v_t refers to a descendant of v whose depth is t-1 higher than v; v_t possesses keyword k_i. Intuitively, farther the v is from v_t, lower rank can be achieved. In case of the multiple matches of certain specific keyword in the result, let 'm' refers keyword matches for user intended keyword k_i, then there might be m values for rank factor $\hat{r}\left(v,k_i\right)$ shown in equation (9).

Let, the obtained rank values be $r_1, r_2, \ldots, r_m$. Thus, the overall rank of r with respect to use keyword k_i would be $\hat{r}\left(v,k_i\right)$. Mathematically, it can be stated as follows:

$$\hat{r}\left(v,k_i\right) = f\left(r_1, r_2, \ldots, r_m\right) \quad (9)$$

Considering above discussion, the rank of v for the complete query can be obtained as $Q = k_1, k_2, \ldots, k_n$ that can be retrieved as shown in equation (10).

$$R\left(v,Q\right) = \left(\sum_{1 \le i \le n} \hat{r}\left(v,k_i\right) \times p\left(v_1, k_1, k_2, \ldots, k_n\right)\right) \quad (10)$$

where $p\left(v_1, k_1, k_2, \ldots, k_n\right)$ refers the proximity of all queries in the retrieved results. In order to represent certain distinct result types to the user, the algorithm proposed in (Dayananda, 2016; Li et al., 2010) returns the cluster set of the SLCA nodes. The two distinct SLCA nodes are paced in the same group only when the paths existing from the XML root to the two SLCA nodes are found to be identical where it estimates individual cluster and returns back the optimally relevant clusters. In addition, the score of

SLCA node 'v' with respect to certain user defined query $Q = k_1, k_2, \ldots, k_n$ can be estimated by the following expression (11).

$$score\left(v,Q\right) = \begin{cases} \sum_{i=1}^{n} weight\left(k_i\right) if dist^*\left(v,k_i\right) = 0 \\ \dfrac{\sum_{i=1}^{n} weight\left(k_i\right)}{\left(\sum_{i=1}^{n} dist^*\left(v,k_i\right) - \theta\right)^y}, \quad otherwise \end{cases} \tag{11}$$

where θ denotes the total iterations or occurrences of the edges iterated on the path existing between v to k_i, γ. The $\text{weight}\left(k_i\right)$ can be estimated employing approaches TF and Inverted Element Frequency (IEF). As stated in the expression, the dist^* function performs for the retrieval of the respective proximity. In this expression, the variable θ is considered for promoting the results where certain keywords possess a deep LCA.

Consider, two results R_1.and R_2 represent the results for query K_1, K_2, K_3. Considering the case where in the result R_1 the LCA of any two query keywords is SLCA, whereas in R_2 the LCA of K_1 and K_2 is deeper in comparison to the SLCA. In this case, the result R2 would possess higher θ and therefore would have higher respective scores.

The aforementioned variable for the distance from v to k_i, dist can be retrieved using the following equation (12).

$$dist^*\left(v,k_i\right) = \begin{cases} dist\left(v,k_i\right), & if dist\left(v,k_i\right) \le avgD \\ avgD + dist,\left(v,k_i\right) - avgD \times \eta otherwise \end{cases} \tag{12}$$

where, avgD represents the average depth of the XML tree with $\eta \in [0, 1]$ that represents the tuning variable. In fact, this formula plays significant role so as to avoid numerous penalties for those node which are located at much far distance from the SLCA node in the hierarchy.

Let the variable avg represent the average number of SLCA nodes in an individual group, then the group score for a group G of s number of SLCA can be obtained by the following equation (13).

$$score\left(G,Q\right) = \begin{cases} \sum_{i=1}^{s} score\left(v_i,Q\right) \times s, & if\, s \le avg \\ \sum_{i=1}^{s} score\left(v_i,Q\right) \times avg, & if\, s > avg \end{cases} \tag{13}$$

In some other approaches, there used to be higher SLCA in a group and these paradigms employ the maximum avg values so as to estimate the respective group score.

4.4 Top-K-Size Keyword Search on Tree Structured Data

An algorithm is proposed for ranking for keyword quires based on LCA size concept and this algorithm is called as top-k algorithm and it is based on stack based on tree structured data. This algorithms computes results for keyword search based on size of LCA and top-k LCA results. In this technique auxiliary index structures are not involved and it can be exploited. Defining layered semantics ranking i.e Tight LCA is done using LCA size. Top- K size LCAs selection is used for filtering semantics in keyword queries. It answers quality problem of others technique, in this it ranks only subset of results using filtering schemes. Top K size LCAs is computed using following steps.

T-LCAsz Threshold Algorithm

This algorithm is based on stack algorithm that returns maximum size order LCAs. Inverted list is given as input for the algorithm for all keywords on XML tree. Size threshold T and keyword Q query. Size of LCAs is estimated based on minimum connecting tree at corresponding LCAs with their size. This algorithm estimates min connecting tree of each LCA and it combines keyword instance into complete or partial LCA in XML tree. Threshold algorithm combines partial LCAs of subset located bottom of XML tree and superset is located higher in tree. Partial LCAs are moved towards ancestors to the size of full LCAs. Size of partial LCA is incremented to infer ancestor node. This approach combines partial LCA of subset to form superset located at higher. The threshold should exceed Maximum size when combination approach is used and size of descendant partial size should not exceed then ancestor, using this concepts. In this framework, the problem of finding all LCAs under a size threshold translates into examining all existing arrangements of keywords, using this calculation of minimum connecting tree of all instance of keyword is avoided.

T-LCAsz Analysis

T-LCAsz algorithm generates LCAs results for query keyword by combining instances of keyword from partial LCA to full LCA. Directional paths are stored in stack where processing of instances are considered for keyword query, where threshold and depth of XML is considered. A instances of keyword enters the lattice from coarseness level(stack level) and reach the last coarseness level(sink stack), depending on size comparisons with partial LCAs. In the worst case, instance will be triggered a partial LCA push action in lattice(stack). Inverted list size is linear on T-LCAsz (Aggeliki et al., 2015). Therefore, T-LCAsz is linear on the size of the inverted lists for a constant k and d. In practice, T-LCAsz performs even better than this worst case scenario. Depth of keyword instance minimum connecting trees limits on width and depth for a threshold. Execution time is reduced by filtering partial LCAs. Lower the threshold the chance filtering partial LCAs is high. Top Keyword LCAsz ranks keyword search results in ascending size order and restrict the answer to the set of LCAs with top-k sizes. The results having same ranks will be having same size with respect to query keyword as per the TLCA semantics. The target aware results is constructed based on Information gain, modifiers, specifiers, and relationships. Query contains return specifiers and modifiers as information type. It is same as return and where clause in an XQuery. The concept of information gain is used to quantify conditional entropy where information is used to define the outcome for a random variable. Using return specifier is used retrieve target entity which is not known to user. Atomicity and Intactness are the two properties that are based on target entity to retrieve relevant answers to define target aware result.

5. CONCLUSION

In this review many solution is given for retrieving relevant information from huge data over internet source or huge XML data source. In the emerging source of XML data, design of efficient and effective keyword search play a major role in Information retrieval system. The detail discussion on various LCA techniques are discussed, Object-Relationship-Attribute (ORA)-Semantics based keyword search and various ranking techniques which includes retrieving top K answer for user intended query are discussed. Annotation path helps us to retrieve matching XML scheme for effective retrieval and finally hybrid logic for XML reference is discussed for increasing reliability and efficiency in the field of IR.

REFERENCES

Aggeliki, D., Dimitri, T., & Timos, S. (2015). *Top-k-size keyword search on tree structured data.* Elsevier Information Systems; doi:10.1016/j.is.2014.07.002

An, A., & Kargar, M. (2011). Keyword search in graphs: Finding r-cliques. *Proceedings of the VLDB Endowment, 4*(10), 681-692. DOI: 10.14778/2021017.2021025

Bao, Z., Chen, Z., Guo, J., Ling, T. W., Lin, X., Wang, W., & Zhou, J. (2012). Fast SLCA and ELCA computation for XML keyword queries based on set intersection. In *IEEE 28th International Conference on Data Engineering* (pp. 905-916). IEEE. DOI: 10.1109/ICDE.2012.75

Bao, Z., Lee, M. L., Ling, T. W., & Zeng, Z. (2013). A semantic approach to keyword search over relational databases. In *International Conference on Conceptual Modeling* (pp. 241-254). Academic Press. DOI: 10.1007/978-3-642-41924-9_21

Bao, Z., Lu, J., Ling, T. W., Wu, H., & Xu, L. (2010). An effective object-level XML keyword search. In *International Conference on Database Systems for Advanced Applications* (pp. 93-109). Academic Press. DOI: 10.1007/978-3-642-12026-8_10

Bao, Z., & Wu, H. (2011). Object-oriented XML keyword search. In *International Conference on Conceptual Modeling* (pp. 402-410). Academic Press. DOI:10.1007/978-3-642-24606-7_30

Bhalotia, G., Chakrabarti, S., Hulgeri, A., Nakhe, C., & Sudarshan, S. (2002). Keyword searching and browsing in databases using BANKS. In *18th International Conference on Data Engineering, 2002. Proceedings,* (pp. 431-440). Academic Press. 10.1109/ICDE.2002.994756

Cao, J., & Nguyen, K. (2010). Exploit keyword query semantics and structure of data for effective xml keyword search. In *Proceedings of the Twenty-First Australasian Conference on Database Technologies* (*vol. 104*, pp. 133-140). Academic Press.

Chan, C. Y., Goenka, A. K., & Chan, C. (2007). Multiwayslca-based keyword search in xml data. In *Proceedings of the 16th international conference on World Wide Web* (pp. 1043-1052). Academic Press. DOI: 10.1145/1242572.1242713

Chatur, P. N., & Lambole, P. R. (2017). Effective XML Keyword Query Processing. *International Conference on Electronics, Communication and Aerospace Technology ICECA 2017*, 523-528.

Chen, L., Liu, X., & Wan, C. (2011). Returning clustered results for keyword search on XML documents. *IEEE Transactions on Knowledge and Data Engineering*, *23*(12), 1811-1825. Doi:10.1109/TKDE.2011.183

Cohen, S., Kanza, Y., Mamou, J., & Sagiv, Y. (2003). XSEarch: A semantic search engine for XML. In *Proceedings of the 29th international conference on Very large data bases* (*vol. 29*, pp. 45-56). Academic Press. 10.1016/B978-012722442-8/50013-6

Combi, C., & Masini, A., Olibon, B. I. B., & Zorzi, M. (2017). A Hybrid Logic for XML Reference Constraints. *Data & Knowledge Engineering*, 1–20. doi:10.1016/j.datak.2018.02.005

Dayananda, P., & Sowmyarani, C. N. (2016). Comprehensive study on Keyword Search in Semi Structured Data. Journal of Knowledge Society Research, 7(3), 1-24. doi:10.4018/IJKSR.2016070105

Ding, B., Lin, X., Qin, L., Wang, S., Yu, J. X., & Zhang, X. (2007). Finding top-k min-cost connected trees in databases. In *2007 IEEE 23rd International Conference on Data Engineering* (pp. 836-845). Academic Press. 10.1109/ICDE.2007.367929

Feng, J., Li, G., Ooi, B. C., Wang, J., & Zhou, L. (2008). EASE: an effective 3-in-1 keyword search method for unstructured, semi-structured and structured data. In *Proceedings of the 2008 ACM SIGMOD international conference on Management of data* (pp. 903-914). ACM. DOI: 10.1145/1376616.1376706

Feng, J., Li, G., Wang, J., & Zhou, L. (2007). Effective keyword search for valuable lcas over xml documents. In *Proceedings of the sixteenth ACM conference on Conference on information and knowledge management* (pp. 31-40). ACM. DOI: 10.1145/1321440.1321447

Golenberg, K., Kimelfeld, B., & Sagiv, Y. (2008). Keyword proximity search in complex data graphs. In *Proceedings of the 2008 ACM SIGMOD international conference on Management of data* (pp. 927-940). ACM. 10.1145/1376616.1376708

Guo, L., Shao, F., & Shanmugasundaram, J. (2003). XRANK: Ranked keyword search over XML documents. In *Proceedings of the ACM SIGMOD International conference on Management of data* (pp. 16-27). ACM. 10.1145/872757.872762

He, H., Wang, H., Yang, J., & Yu, P. S. (2007). BLINKS: ranked keyword searches on graphs. In *Proceedings of the 2007 ACM SIGMOD international conference on Management of data*,(pp. 305-316). ACM. 10.1145/1247480.1247516

Jagadish, H. V., Li, Y., & Yu, C. (2004). Schema-Free XQuery. In *Proc. of VLDB* (*vol. 30*, pp. 72-83). Academic Press.

Köpke, J. (2017). Annotation Paths for Matching XML-Schemas. *Data & Knowledge Engineering*, 1–30. doi:10.1016/j.datak.2017.12.002

Li, J., Liu, C., Wang, W., & Zhou, R. (2010). Suggestion of promising result types for XML keyword search. In *Proceedings of the 13th International Conference on Extending Database Technology* (pp. 561-572). Academic Press. 10.1145/1739041.1739108

Liu, C., Li, J., & Zhou, R. (2010). Fast ELCA computation for keyword queries on XML data. In *Proceedings of the 13th International Conference on Extending Database Technology* (pp. 549-560). Academic Press. DOI: 10.1145/1739041.1739107

Mong, L., Thuy, N. L., TokWang, L., & Zhong, Z. (2016). ORA-Semantics based Keyword Search in XML and Relational Databases. In *International Conference on data engineering* (pp. 157-160). Academic Press. DOI: 10.1109/ICDEW.2016.7495638

Papakonstantinou, Y., & Xu, Y. (2005). Efficient keyword search for smallest LCAs in XML databases. In *Proceedings of the 2005 ACM SIGMOD. International conference on Management of data* (pp. 527-538). ACM. DOI: 10.1145/1066157.1066217

Papakonstantinou, Y., & Xu, Y. (2008). Efficient LCA based keyword search in XML data. In *Proceedings of the 11th International conference on extending database technology: Advances in database technology* (pp. 535-546). ACM. DOI: 10.1145/1353343.1353408

Samer, A., & Siobhán, N. (2017). The Emergence Computation of Overflow in Dynamic X,L Tree Based on Prefix and Interval Labelling Schemes. *The International Conference on Engineering & Technology*, 21-23.

Section 4

New Skills and New Mental Models for the New Era of Education

Chapter 14
Extraction and Annotation of News Topics From TV Streams for Web Video Sharing:
A Contribution to Produce Reliable Online Video News Content

Tarek Zlitni
Sfax University, Tunisia

Walid Mahdi
Taif University, Saudi Arabia

ABSTRACT

Today, with increased internet access, users are often interested in new content-based multimedia applications of high added value such as interactive TV, video on demand (VoD), and catch-up TV services such as YouTube or Dailymotion frameworks. Despite the easy and rapid access to media information of these services, they present the risk of the wide propagation of fake news. As a solution, the authors propose that the input for these services must be from a trustworthy traditional media, precisely TV program content. So, the automatic process of TV program identification and their internal segmentation facilitate the availability of these programs. In this chapter, the major originality of the authors' approach is the use of contextual and operational characteristics of TV production rules as prior knowledge that captures the structure for recurrent TV news program content. The authors validate their approach by experiments conducted using the TRECVID dataset that demonstrate its robustness.

DOI: 10.4018/978-1-5225-7347-0.ch014

INTRODUCTION

Today, the continually increasing volumes of digital video have contributed to conceive new high added value services around television and video platforms, such as Interactive TV and VoD (Video on demand). For example, we notice a rise of Internet Protocol Television (IPTV) and the distribution of multimedia content via the Internet, and especially, a wide availability of services like VoD and catch-up TV. However, in order to increase its audience traditional media benefit from the wide spread of Internet to the characterized by the low cost and the easy access. For theses raisons, the majority of TV channels tends to rebroadcast their productions in alternative media by uploading it on VoD and catch-up TV services. In fact, if the input data for these service it is generated from TV streams content approach can help minimize "fake news" on alternative media (Kai et al., 2017), since this traditional content is considered as a truthful media source.

So that, we assume that one of the best ways to get a reliable video Web content on VoD and catch-up TV services is to get is from traditional Media, essentially TV streams. This challenge needs several content-based video operations (Le et al, 2017) such as representation, segmentation, structuring, indexing and retrieval. A typical scenario consists in identifying particular TV programs from large TV streams and then the segmentation of identified program into consistent and semantic units to be available in a short period after the satellite broadcasting. Thereafter, these units were be annotated for a better use by VoD or catch-up TV services. The program identification and their semantic and temporal internal segmentation needed by VoD or catch-up TV services that must usually be done in very short period of time.

Video segmentation, indexing and retrieval are based on techniques that extract relevant features to structure TV streams into distinct programs. The research on this topic covers several complementary areas: video analysis, video abstraction, and video retrieval (Souza et al., 2014). Video analysis addresses the extraction of low-level visual features, such as color, texture, shape and salient points. Finally, video retrieval is based on the extracted features and allows users to query and search the video database (Rossetto et al., 2017). Although many efforts have been devoted into these three areas, the accuracy of the existing systems is still far from satisfactory. Indeed, a thorough study of the state of the art shows that the proposed works are divided into two classes. The works of the first class are interested in locating programs in TV stream (Hu et al., 2011; Manson et al., 2009; Choroś, 2015b). While the works of Abduraman et al. (2011), Goyal et al. (2009), Misra et al. (2010), Liang et al. (2018), Younessian and Rajan (2015) of the second one carry out the internal structuring of TV programs. These works are suitable only for media types known in advance. However, few works offer a whole approach starting from the inter-segmentation step of programs in long TV streams to the intra-segmentation level and finally the annotation step. The last step dedicated essentially to prepare a reliable processed contend ready to be exploited by online video sharing platforms.

Based on this observation, the goal of our contribution is present a whole approach. The proposed approach aims to cover all the required steps to obtain a structured and reliable TV content well annotated for exporting to VoD and catch-up TV services (Figure 1)

The rest of the paper is structured as follows. In section 2, we present the related works and the motivation and context. Section 3 introduces the prior knowledge of media production and grammar video concept. In Section 4, we introduce the TV program identification in TV streams. Section 5 presents the topic detection of identified news programs. Section 6 introduces the segmented video annotation and preparing for Web exporting. Experiments and evaluation results are presented in Section 7. We conclude with discussions of the proposed solution and an outline of future work.

Figure 1. The proposed approach's overview.

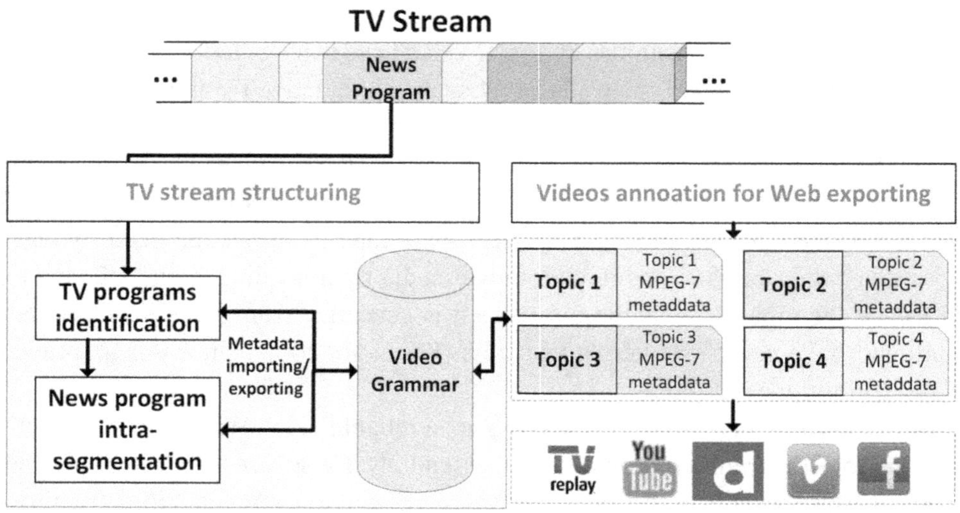

RELATED WORKS

Intensive research efforts were directed to content-based video indexing (Hu et al. 2011; Choroś, 2015b). The video documents are habitually incorporated in large TV streams. Therefore, efficient indexing of TV programs requires firstly their localization and extraction them from the stream. A class of works focuses on locating programs in TV stream (Zlitni and Mahdi 2016), whereas other kind of works is interested on the internal structuring of TV programs. Few efficient contributions have been proposed to structure TV streams. For example, to identify the various programs in TV stream, Naturel and Gros (2008) created hash-table containing key-frame signatures of program references. Then, he used signature similarity between the stream signal and the hash-table elements to locate the programs. The method presented by Manson (2009) aims to detect automatically programs in TV streams. The purpose of its work is to detect the limits of broadcasted programs. The method first detects inter-programs as repeated sequences in the TV stream in order to deduce the programs boundaries. The major drawback of these contributions is that they focus only on programs localization in TV streams without reaching the second level: the internal structuring where we can extract the semantics.

Numerous works have focused on automatic internal structuring of different TV program types. News programs are among the most interesting types of these contributions. According to recent studies, TV news present more than 50% of TV-REPLAY content. Several research teams have been interested in TV news structuring (Abduraman et al., 2011). In general, news programs are produced according to typical production rules that provide important cues for the automatic structuring (Goyal et al., 2009). The majority of the TV news indexing approaches is based on low-level primitives (texture, colors and forms) and/or high-level features such as anchorperson features (Goyal et al., 2009). Misra et al. (2010) segmented TV news into topics by detecting anchorperson shots and text transcripts Liang et al. (2018), Younessian and Rajan (2015). Anchorperson shots were detected based on color feature. Topics were identified by merging neighboring shots from anchorpersons ones based on their visual similarity. Goyal et al. (2009) exploited also anchorperson detection to structure news programs. The authors used a split-and-merge mechanism to detect topic limits. Similarly, Choroś (2015a, 2015b) examines the usefulness of the temporal aggregation method to select report shots and non-report ones in news videos. The work of Dunker et al (2008) provides information to the user according to its specific profile. TV news structuring is based on knowledge like anchorperson shots and interviews. Feature combination deduced from this knowledge is applied to deduce topic transitions. Chaisorn et al. (2006) identified story boundaries using a Hidden Markov Models (HMM) model or inductive rules. Dumont and Quénot (2012) presented an approach based on multimodal features extraction. Their contribution combines audio features like silence segments and visual features as anchorpersons or logos. They used a machine learning method to find the transitions from heterogeneous low-level features. Video analyses discussed in this section as well as in this research are the methods using visual features only. There are also audio-visual approaches analyzing not only visual information but also audio (Asghar et al., 2014; El Khoury et al., 2014). The Machine Learning model required a lot of manually annotated data or features captured from different sources and modalities. In this case, selecting and fusing features available is often a crucial problem. For example, in the work of Younessian and Rajan (2015), textual and visual information comes from different sources of knowledge and consequently different feature spaces, are considered heterogeneous features, and are not directly comparable. The fusion becomes even more difficult when different features have different notions of similarity.

Another limitation is that the internal structuring systems are suitable only for some media types known in advance and pre-extracts of TV streams, and these two conditions are not satisfied in most cases. Currently, there is no work that covers both the inter-segmentation and intra-segmentation steps. Furthermore, a common and very important problem is that in a general setting, the low-level features do not have a direct link to high-level concepts. This raises the question: *how this semantic gap can be bridged*?

The goal of our contribution is to identify and segment news programs incorporated in TV streams. The proposed work uses the rules of TV channel production as prior knowledge. These characteristics are modeled as video grammar able to capture high-level concepts related to the automatic identification and structuring process in the case of TV news programs. Thus, a two-step segmentation of TV stream is required. The first step aims to identify news programs in channel streams and the second one is addressed to detect news program topics.

KNOWLEDGE MODELING OF TV PRODUCTION RULES

Along with such a context, any discussion of knowledge modeling should also include a reference to the kinds of knowledge that can be represented in the knowledge base of an intelligent system, as well as to the basic representation and design techniques. In our case, the kinds of knowledge we are interested refer to the digital TV channel production rules. Actually TV channels broadcast for many hours different programs. Although the resulting streams are long and heterogeneous, inter-programs between these programs are recurrent. Let's take a deeper look into recurrent programs: TV news generic is almost invariant. For many years, TV news start usually with a brief outline of the reports (outline), followed by an alternation of anchorperson's announcement of the upcoming topics and news report (news content). Thus, it would be insignificant to repeat the treatment of its detection at each of its occurrences. So, the basic idea of this concept is to represent the repetition boundaries cues between programs formally according to the well-known principle "*Model once, use multiple times*". More specifically, the approach aims to extract features of a single instance representing other instances of the same unit. The definition of features and inference rules for concepts are described as a video grammar. The video grammar incorporates concept description rules that describe TV Stream content and structure. Algorithms used for the semantic and temporal unit's segmentation of large TV streams are then based on grammar elements.

The main idea is to model recurring visual cues appearing in a stream of a particular TV channel. These cues will be designed only once, and used later, repeatedly in the process of segmentation. A temporal or semantic unit of a particular program is identified in a TV stream if and only if there is an instantiation of one or more grammar rules associated with some features. For example, the identification of anchorperson in news programs is conditioned first by the appearance of a person shot in a TV stream and then validated by other features as the anchorperson face (Le et al, 2017) and studio décor, stored in advance in the grammar. The grammar elements (concepts and rules) are prior knowledge on which our approach of analyzing TV streams to identify typical content structure of TV news program is based.

KNOWLEDGE-BASED GRAMMAR DESIGN

To design a concise grammar with enough structural information, we conduct a study on the production and streaming techniques based on prior knowledge of common TV news program. This study allowed us to extract the most used structure rules, and then exploring them as prior knowledge at the time of TV stream structuring process. Based on this study, we present a set of TV production and streaming knowledge, more specifically concepts relating to locating news programs and intra-segmentation. In the following, we introduce the several concepts and roles proposed for video grammar. This grammar is designed based on descriptive logic rules. This logic allows new concepts and roles to be built using a diversity of constructors (Krötzsch et al., 2012). Concept and role constructors depend on whether concept or role expressions are constructed. For the concepts, we distinguish basic Boolean constructors, role restrictions and enumerations.

DEFINITION OF CONCEPTS

The goal of this step is to define the adequate concepts. Two classes of concepts are defined: generic concepts, applied to any type of program and specific concepts applied to news programs:

- **Segment:** A video segment is a set successive frames in a video stream, characterized with a start and an end time point.
- **Shot:** A series of frames, that runs for an uninterrupted period of time: the moment that the camera starts rolling until the moment it stops. The shot is a continuous sequence between two transitions or cuts.
- **Generic:** Invariant element (segment) that bounds the beginning and the end of a program. This concept is characterized by a set of low-level primitives that differentiates it from other video segments.
- **Program:** A video segment bounded by a starting and an end generics, and having a specific media content. It can be a composite element by other semantic units or a simple element that form entirety the program.

- **News Program (NP):** A particular program, presented by one or two anchorpersons, dedicated to introduce the news (international, sports, economic ...). The NP is defined as a program whose semantic units are topics.
- **Topic:** The fundamental entity that composes news programs. This entity starts with an anchorperson shot that introduce the topic and it is characterized by a specific temporal tempo.
- **AnchorPerson:** The person in charge to present a NP. Generally, only one or maximum two anchorpersons in the same NP.
- **AnchorPersonShot:** The appearance of an anchorperson shot with some conditions (like duration) is generally a thematic (topic) change clue.
- **Decor:** Defines the set of decorative elements and special effects used to identify visually a scene.
- **Signature:** This concept has an entirely technical role to guarantee the uniqueness of the generic instances by encoding their characteristics with visual primitives.
- **Threshold:** Allows representing numerical thresholds used to validate some concepts. As the instances of this concept are used for defining other concepts (e.g. topic), this concept is defined by the constructor Enumeration (O) containing the set of used thresholds values.

DEFINITION OF ROLES

The roles are defined to model the relationships between concepts. They represent the links between instances of concepts. The roles can be interpreted as binary relations.

Starts_with: *A temporal relationship between two segment instances A, B . The segment B is the first part of segment A.*
Ends_with: *reciprocal relationship of* ***starts_with****. The segment B is the last part of segment A.*
follwed_by: *A relationship that expresses the temporal direct adjacency between two segment instances.*
Duration: *Determines the duration of a video segment.*

Thus, we get the following concepts formalization:

$$\text{Program} \equiv \text{Segment} \sqcap \text{starts\_with .Generic.} \top \sqcap \text{ends\_with .Generic.} \top$$

$$\text{NP} \equiv \text{Program} \sqcap \text{starts\_with: GENERIC\_START\_NP} \sqcap \text{ends\_with: GENERIC\_END\_NP} \sqcap \exists \text{TOPIC.} \top \sqcap \exists \text{Duration.} \geq \mathit{THRE\_NP}$$

$$\text{Topic} \equiv \text{Segment} \sqcap \exists \text{Duration.} \geq \mathit{THRE\_TOPIC} \sqcap \text{starts\_with . AnchorPersonShot} \sqcap \text{follwed\_by .Report}$$

$$\text{Report} \equiv \text{Segment} \sqcap \exists \text{Shot.} \top \sqcap \text{follwed\_by . AnchorPersonShot}$$

$$\text{AnchorPersonShot} \equiv \text{Shot} \sqcap \exists \text{ AnchorPerson.} \top \sqcap \text{ Duration.} \geq \mathit{THRE}\text{\_ANCHORPERSON}$$

$$\text{AnchorPerson} \equiv \exists \text{decor: STUDIO} \sqcap \leq 2 \text{Person} \sqcap \text{Person.vue:FACE}$$

Signature ≡ ∃VisualDescriptor:COLOR⊓∃ VisualDescriptor:POI

THRESHOLD ≡ {*THRE_TOPIC*,THRE_*ANCHORPERSON*, *THRE_NP*}

With ***ABox*** contains:

{Generic (GENERIC_START_NP), Generic(GENERIC_END_NP), Decor(STUDIO), VisualDescriptor (COLOR), VisualDescriptor (POI)}

OVERVIEW OF TV STREAM STRUCTURING PROCESS

Our contribution is based on prior knowledge modeled in video grammar and governing both TV channels rules of production and streaming process. This knowledge is modeled once, but exploited several times during the process of TV stream structuring and segmentation. Based on this grammar, we propose a process of inter-segmentation to identify given news programs (***NP***) within large TV streams. This process relies on matching visual feature signatures already stored in the grammar and those captured from an input TV streams.

After the inter-segmentation step, the second step proposes to identify the internal structure of the identified programs by the first step. As already mentioned, we focus, at this step, on the case of ***NP***. ***NPs*** are characterized by a large audience and almost invariable syntactic structure. This contribution takes its originality from the exploitation of structural characteristics that govern the organization of the ***NP*** content.

TV STREAM SEGMENTATION: CASE OF NEWS PROGRAM IDENTIFICATION

NPs identification proceeds mainly in two steps. The first one aims to pile a video grammar by concepts that help to identify typical structures of TV news. This step relies on a TV news graphic design especially ***NP*** visual jingles. The grammar is expressed by feature set deduced from the production rules used by each ***NP***. For example, for TV news grammars, the segmentation of a ***NP*** content requires the apparition of the features of anchorperson's shots (decor, appearance of anchorperson) to detect the topics. The second step identifies news program occurrences in the TV stream based on the similarity measurement of grammar signatures.

GRAMMAR SIGNATURE FOR NP JINGLE

NP jingles are visual markers whose principal role is to visually identify their boundaries during a long period. As all instances of ***NP*** have the same visual markers, we represent the visual characteristics of an instance to identify the other ones afterwards. We propose a new spatial-temporal method to create a signature for programs identification (1).

$$Vsig = \left(\left[\overrightarrow{SigF}\right], N_{frame}, T_{step}\right) \tag{1}$$

The signature creation process starts by selecting a set of frames and a temporal step (T_{step}) which separates them. For each selected frame, we compute ***SigF***, the visual features extracted from this frame. Numerous techniques were proposed to create image signature (Athanasakos et al., 2007; Lakshmi et al., 2008). To ensure the signature robustness, we opt to exploit a colorimetric feature and an interest point descriptor.

Colorimetric Feature

Color histograms are popular colorimetric features used to represent images in several multimedia applications. However, histograms do not consider spatial information, thus images with very different appearance can have similar histograms. So, we opt to use Color Coherence Vector (CCV): a histogram-based descriptor that incorporates spatial information. The CCV histogram buckets are subdivided according to spatial color coherence (Ravani et al., 2010). Each image pixel is classified as coherent if it is part of a large similarly-homogeneous color region or incoherent otherwise. CCV stores the number of coherent versus incoherent pixels for each color.

The color coherency threshold is fixed to 1% of the image size. The CCV computing is composed essentially of three steps. The first step eliminates small variations between adjacent pixels using a mean filter. Then, we proceed to a discretization of color space, to obtain only N_{CCV} distinct colors to decrease the luminosity variation, and reduce the size of frame signature. For this segmentation, we adopt the Statistical Region Merging (SRM) method (Nock and Nielsen, 2004). The advantages of this method are simplicity and performance without the use of color space transformations. The image is segmented according to CCV buckets. After the image segmentation process, each pixel belongs only to a single region. A pixel is coherent if the size of the region that it belongs to exceed a coherency threshold, otherwise, the pixel is incoherent. αj and βj are respectively the numbers of coherent and incoherent pixels of the jth color. CCV consists of $((\alpha 1,\beta 1),\ldots,(\alpha Nc,\beta Nc))$. So, the set of bins represents the CCV frame signature (2).

$$SigF_{CCV} = \left\{\overrightarrow{CCV_i}\left(\alpha_i, \beta_i\right), i \in \left[1..N_{CCV}\right]\right\} \tag{2}$$

Points of Interest (POI) Feature

The Point Of Interest (*POI*) is a point in an image that has outstanding properties in comparison to its adjacent points. Also, *POIs* are characterized by the robustness to blur, luminosity variation and geometric transformations. Thus, using a *POI* descriptor provides a rich and compact signature at once. Furthermore, studies of evaluation of different *POI* detectors under different conditions like luminosity variation or image resizing proved that better results are obtained by the *Scale Invariant Feature Transform* (SIFT) detector (Lowe, 2004). The POI's frame signature is expressed as follows (3):

$$SigF_{POI} = \left\{\overrightarrow{POI_k}, k \in \left[1..N_{POI}\right]\right\} \quad (3)$$

NEWS PROGRAMS LOCATING

For locating news programs we propose two steps. First we use feature weights to assign relative importance to each feature within the grammar. Second we proceed by locating ***NPs*** based on a visual similarity comparison of the features captured at time t from TV stream and those saved in video grammar. More details on these two steps are given in the following sections.

Weights of Grammar Features

Since a frame has a composed-signature ($Sigf_{CCV}$ and $Sigf_{POI}$), we combine their similarities to obtain a single decision value (4). This combination aims at increasing the identification rates. The features similarity combination is calculated using the average rule, following normalizing coefficients of two signature features.

$$simSigF\left(k,t\right) = \frac{w_1 \times simSigF_{CCV}\left(k,t\right) + w_2 \times simSigF_{POI}\left(k,t\right)}{w_1 + w_2} \quad (4)$$

According to our experiments, we conclude that a discriminative descriptor d differs from a channel to another. This difference is explained mainly by the fact that a visual jingles program exploits graphic compositions rich in a particular descriptor rather than another. Therefore, the weighted coefficient w_d of the descriptor d must be relative to its efficiency of programs identification to indicate the relevance of d. The importance of the descriptor d is evaluated by its capacity to identify the maximum number of correct identifications. d is significant if both recall and precision have high values at once. So, w_d is a combination formula of these two metrics (5).

$$w_d = F1 = \frac{2 \times Recall \times Precision}{Recall + Precision} \quad (5)$$

To select the optimal weight value for the descriptor d of each TV program/channel, we conduct a training stage. This training is carried out on ground truth composed of a set of visual jingles segments, with an average duration equal to 35 seconds. The ground-truth dataset were randomly selected from a subset of the news programs.

Table 1 summarizes the weight values for various TV channels.

Table 1. W_1 and w_2 values for each channel.

TV Channel	Recall (R)		Precision (P)		w_d (F1)	
	d_1	d_2	d_1	d_2	w_1	w_2
M6	0.8	0.91	1	0.92	0.89	0.91
RTV	0.95	0.76	0.75	0.87	0.83	0.81
LCI	0.99	0.73	0.98	0.9	0.98	0.80
Itele	0.83	0.71	0.67	0.81	0.74	0.75
France 24	1	0.66	0.84	0.84	0.91	0.74

News Program Identification

NP identification relies on detecting the start time of this program in TV streams. The basic idea of localization consists in comparing the visual similarity of the video signal at time ***t*** in a TV stream and the ***NPs*** signatures in video grammar. To identify the video type at the current time ***t***, we measure the similarity between the signature created at this time (current frame) and all ***NPs*** signatures in video grammar. Locating ***NP*** is based on the measurement of signature similarity respectively between the $\boldsymbol{SigF_{CCV}}$ and $\boldsymbol{SigF_{POI}}$ features in video grammar and the TV stream signal. For $\boldsymbol{SigF_{CCV}}$, we calculate a normalized average of *CCV* bins similarities defined in frame signature (6).

$$SimSigF_{CCV}\left(k,j\right)=\frac{\Sigma\left(\overrightarrow{CCV_k},\overrightarrow{CCV_j}\right)-\left(\sum_{i=1}^{N_{CCV}}\left|\alpha_k\left(i\right)-\alpha_j\left(i\right)\right|+\left|\beta_k\left(i\right)-\beta\alpha_j\left(i\right)\right|\right)}{\Sigma\left(\overrightarrow{CCV_k},\overrightarrow{CCV_j}\right)} \quad (6)$$

The $SigF_{POI}$ signature similarity measurement (7), is calculated by an average distance: *DistSIFT* (FS,GS), proposed by Hua et al. (2012), comparing SIFT features for stream frames (FS) and the ones extracted from the video grammar (GS). This measurement is suitable for the case of resized images. Thus, we choose this measurement since we only treat the case of resizing without orientation variation.

$$Sim\left(\overrightarrow{POI_k},\overrightarrow{POI_j}\right)=\begin{cases}1, & if\ Dist_{SIFT}\leq Thre_{SIFT}\\0, & otherwise\end{cases} \quad (7)$$

with:

$$Dist_{SIFT}\left(FS,GS\right)=\frac{1}{m}\sum_{i=1}^{m}\left(\min_{1\leq j\leq n}\sqrt{\sum_{k=1}^{128}\left(FS_{ik}-GS_{jk}\right)^2}\right)$$

where

- *m*: the number of SIFT features of *FS,*
- *n*: the number of SIFT features of *GS,*
- FS_{ik}: the kth element of the ith feature vector of F*S,*
- GS_{jk}: the kth element of the jth feature vector of *GS*.

TV NEWS PROGRAMS INTRA-SEGMENTATION

The choice to structure ***NPs*** is essentially justified by two reasons. On the one hand, they are produced for and watched by a large public of TV viewers thanks to its pertinent information. On the other hand, despite the style diversities from a TV channel to another, they are generally produced according to relatively regular rules of production that offers significant cues for automatic content structuring (Jacobs et al., 2004; Misra et al., 2010).

NP Production Knowledge

According to production knowledge of TV channels, a news program has an operational structure defined by three main characteristics that manage the intra-content structure of ***NP***. We note that these characteristics are not part of the grammar video rules, but rather are used to identify key concepts of the grammar:

- **Structural Characteristic:** Based on the studio/topic concept where the anchorperson is the principal actor that plays the role to announce the topics changes. Each time a new topic is started, the anchorperson appears to introduce it.
- **Framing Characteristic:** Highly related to the structural one and it conditions the framing of the anchorperson shot according to the golden number principle. The golden principle creates imaginary lines, called strength lines that slice the image into three equal horizontal and vertical regions (David, 1991). The intersections of these 4 lines emphasize the key points of the shot: the points within that the spectator eye is attracted, therefore areas where the anchorperson must be framed.
- **Design Characteristic:** The studio décor of each TV channel plays an essential role to transmit to the spectators the concept of visual identity referring to particular TV news. This identity is created by the design of appropriate graphics for a long duration.

Anchorperson Shots Modeling

The Anchorperson shot is an important cue to extract the structure of ***NPs***. We propose in this section our approach composed of five modules for both constructing and characterizing models for each kind of anchorperson in different news programs.

Shot Detection and Face Shot Filtering

We proceed, at first, by the detection of all shots. In most cases of ***NPs***, the type of shot boundaries is "*hard cut*". In our work, we opted for the ECR technique thanks to its performance for this type of shots boundary. This technique relies on the fact that the edges of the objects within the frames would definitely change across a boundary. In other words, temporal visual discontinuity always comes with structural discontinuity. The metrics *TnearFar* and *TfarLastFar* used in ECR are obtained by a ratio of pixels numbers of two adjacent frames. Therefore, if the frames belong to the same shot, their values are near to 1.

According to our experiments, we adopt this technique to fix *TnearFar* and *TfarLastFar* thresholds to 2 and 3.1 respectively. Thus, *TnearFar* and *TfarLastFar* values that exceed these two thresholds indicate a shot detection.

To filter face shots, we used the AdaBoost algorithm based on Haar feature. These features are functions allowing to set the difference in contrast between several contiguous rectangular areas in an image. Fourteen filters of rectangular forms were used to code existing contrasts in a face and the space relations by computing the difference between the sum of the pixels in the white regions and the sum of the black ones. Thus, robust classifiers were made-up with several weak classifiers by the AdaBoost algorithm. The relevant classifiers can be combined in a cascade structure. A positive result of the first classifier started the evaluation by a second classifier that in turn started the following classifier. A negative detection of a classifier involved a rejection of the window.

Anchorperson Shot Identification

The anchorperson shots detection is based on the identifying of the shots where their faces appear. For this, we proceeded by a bi-clusters classification of all faces according to a similarity measurement. The goal is to classify similar faces into two clusters: C_1 for anchorperson faces and C_2 for other ones. Thus, the anchorperson face belongs to the large cluster. Each face image is considered as a vector in a high-dimensional space equal to the number of image pixels. So we had to deal with a problem of pattern comparison in high-dimensional spaces. Significant improvements are achieved in this case by using a mapping of all faces into a lower-dimensionality space based on the Principal Component Analysis (PCA) to decrease the dimensionality and retain as much as possible of the variation of the original dataset.

The best low-dimensional space W_k is determined from the covariance matrix $A \times A^T$ (8).

$$A \times A^T = \left[X_1 - \bar{X}, \ldots, X_n - \bar{X}\right] \rightarrow W_K = \left[U_1, \ldots, U_K\right] \tag{8}$$

After reducing the dimensionality, we projected each face image in this space w_k (9).

$$X^{proj} = W_K^T * \left(X - \bar{X}\right) \tag{9}$$

Then, we compute the reconstructed face image from the original space (10).

$$X^{recons} = X + W_K^T * X^{proj} \tag{10}$$

The reconstruction error (11) is the distance between the original face image and the reconstructed face image.

$$E^{recons}\left(X\right)=\sqrt{\sum X\left(p\right)-X^{recons}\left(p\right)} \tag{11}$$

To segregate elements of C_1 and C_2, we exploited the E^{recons}. Indeed, the images corresponding to the presenter faces were logically those that had a low E^{recons} value. So, the best E^{recons} is the value that separates$C_1$and C_2.

Experimental results proved that all faces having E^{recons} inferior than a threshold are anchorperson faces.

Background Detection

The anchorperson shots show particular characteristics. There is no movement of camera, framing is always similar and generally, the background is fixed or varies very little. Accordingly, the structuring of anchorperson shot requires initially a stage of background detection and then the selection of the interest zones. Thus, we suggest describing the space structure of the anchorperson shot through the automatic selection of certain interest zones of the background by applying an anchorperson pattern. To locate the anchorperson body, we first start with the location of his face. Thereafter, we proceed to locate the rest of the body. The pattern is deduced following a study of Bogin and Varela-Silva (2010) for the Human Body proportion where the size of the shoulders is between 2 and 2 + 1/3 of the width of the head (Figure 2.a).Background zones are selected with a little shift distance to the body pattern (Figure 2.d), to avoid some noise due to the little motion of anchorperson and its hair area.

Figure 2. Anchorperson pattern and studio background detection.

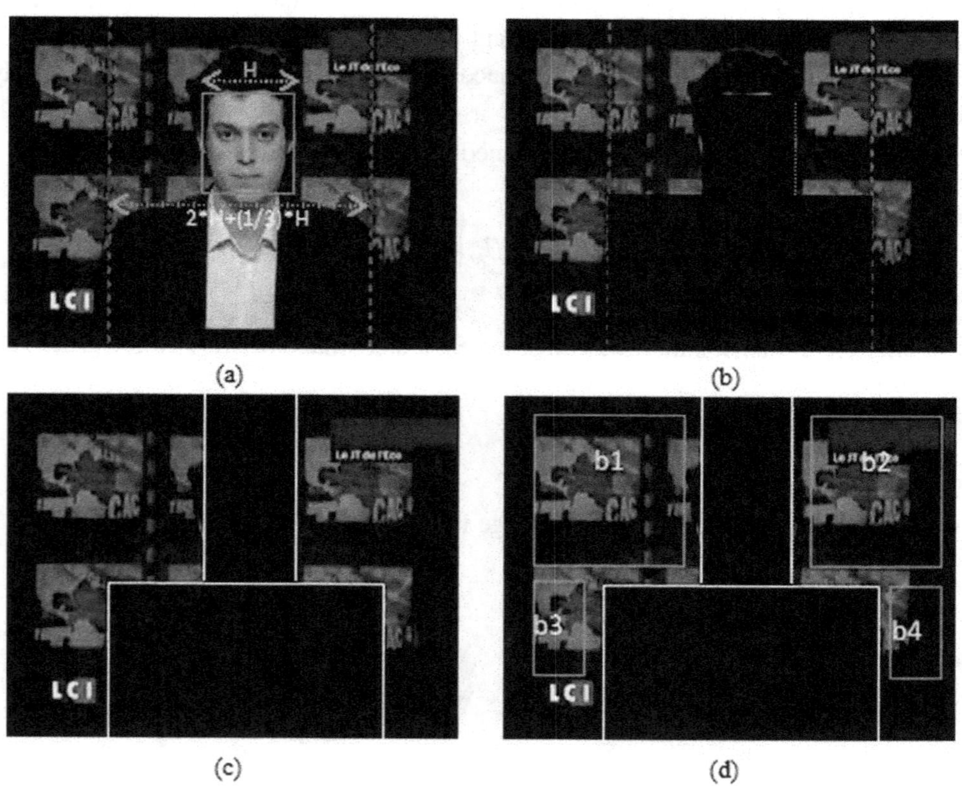

Background Characterization

Once located, each background zone b_j is characterized by five texture features. A co-occurrence matrix, detailed by Arvis et al. (2004), is calculated for each zone using 5 metrics: average (12), energy (13), inertia (14), entropy (15) and differential inverse moment (16).

$$average = \sum_x \sum_y coo(x,y) \quad (12)$$

$$energy = \sum_x \sum_y coo(x,y)^2 \quad (13)$$

$$inertia = \sum_x \sum_y (x-y)^2 \, coo(x,y) \quad (14)$$

$$Entropy = -\sum_x \sum_y coo(x,y) \log\left(coo(x,y)\right) \quad (15)$$

$$differential\,inverse\,Moment = \sum_x \sum_y \frac{1}{1+(x-y)^2} coo(x,y) \quad (16)$$

To highlight the importance of the descriptor f_{ij}, we pondered it with a weight w_{fij} (17).

$$w_{f_{ij}} = \frac{\left(1 - \frac{\delta_{f_{ij}}}{\sum_N \sigma_{f_{ij}}}\right)}{\sum_N \left(1 - \frac{\delta_{f_{ij}}}{\sum_N \sigma_{f_{ij}}}\right)} \quad (17)$$

where δf_{ij} is the standard deviation measuring dispersion, around the average, of N computed values for the descriptor f_{ij} from N anchorperson shot. Thus, the larger the standard deviation value is, the less important the descriptor f_{ij} is and thus the more insignificant the value of the weight W_{fij} is. Accordingly, the weights W_{fj} should satisfy the following constraint (18).

$$\sum_{f_{ij}} w_{f_{ij}} = 1 \tag{18}$$

Specific NPs: Case of 2 Anchorpersons' Programs

Actually, a new style of TV news appears. *NPs* of this style are presented by two anchorpersons. In this case, all faces within the program are grouped in three clusters, one cluster for each anchorperson and the third for the rest of faces in the program.

To treat this particular case, we proceed as follows. After the identification of the first anchorperson, all the faces belonging to its cluster are removed from the whole set of faces $\left(F = F - \left\{\forall f_i \mid f_i \in P_1\right\}\right)$ and the classification process is run again. Following the second classification's iteration, two cases could be obtained. If there is another class, it means that the news program includes a second anchorperson. Otherwise, the classification process is stopped.

Topic Segmentation Improvement

Even though the program is presented by two anchorpersons, the topic details are presented by a large shot including only the anchorperson who is talking. According to this assumption, adjacent anchorperson shots are merged. In fact, after identifying all anchorperson shots, we sort them according to their start's time.

All adjacent shots are compared to a temporal threshold, fixed at 25 seconds according to the study of Misra et al. (2010). Each short ***shot i*** whose duration is inferior to a certain threshold, is merged to its adjacent ***shot i+1***. So, we obtain a single shot that describes a specific topic.

The content structuring of the ***NP*** is carried out according to two modes, the *stand-alone* mode or the *rewire* mode.

The *stand-alone* mode is exploited *off-line* on the first input of the program to our system. In this case, all the steps of the structure learning stage are carried out. The topic segmentation is consequently a direct exploitation of the results provided in output of the identification stage of the anchorperson shots.

Videos Annotation and Preparing Web Exporting

After the achievement of TV programs identification step and then the TV news structuring stage, we obtain a separated the segmented news topics. Despite these ones are extracted from a trustworthy source, it will be preferable an additional annotations. Thus, for Web exportation, we opt to add a metadata information. These metadata is very important, mainly for improving the quality of indexing by search engines on the Web. Consequently, it leads to an improvement of the ranking of these videos in the results of research on the Web and decrease also the risk of fake news.

In our approach, the metadata are extracted from the video grammar instance. To standardize the descriptions of these metadata, we use the MPEG-7, a standard that facilitates indexing and searching by the content of multimedia documents. This standard contains a rich set of descriptors. MPEG7 proposes also an architecture that combines descriptors related to the structure of the image and the semantic entities that compose it. The standard includes a set of descriptors suitable for multimedia applications

Figure 3. Example of a generated MPEG-7 metadata file

```xml
<?xml version="1.0" encoding="UTF-8"?>
<Grammar xmlns:mpeg7="urn:mpeg:mpeg7:schema:2001"
chame="TV5" date="14-4-2017"
xmlns:xsi="http://www.w3.org/2001/XMLSchema-instance"
SchemaLocation="schemaGrammar.xsd">
 <Programs>
 <Program type="News Story">
 <Generic type="GENERIC_START_NP">
        <MediaTime>
        <MediaTimePoint>T00:00:00</MediaTimePoint>
        <MediaDuration>PT18S</MediaDuration>
        </MediaTime>
        <MediaProfile>
        <MediaInstance>
        <InstanceIdentifier>starting generic of NP of TV5
        </InstanceIdentifier>
        <MediaLocator>
        <MediaUri>start_NP_tv5.avi</MediaUri>
        </MediaLocator>
        </MediaInstance>
        </MediaProfile>
        <Signature>
        <VisualDescriptor type="POI">
        ...
        </VisualDescriptor>
        <VisualDescriptor type="CCV">
        ....
        </VisualDescriptor>
        </Signature>
 </Generic>

 <Anchorpersons>
 <Anchorperson>
        <Name>
        <GivenName>Laurent</GivenName>
        <FamilyName>DEBESSE</FamilyName>
        </Name>
        <Face>
        <ObjectList components="visual">
        <VisualSummaryComponent>
        <VideoSourceLocator>
        <MediaTime>
        <MediaTimePoint>T00:00:10:001F1000</MediaTimePoint>

        <MediaDuration>PT1S8N10F</MediaDuration>
        </MediaTime>
        </VideoSourceLocator>
        <Region>
        <Box mpeg7:dim="2 4">46 46 46 77 85 77 85 46
        </Box>
              </Region>
              </VisualSummaryComponent>
              </ObjectList>
              </Face>
        </Anchorperson>
        </Anchorpersons>
        </Program>
        </Programs>
        </Grammar>
```

content-based indexing and retrieval. This is ensured by a set of descriptors that encode visual characteristics such as shape, texture, color and object motion.

The generated MPEG-7 instances are structured in tow levels (Figure 3). The first one describes the channel (Channel Name, Description, date of Stream, list of Programs …). The second level describes the News story (generic file, visual descriptors, duration, anchorperson(s), anchorperson 'face regions ...).

EXPERIMENTAL STUDIES

In this section we have chosen to follow a three-round description of experiments we conducted. The first one addresses results obtained for inter-segmentation process and the second one is dedicated to the intra-segmentation process.

Experiments of TV Stream Inter-Segmentation

To evaluate the TV stream segmentation, we used a varied dataset composed of streams from 5 different TV channels. This dataset was collected from several days of streaming. A variety of programs and inter-programs (news, sport, varieties, documentaries, pubs…) are contained in these streams. The dataset is composed of SDTV quality videos with a frame rate equal to 25. Videos resolutions are varied (512X288, 320X240 and 640X360). Table 2 recapitulates the dataset description and the identification results of TV stream segmentation level.

Table 2. Experimental results of TV stream segmentation.

TV Channel	#programs	#identifications	#correct identifications
M6	48	46	45
RTV	52	44	43
LCI	112	114	101
Itele	78	75	65
France 24	66	70	63
All channels	356	349	317

First, we experimented the inter-segmentation level using each descriptor separately. We note that CCV descriptor offers better recall rates in most cases than POI descriptor. Though, the precision rates are better using the POI descriptor (Figure 4).

To improve the stage of locating programs, we conducted a second experiment combining both features. As shown in Figure 4, results prove that, for the most channels (except LCI Channel), best identification rates are obtained with a weighted fusion of POI and CCV features. We note that the CCV descriptor gives better recall rate than the POI one, in most cases. However, the precision rate is higher using the POI descriptor. According to this observation, we opt to weight the two features using the metric F1. This metric takes, at once, advantage of the performance of precision and recall rates. For a better evaluation of the relevance of the proposed features for locating programs, we opted to use the ROC curve method. This method is defined specially for two classes (positive and negative) classification systems. Thus, it determinates the ability of a particular system to separate positive and negative elements. In our work's

Figure 4. Comparison between rates of POI and CCV features

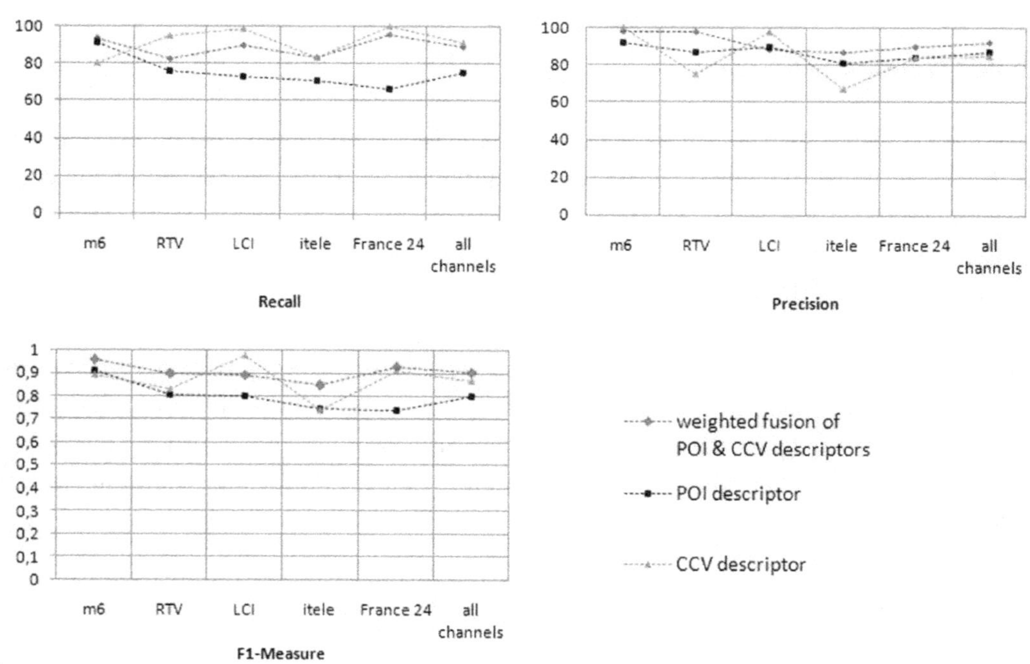

context, a positive element corresponds to a program which has been identified correctly, otherwise, is considered as a negative element. So, we use the ROC curve to represent graphically the performance of our system of TV programs locating. The ROC method is essentially based on four metrics:

- *Correct Accepting Rate (CAR)*
- *Correct Rejection Rate (CRR)*
- *False Rejection Rate (FRR).*
- *False Acceptance Rate (FAR)*

The curve represents, then, the evolution of CAR values according to the FAR values (Figure 5a).

The results revealed from the ROC curve justify the effectiveness of POI and CCV features to identify the TV programs types from a threshold value equal to 0.57. These results confirm also that a better program type's identification is achieved using a weighted combination of the two features. Global results of programs identification are more detailed in Figure 5b.

In this experimental results, although the high precision value (92, 16%), we conclude that the recall was rather satisfactory (89, 08%). Meanwhile, it could be ameliorated. The degradation of the recall was essentially due to some missed identifications. The main reason behind the missed cases was mainly due to the low quality of the TV streams, such as a blurred signal.

Experiments of News Programs Intra-Segmentation

For this level, we conducted two experiments. The first one was conducted on the same corpus of the first level experiments, ***NPs*** already extracted from the inter-segmentation level. To get comparative results to similar works, we conducted a second experiment on the TRECVID dataset.

Results on Long TV Stream Programs

On the first experiments, we used 10 recordings of *NPs* extracted from the same streams used in inter-level-segmentation experimentation. We used two recordings for each channel's *NP*: one for the segmentation

Figure 5. Results of programs types' identification
a. ROC curve, b. Recall and precision rates

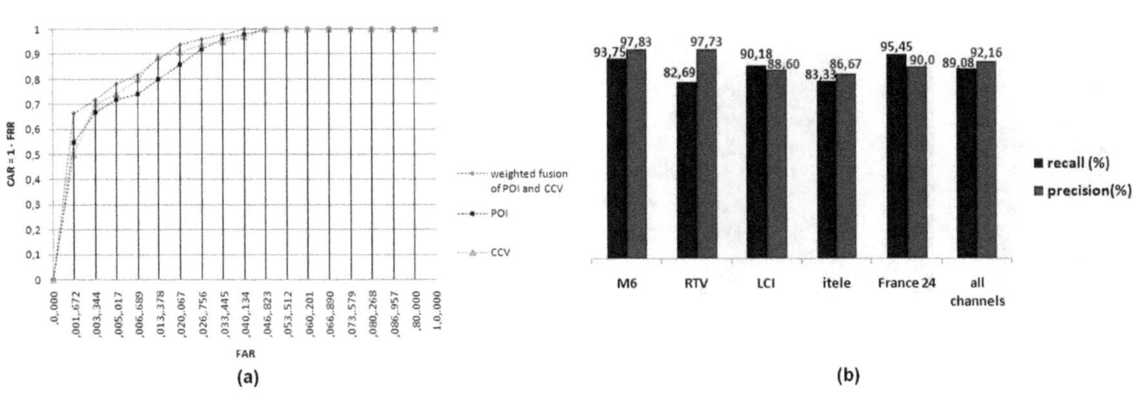

in *stand-alone* mode and the other for the *rewire* one. We also created a truth ground that contained, for each *NP*, the numbers of shots, face shots and anchorperson shots (Table 3).

First, we carried out the detection of all shots and the identification of face ones. High rates for faces shot (Figure 6a) improved anchorperson's shot detection and consequently topics detection.

Then, we evaluated our approach in both stand-alone and rewire modes. For the rewire mode we conducted two experiments. The first experiment used frame's background features of the news studio and the second one did not consider these features. This second choice was made for few ***NPs*** types that did not have a regular background.

Result on Varying Studio Background

Currently, a new trend of NPs appeared using a totally or partially variant studio's background. These programs use a background with a complex graphic composition.

To measure the performance of rewire mode for this type of ***NPs***, we conducted experiments regardless of the features of the studio background. Topics detection rates are shown in Figure 6b.

The obtained results confirm satisfactory performances in both modes with an average of topics detection rate of about 94% for the stand-alone mode and more than 80% for the rewire one. These rates prove that rewire mode results depend strongly on the stand-alone mode ones. Missed detections are explained on the one hand by the missed face shots, and on the other hand by the detections never obtained in the case of the France 24's ***NPs*** in stand-alone mode. This case is due to the fact that the

Table 3. Ground truth of TV news segmentation

	M6		RTV		LCI		Itele		France 24	
	NP1	*NP2*	*NP1*	*NP2*	*NP1*	*NP2*	*NP1*	*NP2*	*NP1*	*NP2*
# Shot	499	494	369	302	345	322	351	280	414	370
#face Shot	284	279	219	135	183	143	225	145	231	198
#Anchor Shot	81	78	36	33	54	39	60	38	117	106

Figure 6. Results on long TV stream News programs
a. Shot detection rates. b. Topics detection rates

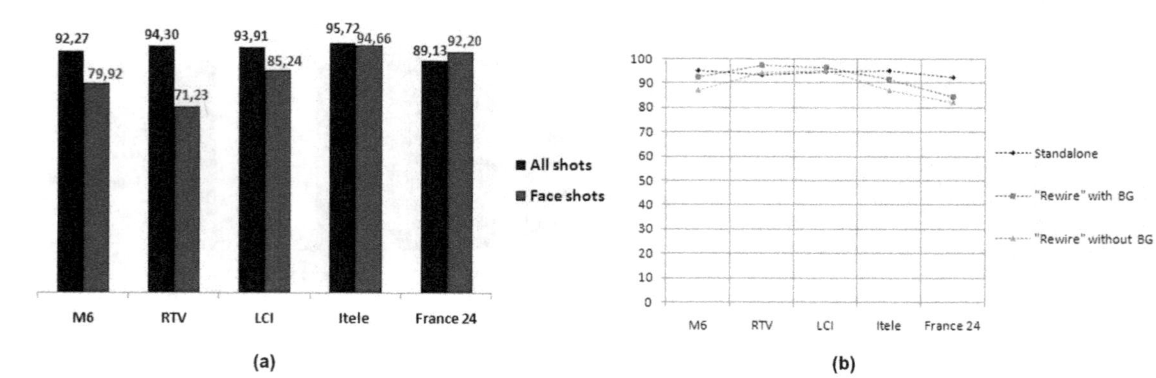

anchorperson in ***NP2*** was different from the one in ***NP1***. Therefore, the method of pairing between the signature from video grammar and the other one extracted from ***NP2*** is nondeterministic. The immediate solution to overcome this problem is to consider several anchorperson faces per particular channel's ***NP*** in the structure suggested in this grammar.

Results on the TRECVID Dataset Result

The second experiment was conducted on TRECVID 2003 dataset (Smeaton et al., 2003). We chose this dataset, widely used by the community, in order to evaluate our results to recent works. The TRECVID 2003 collection contained more than 2900 story boundaries. TV news programs of this dataset were collected from different channels. However, this dataset presents some drawbacks. The main disadvantage is that it contains only old style TV news. New styles of TV news, especially ones that having animated background and more than one anchorperson are not included. So, conducting a comparative evaluation of these new types of news is a real difficulty.

In these experiments, previous approaches (Dumont and Quénot, 2012; Goyal et al., 2009; Misra et al.; 2010) were selected for comparison according to two criteria. First, these approaches are chosen because they are based on the same assumption to segment news broadcasts: anchorperson shots detection as the starting point for detecting news topics. The second criterion was that all approaches were experimented on the same standard TRECVID 2003 dataset. To reach an accurate comparison with other works, we used same metrics as defined in (Dumont and Quénot, 2012). So, we evaluated the news segmentation efficiency using the precision, recall and F1 measures. Comparative results are presented in Figure 7.

These results prove the significance performance of our approach. This performance is most evident with the rates obtained by the stand-alone and rewire (with BG) modes. In the rewire mode, we performed an experiment based on the direct match between features of the studio decor stored in the grammar regardless of anchorperson faces. Another experiment was carried out without considering the studio décor features. This second choice is mainly justified by the fact that there were cases of ***NPs*** with various backgrounds for each topic. So, although the studio background features increased significantly the detection rate that confirms the importance of prior knowledge, it presents some problems for this particular type of TV news. Besides, in spite of the complexity and the animation effects, which characterized the background of some ***NPs***, we obtained good rates. This confirms the importance of the weighted attribute to the various background features.

Figure 7. Comparative results of experiments

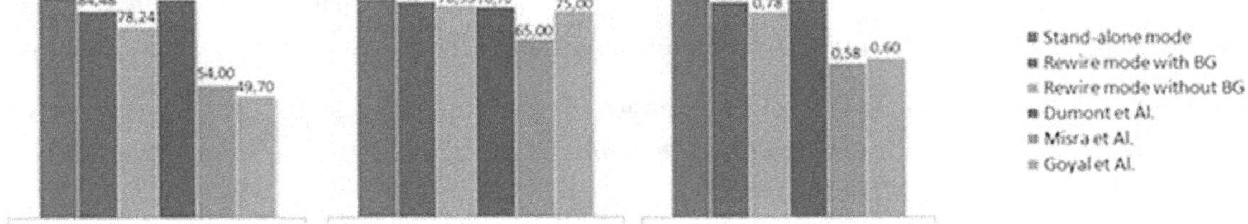

DISCUSSION AND FUTURE WORKS

In this paper, we presented a novel approach dedicated to generate News programs from large-scale TV streams for VoD and catch-up TV services. The goal of the proposed approach is to structure compact large TV streams and then generate a reliable Web content on alternative media. Thus, generating the input of online video sharing platforms from TV channels content has two main benefits. The first one is the increasing of the audience users seeing the access to these platforms is more rapid and less expensive. The second benefit concern the authenticity of content. In fact the content generated from TV streams to web media is reliable and consequently it's a preventive solution to reduce the risk of *fake news*. The main originality of our approach is the use of the contextual and operational characteristics of TV production rules as prior knowledge. Based on inferred grammar, our approach proposes three steps. The intra-segmentation step aims to identify news programs in TV stream according to the visual similarity between the visual features directly calculated from the TV stream and those involved by inferred grammar. The second step, called the intra-segmentation, is addressed to determine the internal structure of the identified news programs. The last stage is dedicated to annotate segmented videos using MPEG-7 standard. This stage aims to enrich the segmented topics by additional metadata. These metadata will enhance the quality of indexing of the exported media on the Web and ameliorate its ranking on video sharing platforms.

The performance of our approach was investigated by experiments conducted using TRECVID 2003 dataset, and we proved its effectiveness.

Further investigation shall focus on the adaptation of this approach to treat other types of programs. We also intend to enrich the grammar by other features like audio features.

We would also to develop the proposed approach by adding an automatic hashtags generator. This component aims to facilitate the work of content manager which generally do this task manually. Also, the hashtags generation enhance the media audience ranking.

REFERENCES

Abduraman, A. E., Berrani, S. A., & Mérialdo B. (2011). TV program structuring techniques: A review. In *TV Content Analysis: Techniques and Applications*. Academic Press.

Arvis, V., & Debain, C., Berducat, M., & Benassi, A. (2004). Generalization of the Co-occurence Matrix for Color Images: Application to Color Textures Classification. *Image Analysis & Stereology*.

Asghar, M. N., Hussain, F., & Manton, R. (2014). Video indexing: A survey. *Int. J. of Computer and Information Technology*. doi:10.1007/978-3-319-24306-1_2

Athanasakos, K. C., Doulamis, A. D., & Karanikolas, N. N. (2007). A Signature Tree Content-based Image Retrieval System. *10th International Conference on Computer Graphics and Artificial Intelligence*.

Bogin, B., & Varela-Silva, M. I. (2010). Leg Length, Body Proportion, and Health: A Review with a Note on Beauty. *International Journal of Environmental Research and Public Health*, *7*(3), 1047–1075. doi:10.3390/ijerph7031047 PMID:20617018

Chaisorn, L., & Chua, T. S. (2006). *Story boundary detection in news video using global rule induction technique. Procs of the IEEE Inte. Conf. on Multimedia and Expo.*

Choroś, K. (2015a). Automatic fast detection of anchorperson shots in temporally aggregated TV news videos. In *ACIIDS 2015. LNCS*. Heidelberg, Germany: Springer. doi:10.1007/978-3-319-15705-4_33

Choroś, K. (2015b). Automatic Categorization of Shots in News Videos Based on the Temporal Relations. *Computational Collective Intelligence, 7th International Conference, ICCCI 2015.*

David, C. (1991). *Practice of the video.* Edition LAROUSSE.

Dumont, E., & Quénot, G. (2012). Automatic Story Segmentation for TV News Video Using Multiple Modalities. *International Journal of Digital Multimedia Broadcasting*, *2012*, 1–11. doi:10.1155/2012/732514

Dunker, P., Gruhne, M., & Sturtz, S. (2008). Personal Television: A Cross modal Analysis Approach. *IEEE International Symposium on Consumer Electronics.*

El Khoury, E., Sénac, C., & Joly, P. (2014). Audiovisual diarization of people in video content. *Multimedia Tools and Applications Journal, 68*(3).

Goyal, A., Punitha, P., Hopfgartner, F., & Jose, J. M. (2009). Split and merge based story segmentation in news videos. *Proc. of the 31th European Conference on IR Research on Advances in Information Retrieval.* 10.1007/978-3-642-00958-7_82

Hu, W., Xie, N., Li, L., & Zeng, X. (2011). A Survey on Visual Content-Based Video Indexing and Retrieval. *IEEE Transactions on Systems, Man and Cybernetics. Part C, Applications and Reviews*, *41*(6).

Hua, S., Chen, G., Wei, H., & Jiang, Q. (2012). Similarity measure for image resizing using SIFT feature. *EURASIP Journal on Image and Video Processing.*

Jacobs, A., Miene, A., Ioannidis, G. T., & Herzog, O. (2004). Automatic shot boundary detection combining color, edge, and motion features of adjacent frames. *TRECVID Workshop Notebook Papers.* doi:10.1186/1687-5281-2012-6

Kai, S., Amy, S., Suhang, W., Jiliang, T., & Huan, L. (2017). Fake News Detection on Social Media: A Data Mining Perspective. *ACM SIGKDD Explorations Newsletter*, *19*(1), 22–36. doi:10.1145/3137597.3137600

Krötzsch, M., Simancik, F., & Horrocks, I. (2012). *A description Logic.* Primer Technical Report, arXiv.org, volume CoRR abs/1201.4089

Lakshmi, D., Damodaram, A., Sreenivasa, M., & Lal, J. (2008). Content based image retrieval using signature based similarity search. *Indian Journal of Science and Technology*, *1*(5).

Le, N., Bredin, H., Sargent, G., India, M., Lopez-Otero, P., & Barras, C. (2017). Towards large scale multimedia indexing: A case study on person discovery in broadcast news. *International Workshop on Content-Based Multimedia Indexing - CBMI 2017.* 10.1145/3095713.3095732

Liang, H., Chen, L. H., & Chen, W. S. (2018). Video Caption Extraction Using Spatio-Temporal Slices. *International Journal of Image and Graphics*, *18*(2).

Lowe, D. (2004). Distinctive Image Features from Scale-Invariant Key points. *International Journal of Computer Vision*, *60*(2), 91–110. doi:10.1023/B:VISI.0000029664.99615.94

Mahdi, W., & Zlitni, T. (2016). Navigating News Programs in Large TV Streams: A Knowledge-Based Approach. *International Journal of Knowledge Society Research*, *7*(1), 80–98. doi:10.4018/IJKSR.2016010106

Manson, G., Naturel, X., & Berrani, S. A. (2009). Automatic Program Extraction From TV Streams. *European Interactive TV Conference, EuroITV'09*.

Misra, H., Hopfgartner, F., Goyal, A., Punitha, P., & Jose, J. (2010). TV News Story Based Segmentation one Semantic Coherence and Content Similarity. *Proc. of 16th International Conference one Multimedia Modeling*.

Naturel, X., & Gros, P. (2008). Detecting repeats for video structuring. *Multimedia Tools and Applications*, 38.

Nock, R., & Nielsen, F. (2004). Statistical Region Merging. *Trans. Pattern Anal. Mach. Intell.*

Ravani, R., Mirali, R. M., & Baniasadi, M. (2010). Parallel CBIR System Based on Color Coherence Vector. *Proc. of International Conference on Systems, Signals and Image*.

Rossetto, L., Giangreco, I., Tănase, C., & Schuldt, H. (2017). Multimodal video retrieval with the 2017 IMOTION system. *Procs of the 2017 ACM International Conference on Multimedia Retrieval (ICMR 2017),* 457–460.

Smeaton, A. F., Kraaij, W., & Over, P. (2003). TRECVID 2003 - an overview. *TRECVID 2003 - Text REtrieval Conference TRECVID Workshop*.

Souza, C. L., Pádua, F. L. C., Nunes, C. F. G., Assis, G. T., & Silva, G. D. (2014). *A unified approach to content-based indexing and retrieval of digital videos from television archives*. Artificial Intelligence Research.

Younessian, E., & Rajan, D. (2015). Multi-modal fusion for associated news story retrieval. *Multimedia Tools and Applications Journal, 74*.

Chapter 15
A Conceptual Model for Cloud-Based E-Training in Nursing Education

Halima E. Samra
La Trobe University, Australia

Alice S. Li
La Trobe University, Australia

Ben Soh
La Trobe University, Australia

Mohammed A. AlZain
Taif University, Saudi Arabia

ABSTRACT

In specific fields, such as e-training in nursing, involving computer-intensive training scenarios, there is an increased demand to deliver training services to a larger number of learners, and with it, the need for cloud services. However, to date there has been a lack of a formalized framework relating to the use of cloud computing for on-demand interactive e-training resources in nursing education. To this end, this chapter formalizes a conceptual framework for a cloud-based e-training system in nursing education. The conceptualization takes into consideration nursing e-training system requirements, with a focus on applying cloud computing technologies to ensure the dynamic scalability of virtual distributed services and computing power while maintaining QoS and security.

INTRODUCTION

Nursing education has recently undertaken a technological advancement and has employed effective teaching and training methods to address learning requirements and assist learners to build life-long skills, develop their problem-solving abilities and employ resources to obtain the best benefit from technology. Internationally, there is a growing trend toward the need for effective clinical skills education that is flex-

DOI: 10.4018/978-1-5225-7347-0.ch015

ible and suits different learning expectations (Bloomfield et al. 2013). Therefore, to accommodate the intensive practical needs of nursing students and guarantee an effective training processes, it is essential that the most effective teaching methods are utilized, supported by technologies that offer an effective intelligent adaptive training environment to develop learning and to promote skills. Simulation-based learning is the next era in the history of clinical practical education. The use of simulation within clinical skills education is an emerging educational technique gaining global recognition as a valuable pedagogic approach and an effective way to ensure practical skills in a safe simulated clinical environment. Generally, for e-training in nursing to be effective, important requirements must be satisfied, including that the e-training environment should provide dynamic scalability of virtual distributed services, computing power, QoS and security.

E-training is widely regarded as a valuable mechanism for the acquisition of clinical skills in nursing education where flexible access to e-training resources and the opportunity to engage in independent learning enables students to practice skills at a time of their own choosing and at their own pace(Bloomfield and Jones 2013). One of the benefits of utilizing the powerful features of e-training tools is that it provides valuable feedback to the learners throughout the training process, it allows them to select the learning content; and it enables them to engage in self-assessment and to evaluate the results of their learning. However, resources need to be developed to provide an online and realistic training environment. Therefore, it is important to be able to access virtual resources, such as online simulations and virtual lab repositories, to provide on-demand up-to-date training.

A promising solution to the limitations facing e-training in nursing in relation to scalability to enable it to cope with the increasing number of users and resources is cloud computing, which can provide educational institutions with a distance computing infrastructure and data as a service on-demand over the Internet (Fasihuddin et al. 2012). It also offers educational platforms and services, and virtualization by combining all resources and centralized data storage (Ghazizadeh 2012).

Particularly, for nursing clinical skills learning where services such as virtual labs, simulations, and multimedia provision are computer-intensive and should be offered in a highly scalable way, the cloud-based environment can enable both students and their instructors to access computing resources on-demand for lectures and labs, according to their learning needs (Gonzalez-Martínez et al. 2015).

However, many challenges associated with cloud deployment may arise, such as data privacy, security, availability, consistency, and transmission. Many research studies have proposed solutions to tackle these problems. One solution for educational institutions is for them to build their own private cloud in which to place their sensitive data under their own management. Other research suggests using a cloud backup for important data and information to ensure the requirements of the Service Level Agreement, which specifies the terms of the contract, are met in relation to the provision of the cloud services (Fasihuddin et al. 2012).

Despite the effective solutions which have been applied to cloud models to fit e-training environments, there is still a growing need for a customized e-training cloud platform that meets the specific needs of training environments. For this reason, the aim of this chapter is to enhance the conceptual model developed in (Samra, Li, Soh, & AlZain, 2017) for a cloud-based e-training system in nursing, taking into consideration e-training requirements and utilizing the capabilities of cloud computing to address the challenges, such as QoS in terms of resource provisioning and security in terms of data protection in transmission and storage.

This chapter is organized as follows. Firstly we present an overview of nursing education and cloud computing concepts, respectively. We then discuss some related work in cloud-based e-training, and describe the proposed conceptual model. Subsequently we conclude the chapter and summarize the contribution in the area of cloud-based e-training in nursing education.

NURSING EDUCATION OVERVIEW

Nursing education aims to prepare nurses educationally and operationally in their fields by applying theoretical fundamentals and simulating the actual practice of the clinical skills. In traditional nursing education, classrooms deliver the theoretical aspects of the skills after which the technical skills are demonstrated in special sessions individually or in groups under the supervision of the nursing educators (Majima, Maekawa, Soga, & Sakoda, 2013). Although the theoretical component adequately covered the foundation level, it was well recognized that there was a gap between the theory taught in the classroom, and the real practice in clinical environments, predominately because the theoretical and practical components were delivered by two different lecturers, resulting in dissatisfaction, confusion, and a lack of confidence among students, leading to frustration and a failure to complete their courses (ALLAN, 1989). Therefore, it became the goal of nursing educators to minimize this gap and develop nursing programs which effectively utilized technological advances, such as deploying e-training, self-directed learning and simulation-based learning which depend on student-centred strategies that provide appropriate links between the theoretical and clinical components of the teaching plan.

To convey practical experience and impart clinical skills to the leaner in nursing education, online space appears to be well suited. Firstly, e-training, when implemented as a fully online learning mode to replace traditional classes, is highly dependent on textual material as teaching resources and learning activities are quite limited, resulting in unsatisfactory learning outcomes. Blended learning, also referred to as "hybrid" learning (Garrison, 2011), is favoured by the training-based education community, as it provides learners with the opportunity to benefit from the best of both worlds. Nursing education planners have explored e-training technological capabilities and services in their student-centred strategies represented in simulation-based learning and virtual laboratories. E-training offers solutions to nursing education, with its accessible and convenient learning setting which allow flexible and self-directed practise; and independent and interactive learning environments (Abdelaziz, Kamel, Karam, & Abdelrahman, 2011) (Al-Huneiti, Hunaiti, & Al-Masaeed, 2018). Figure 1 depicts an example of e-training in nursing.

CLOUD COMPUTING OVERVIEW

The cloud computing paradigm is a promising solution for the next generation of e-training methods, such as nursing education. It provides efficient on-demand network access to centralized resource sharing, such as networks, storage, servers, applications, and services (Höfer and Karagiannis 2011). The cloud environment is a scalable infrastructure that supports and interconnects several cloud computing services. This new environment is based on the concept of "dynamic provisioning" which means presenting a group of interconnected and virtualized computers as one or more unified computing resource(s)

(Rajkumar et al. 2013). Generally, the definition of cloud computing depends on an understanding of its characteristics and the services that can be deployed as a cloud service. The definition of cloud computing as proposed by the National Institute of Standards and Technology (NTIS) encompasses three main components: key cloud characteristics, deployment models, and service models (Sosinsky 2010).

Cloud Computing Characteristics

The NIST definition of cloud computing outlines the following five key characteristics that every cloud should offer (Rountree and Castrillo 2013):

- **On-Demand Self-Service:** The client can request and receive access to a service in an automated process without the need for human interaction with the cloud service providers (CSPs).
- **Broad Network Access:** The client only needs a basic network connection in order to access the services and applications in the cloud.
- **Resource Pooling:** CSPs pool their computing resources to serve multiple clients and dynamically assign and reassign both physical and virtual resources to meet their clients' changing requirements. By exploiting virtualization, CSPs can increase their system density.
- **Rapid Elasticity:** The cloud resources can be elastically provisioned upon user demand and needs. This can be achieved by using system scalability to make resources available (Sosinsky 2010).
- **Measured Service:** Cloud service usage should be measured using metrics, such as the storage size used by the client, the network usage, etc. and should support the "pay as you go" feature.

Cloud Deployment Models

A deployment model refers to the location and management of the cloud's infrastructure (Sosinsky 2010). There are four different types of clouds that a user can subscribe to, depending on his requirements:

- **Public Cloud:** The public cloud infrastructure is available for public use and CSPs are responsible for the management and administration of the systems that are used to deliver the service (Rountree and Castrillo 2013). The client needs to connect to the Internet to access the cloud.
- **Private Cloud:** The private cloud infrastructure is customized for an organization's specific use and can be managed by the organization or a third party (Sosinsky 2010). The client can access this cloud through a local area network, or a wide area network, but in remote cases, clients can be offered a virtual private network (Rountree and Castrillo 2013).
- **Community Cloud:** A community cloud is where a cloud has been organized to serve a specific group of organizations which share similar requirements (Sosinsky 2010). The group can obtain more privacy than is available on a public cloud but the group members share the management and administration responsibilities (Rountree and Castrillo 2013).
- **Hybrid Cloud:** A hybrid cloud model is a combination of two or more cloud models (Rountree and Castrillo 2013). Every cloud retains its own characteristics, but they are bound together as one unit. It is a complex environment, but to address this problem, a hybrid cloud may offer standardized and proprietary access to the data or the application (Sosinsky 2010).

Cloud Service Models

A service model presents a set of definitions that describe the service characteristics and criteria (Sosinsky 2010). Cloud computing delivers infrastructure, platforms, and software as services. Therefore, these services are generally classified into three classes known as cloud service models:

1. **Infrastructure-as-a-Service (IaaS) Model:** The IaaS model provides clients with virtualization platforms, including processing and storage resources (virtual machine, virtual storage, and networking). However, running and maintaining the operating system and the software in these resources is the client's responsibility (Rountree and Castrillo 2013).
2. **Platform-as-a-Service (PaaS) Model:** The PaaS model provides an operating system, a development platform and software and hardware tools which are delivered by CSPs to the client. A client is responsible for installing and managing their deployed application (Sosinsky 2010).
3. **Software-as-a-Service (SaaS) Model:** Complete software applications are delivered by the CSPs over the Internet and are accessed via a web browser so the customer does not need to purchase a software license or worry about maintenance (Sosinsky 2010).

Cloud Computing Technologies

In the past, technological limitations made it difficult for CSPs to deliver a viable service and cope with growing customer demands. A lack of components, inadequate and expensive servers, and the limited capabilities of applications were obstacles that hindered the development of cloud computing (Rountree and Castrillo 2013). Now cloud computing has embodied technological advances, such as distributed systems, virtualizations, Web 2.0, and service-oriented solutions to cloud difficulties (Rajkumar et al. 2013).

- **Distributed Systems:** Clouds are large distributed computing components which provide services on demand to clients as a single coherent system. Sharing and utilizing cloud computing resources (infrastructure, runtime environments, and services) has resulted in clouds being characterized in terms of scalability, concurrency, and ubiquity (Rajkumar et al. 2013).
- **Virtualizations:** Clouds have a virtually unlimited capacity by dividing physical resources (hardware, storage, runtime environments, programming language, and networking) into several virtual machines (VMs), using software called hypervisors that create separate virtual machine instances. As a result, it provides clouds with application isolation, quality of service, and protection (Rajkumar et al. 2013).
- **Web 2.0:** At present, the primary web interface that clouds use to deliver their services encompasses new technologies and services which bring interactivity and flexibility to Web pages, making cloud computing an attractive solution for constructing computing systems(Rajkumar et al. 2013).
- **Service Orientation:** Service orientation is the core reference model for cloud computing systems with loosely coupled features where the service is reusable and serves different scenarios (Rajkumar et al. 2013).

RELATED WORK

The development of nursing education is tied to the strong demand for highly skilled nurses who can manage complex health care environments which require contact between the nurse and the patient in a safe manner. These requirements have challenged nursing education developers to advance student-centred learning to overcome the continuing obstacles that hinder nursing practice. Therefore, there is a need to ensure future nurses have the ability to engage in long-life learning to develop their knowledge and experience over time, even after graduation (Cadorin, et al., 2015). E-learning promotes self-directed learning, allowing students to choose the learning content, resources, tools that support their different characteristics, requirements, and level of performance (Button, Harrington, & Belan, 2014). However, for self-directed learning to be incorporated in nursing education with its complex clinical environment, there is a need for tutorial strategies to guide and analyse the learning process for better student performance (Cadorin, et al., 2015) ; for instance, to strengthen the self-directed learning setting, the integration of clinical skills education systems and intelligent tutoring systems (ITSs) which focus on monitoring and supervising the learning process with the provision of appropriate advice to the student when needed, has been shown to have a great advantage (Sorensen, Ramachandran, & Salvendy, 2007) (Tshiamo, Kgositau, & Magowe, 2017).

Clinical skills are one of the most essential components of modern nursing education. A clinical environment is a setting where theoretical knowledge can be applied in a simulated a practical session so nurses can develop and learn skills in a safe environment (Khatiban & Sangestani, 2014). In conventional nursing education, clinical experience is valued as a fundamental part of the nursing curriculum. However, nursing students were treated as handmaidens or 'an extra pair of hands' and they often passed their exams without having acquired the necessary skills, resulting in a lack of confidence where nurses felt their training was insufficient to prepare them to work in a clinical field (Alty, 1994). Facing these challenges, nursing educators tried to narrow the gap between the theoretical and practical components of the curriculums by implementing technologies which have had an immense impact on enhancing the provision of clinical skills in a risk-free environment, enabling students to develop and acquire the skills required to provide safe and high quality care (Button, Harrington, & Belan, 2014). Consequently, new clinical models and simulations are required to enhance practical training in nursing education.

Cloud-based technologies play a significant role in the technology-enhanced learning domain. Many research studies have contributed to the deployment of cloud technologies in the field of education, in general. As a first step, most research which has been conducted discussed the beneficial aspects of cloud deployment in the field of education (Ghazizadeh 2012), whereas others tried to utilize cloud computing by introducing services such as TLaaS (Teaching and Learning as a Service) (Radhakrishnan et al. 2012). A pedagogical cloud framework was presented by Chaabouni and Laroussi in (Uden et al. 2014) using the modeling of the pedagogical indicators in a collaborative and cooperative way on a cloud with the SaaS model. The aim was to analyse learner usage and manage indicators on the execution of educational scenarios. Several of the proposed cloud architectures focused on the management of the cloud infrastructure. The BlueSky cloud framework for e-training was proposed by Dong et al. In (Jaatun et al. 2009). They combine traditional middleware functions, such as load balancing and data caching in their architecture to deliver a reliable, scalable, and cost-effective service to e-training systems. CloudIA is a project conducted by Doelitzscher et al. (Jaatun et al. 2009) to design and build a private cloud for on-demand computing resources, and running e-training applications and collaboration software. Snow Leopard, developed by Cayirci et al. (Jaatun et al. 2009), is a private cloud for educational purposes in the military domain. It resolves security challenges in its design, including some mechanisms such as privacy, anonymity and traffic analysis. The Virtual Computing Laboratory (VCL) is a training environment that is attracting increased research interest in the cloud domain. A proposed Virtual Laboratories Cloud System (VLCS) for linking e-training environments with VCLs offers one possible way to improve the learning process by using cloud computing to offer Virtual Laboratories as a Service (VLaaS) (Rădulescu 2014). The CLEM project, Cloud E-learning for Mechatronics, was developed by Chao et al. (Chaoa et al. 2015) for remote laboratories in the cloud. They developed an ecosystem for learners and educators to share a platform that allows a large number of distributed mechatronic devices to be pooled and used for e-training.

Although there is a considerable shift toward research in the use of cloud computing in an educational context from different perspectives as discussed above, there is a lack of a well-defined cloud framework that suits training activities and handles the requirements and services for an e-training environment in nursing education. Therefore, our aim is to contribute to the technology-enhanced learning research area with a conceptual model for a private cloud for nursing e-training systems, which can support computing-intensive learning scenarios by using technological features of clouds, such as virtualization, on-demand resource provisioning, scalability, security, reliability and quality of service.

CONCEPTUAL MODEL FOR CLOUD-BASED E-TRAINING IN NURSING

E-Training System Description

Online education in nursing encompasses practical lectures, using resources such as virtual labs and online simulations. The online education also incorporates an e-learning Management System (LMS) for enrolments and virtual classes where lectures are posted as PowerPoint slides, video clips, and e-library.

E-Training Users' Requirements

There are three categories of users involved in the actual e-training systems in nursing:

Students can enrol in a course, login to their account, attend virtual classes, use the material available in LMS, reserve a virtual lab or simulation for a practical session in nursing (see Figure 1).

Lecturers can login to their account, create a course, utilize material available in LMS for the course, reserve a virtual class, reserve virtual lab or participate in a simulation

Institution Management is responsible for the management of the learning process and for configuring the whole cloud platform.

Cloud-Based E-Training System Requirements

In line with e-training system specifications, we propose a private cloud for an e-training system in nursing which supports both the individual trainee and/or collaborative work practices through courses prepared by instructors. The characteristics of such an environment vary from the long-term predication for scheduling regular courses to the periodic nature of an educational event, such as enrolment and examination periods. Nursing e-training activities are computer-intensive and require a high-scale mechanism and full resource provisions to respond to sudden demand variation (Gonzalez-Martínez et al.

Figure 1. E-training in nursing

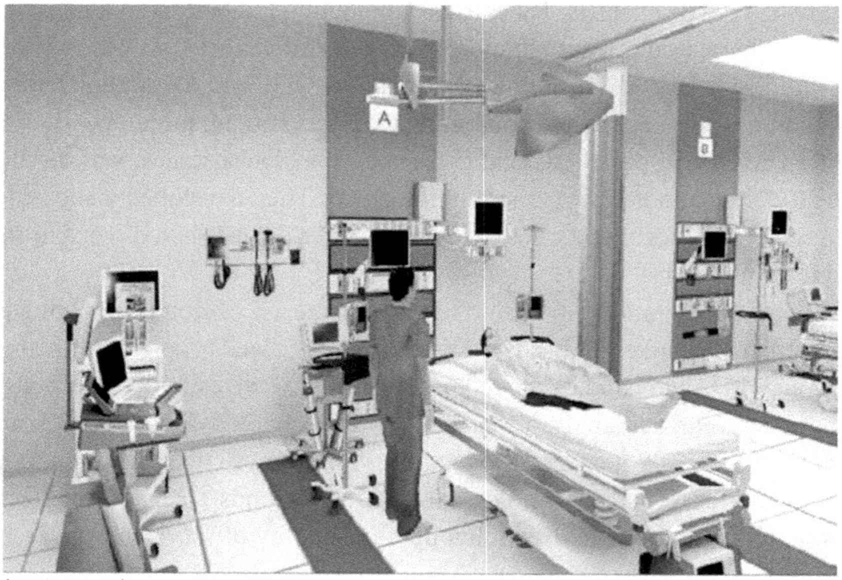

http://www.medscape.com/viewarticle/780819_4

2015). Therefore, by utilizing cloud computing features, such as virtualization and on-demand resource provisioning, our design will support scalability and elasticity where resources can be dynamically added or removed according to the user requirements while maintaining QoS (Rajkumar et al. 2013).

Virtualization

Virtualization is one of the most important underlying technologies that supports cloud models. It abstracts the hardware from the operating system (OS) using a hypervisor or Virtual Machine Monitor (VMM) (Younge et al. 2010). Virtualization is the key management feature of the infrastructure layer.

Infrastructure Management

The lower level of the cloud construction involves resource deployment and the management of the physical resources (computer servers, storage block, desktop, and networks). Virtualization forms the upper level in the infrastructure layer by virtually relocating physical resources to multiple virtual machine instances.

Server Virtualization

One physical server (CPU, memory, desks) can run multiple computing environments where every VM can run multiple operating systems and applications in an isolated environment. This helps with the management of a computing-intensive environment by creating a scalable system of different distributed computing devices and increases system performance. In addition, ensuring each VM is isolated will address the security concerns (Jaatun et al. 2009).

Storage Virtualization

Storage virtualization provides cloud systems with **storage pools** where all the computing resources from various storage facilities can be viewed as a single resource which can be shared among virtual instances. Storage virtualization is a key component in data management because it helps with resource allocation to users (Jaatun et al. 2009).

Desktop Virtualization

Desktop virtualization creates a desktop image as a virtual machine and provides a comprehensive desktop environment for users. It also reduces the risk of data loss and minimizes security concerns as none of the data stored in the data center will be affected by hardware failure (Portnoy 2012).

Network Virtualization

Virtual machines can be configured with one or more network interface cards (NICs). The hypervisor supports the creation of a virtual network where the vNICs can be interconnected via virtual switches that provide segmentation and isolation for network traffic, ensuring security and data integrity (Portnoy 2012).

On-demand Resource Provisioning

Resource provisioning enables the dynamic provision of on-demand resources which can be achieved by easy scheduling and reservation of computing resources (Rajkumar et al. 2013). In an e-training system, the middleware of the cloud should be designed to support the ability to schedule and reserve virtual labs and simulations in nursing.

Middleware Management

The efficient management of a computing platform is essential to provide a good nursing e-training environment for managing distributed applications. In terms of on-demand resource allocation, the computing platform includes three modules: the *user management* module, the *resource allocation* module and the *access management* module (Tian et al. 2010). In addition to the main controller, there is also a *management node* which is in charge of resource scheduling, security, monitoring and virtual network management.

User Management

The user module consists of several sub-modules, such as the access management module, the login and authentication module and basic information management that update the user database. The database server manages and maintains the user information, such as authentication, resource availability information, platform state information, and connection information (Tian et al. 2010).

Resource Allocation Management

The core of a platform management system is resource allocation which makes scheduling and reserving a resource easy through the collaboration of the three subsystems:

- Resource usage (reservation) which manages the current users' requirements and reserves resources for prospective ones.
- Resources status (analysis) which manages the maintenance of the status of all resources.
- Resources renewal (runtime environment) which manages the reuse of the resource for the same user.

Access Management

This manages the user's access to resources, including distance access and connection management.

Management Node

The management node is the core of the computing platform that works as an intermediate component between users and their requested services. The management node first verifies the user's account before receiving the user's requests, and then allocates the available real and virtual services to the user for a specified time (Tian et al. 2010). The management node controls the cloud platform, offering efficient mechanisms for *scheduling*, *monitoring*, and *security* management, as shown in Figure 2.

Scheduling

Specific APIs are developed to deal with the scheduling and reservation tasks using the task assignment and execution mechanism. The MapReduce programming model is used for scheduling and executing tasks where the application task is automatically divided into many multiple tasks, each of which may be executed on any node in the cluster (Jaatun et al. 2009).

Monitoring

An e-training system depends on high reliability and service provisioning to ensure the effectiveness of the training sessions. A heartbeat monitoring mechanism is used where the active system and the stand-by system monitor each other's heartbeat and any system that shows no heartbeat for a period of time is considered to be in shut-down mode. Then the current operation can be transferred to another host without interrupting the operation (Matsumoto and Yutaka 2011).

Security Management

For an e-training cloud which is a private cloud, security management issues are lessened due to the proprietary of the infrastructure and the fact that the cloud is managed by the institution. However, data privacy and security still need to be managed by maintaining the e-training environment requirements, such as ensuring that confidential data, such as information on training sessions, examinations and stu-

Figure 2. Proposed cloud-based e-training framework and management node

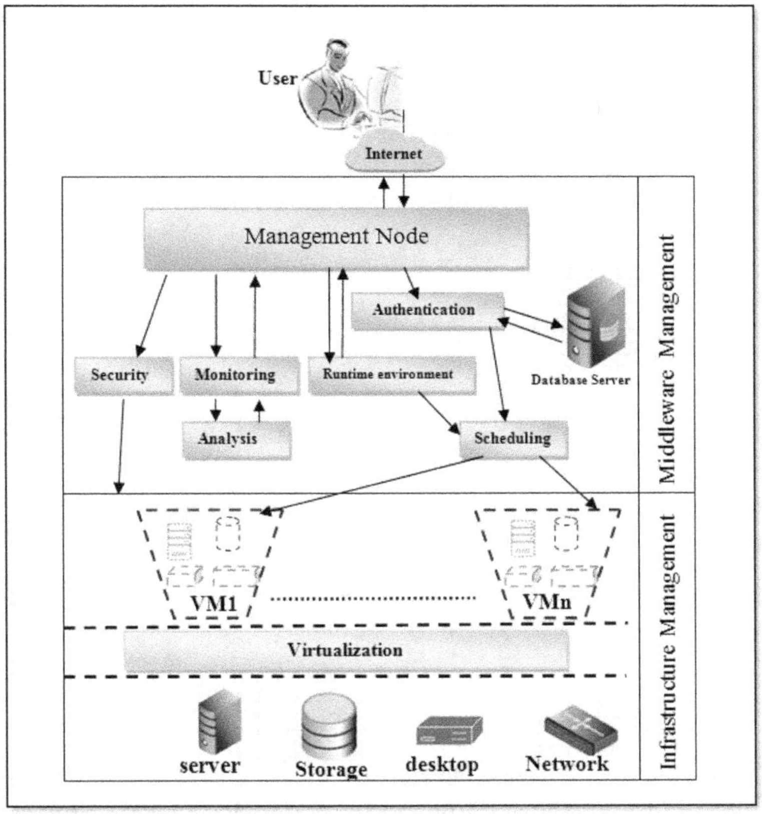

dent records are kept private by limiting access to specific users. Also, data traffic must also be secure. Therefore, to provide a secure, controlled execution environment, a reliable flow control mechanism which can be achieved by labelling each data, service, and user with a security classification and authentication level, will ensure that only users with the appropriate security clearance can access the data (Jaatun et al. 2009). Security management for the cloud infrastructure can be done using hypervisors at the hardware level to reinforce the isolation between VMs to prevent data access and provide a secure, multi-tenant environment. The encryption and authentication of communication is essential to avoid data being compromised during transmission. Therefore, access to services can be through secure remote connections, such as the open Secure Shell protocol (OpenSSH) (Tian et al. 2010).

Architecture of the Cloud-Based E-Training System

To address the nursing e-training requirements, the proposed conceptual model includes an infrastructure layer, virtualization layer, platform layer, application layer, and user interface layer in addition to the framework implementation mechanisms and methods for efficiently managing all layers and services, as shown in Figure 3.

Figure 3. Proposed cloud-based e-training architecture

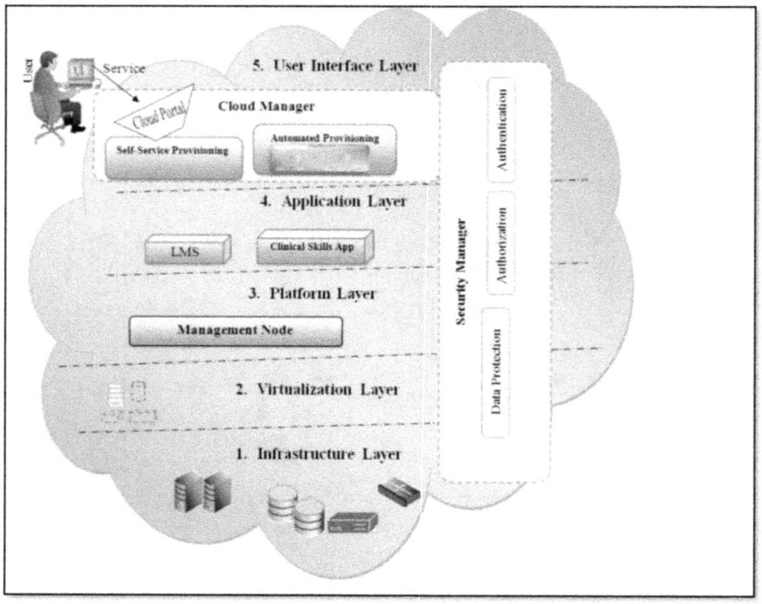

Infrastructure Layer

The infrastructure layer comprises physical machines in the lower level which that provides scalable and elastic computing foundation components, which support training sessions with powerful CPUs which constitute the lower layer for the processing layers of the application. The data storage where nursing teaching and training resource materials, being elastic, can dynamically scale up and down to meet the application storage needs. Also, a scalable file storage component constitutes the foundation for the higher level storage component. Finally, an API for cross-integration enables collaboration between inter-component and inter-application (Zaigham 2013).

Virtualization Layer

Virtualization will support our architecture design at the hardware level, operating system level, and network level. Using virtualization, the physical machines in the hardware level can replicated and pooled and dynamically provisioned, based on the current workloads of various applications to achieve on-demand services. Also, it is possible to open more than one operating system platform in a single physical machine for efficient resource sharing and to serve more users (Tian et al. 2010).

Platform Layer

The platform layer provides an elastic and scalable application development platform through an API. This layer provide an integration platform where various resources that belongs to different platforms in the infrastructure layer (Fasihuddin et al. 2012). In addition to the provision of services such as scheduling and reservation, data services and multi-tenancy support which helps to build scalable cloud-based e-training simulation, and an e-learning LMS framework in nursing education.

Application Layer

The application layer provides different types of applications that support the e-training process where the instructor can use these applications to build their own courses by collecting materials from the pool available in the infrastructure layer. Also, it supports multi-users as the application allows many trainees to access the same shared pool of resources.

User Interface Layer

The interface layer provides various entry points to the nursing e-training cloud, secure access to the functionality of the system, in addition to web-based UIs.

Nursing E-Training Cloud Components

The workflow in the nursing e-training private cloud relies on two core components that manage the overall life cycle of the system, viz E-training Cloud Manager and E-Training Security Manager.

E-Training Cloud Manager

Automated Provisioning Manager

This is a dashboard function which monitors and reports the usage of the CPU, memory, and the disk I/O frequency (Matsumoto and Yutaka 2011). It is used to visualize resource management by applying important metrics, such as the current number of trainees using the cloud and their requested services to automatically scale the resources and provide services.

E-Training Cloud Portal

The e-training cloud portal provides a template interface for users to access all management-related interactions through self-service provisioning capabilities (Zaigham 2013). It also enables the administration and monitoring of services. On the back end, the portal will interface with the APIs to easily interact with the other applications and components of the system (Rountree and Castrillo 2013).

E-Training Cloud Security Manager

The e-training cloud security manager provides security and privacy in terms of a secure authentication and authorization process. User authentication is verified upon login to the cloud using multifactor authentication. Data confidentiality and integrity can be achieved by applying a reliable flow control mechanism using a labelling technique to protect data from unauthorised access or modification. In addition, the security manager will enforce isolation while virtualizing resources to ensure every virtual machine instance operates in an isolated environment. Finally, data can be protected during transmission by applying the OpenSSH protocol.

CONTRIBUTION

To date there has been a lack of a formalized framework relating to the use of cloud computing for on-demand interactive e-training resources in nursing education. To that end we propose a conceptual model for cloud-based e-training in nursing, which is the major contribution of this chapter. We detail the design and requirement issues in regard to nursing e-training. We also explore the features of cloud computing to enable the effectiveness of virtualization, on-demand resource provisioning, delivering scalable services, achieving QoS and maintaining data security, which constitute essential nursing e-training system requirements. Other findings in this chapter include:

- The features of cloud computing, such as dynamic scalability and effective usage of resources, make cloud deployment a logical option for educational environments.
- The utilization of the powerful virtualization technology enable clouds to auto scale for their computing resources according to the services requested by the user.
- In the e-training environment, there is a high demand for computing resources during the learning life cycle, and particularly during peak periods, such as enrolment, assignment submission and examination periods.

In conclusion, cloud-based technologies play a significant role in the technology-enhanced learning domain. The adoption of cloud technologies in the educational environment has a positive impact on the learning process by offering new tools and services to improve and support the learning life cycle, including interactivity. In specific fields, such as nursing training, that involve computer-intensive training scenarios, there is an increased demand to deliver training services to a larger number of learners, therefore the need for cloud services.

REFERENCES

Abdelaziz, M., Kamel, S., Karam, O., & Abdelrahman, A. (2011). Evaluation of E-learning program versus traditional lecture instruction for undergraduate nursing students in a faculty of nursing. *Teaching and Learning in Nursing*, *6*(2), 50–58. doi:10.1016/j.teln.2010.10.003

Aity, A. (1994). Seclusion and nurse education. In Seclusion and mental health (pp. 110-126). London: Chapman & Hall.

Al-Huneiti, R., Hunaiti, Z., & Al-Masaeed, S. (2018). Nurses' Attitudes Towards E-Learning for E-Health Education. In I. R. Association (Ed.), *Nursing Education, Administration, and Informatics: Breakthroughs in Research and Practice* (pp. 67–74). IGI Global. doi:10.4018/978-1-5225-5490-5.ch005

Allan, P. (1989). Nursing education - a luxury or necessity? Current Issues in Nursing.

Bloomfield, J., Cornish, J., Parry, A., Pegram, A., & Moore, J. (2013). Clinical skills education for graduate-entry nursing students: Enhancing learning using a multimodal approach. *Nurse Education Today*, *33*(3), 247–252. doi:10.1016/j.nedt.2011.11.009 PMID:22178595

Bloomfield, J., Cornish, J., Parry, A., Pegram, A., & Moore, J. (2013). Clinical skills education for graduate-entry nursing students: Enhancing learning using a multimodal approach. *Nurse Education Today*, *33*(3), 247–252. doi:10.1016/j.nedt.2011.11.009 PMID:22178595

Bloomfield, J., & Jones, A. (2013). Using e-learning to support clinical skills acquisition: Exploring the experiences and perceptions of graduate first-year pre-registration nursing students — A mixed method study. *Nurse Education Today*, *33*(12), 1605–1611. doi:10.1016/j.nedt.2013.01.024 PMID:23473860

Button, D., Harrington, A., & Belan, I. (2014). E-learning & information communication technology (ICT) in nursing education: A review of the literature. *Nurse Education Today*, *34*(10), 1311–1323. doi:10.1016/j.nedt.2013.05.002 PMID:23786869

Cadorin, L., Rei, A., Dante, A., Bulfone, T., Viera, G., & Palese, A. (2015). Enhancing self-directed learning among Italian nursing students: A pre- and post-intervention study. *Nurse Education Today*, *35*(6), 746–753. doi:10.1016/j.nedt.2015.02.004 PMID:25735910

Chaoa, K., Jamesa, A., Nanosa, A., Chena, J., Stan, S., Munteanb, I., ... Capelle, J. (2015). Cloud E-learning for Mechatronics: CLEM. *Future Generation Computer Systems*, *48*, 46–59. doi:10.1016/j.future.2014.10.033

Fasihuddin, H., Skinner, G., & Athauda, R. (2012). A Holistic Review of Cloud-Based E-learning System. *IEEE International Conference on Teaching, Assessment, and Learning for Engineering (TALE)*. 10.1109/TALE.2012.6360325

Garrison, R. (2011). *E-Learning in the 21st Century: A Framework for Research and Practice*. Hoboken, NJ: Taylor and Francis.

Ghazizadeh, A. (2012). Cloud Computing Benefits And Architecture In E-Learning. *IEEE International Conference on Wireless, Mobile and Ubiquitous Technology in Education*. 10.1109/WMUTE.2012.46

Gonzalez-Martínez, J., Bote-Lorenzo, M., Gomez-Sanchez, E., & Cano-Parra, R. (2015). Cloud computing and education: A state-of-the-art survey. *Computers & Education*, *80*, 132–151. doi:10.1016/j.compedu.2014.08.017

Höfer, C., & Karagiannis, G. (2011). Cloud computing services: Taxonomy and comparison. *Journal of Internet Services and Applications*, *2*(2), 81–94. doi:10.100713174-011-0027-x

Jaatun, M., Zhao, G., & Rong, C. (2009). *Cloud Computing, First International CloudCom 2009*. Berlin: Springer-Verlag.

Khatiban, M., & Sangestani, G. (2014). The effects of using problem-based learning in the clinical nursing education on the students' outcomes in Iran: A quasi-experimental study. *Nurse Education in Practice*, *14*(6), 698–703. doi:10.1016/j.nepr.2014.10.002 PMID:25458229

MacKenzie, S. (2012). *Human-Computer Interaction: An Empirical Research Perspective*. Burlington, MA: Elsevier Science.

Majima, Y., Maekawa, Y., Soga, M., & Sakoda, M. (2013). *A Proposal of the New System Model for Nursing Skill Learning Based on Cognition and Technique*. Berlin: Human-Computer Interaction, Part II. doi:10.1007/978-3-642-39262-7_16

Matsumoto, H., & Yutaka, E. (2011). *Dynamic resource management in cloud environment*. Fujitsu Sci. Tech.

Portnoy, M. (2012). *Virtualization Essentials*. Hoboken, NJ: Wiley.

Radhakrishnan, N., Poorna Chelvan, N., & Ramkumar, D. (2012). *Utilization of Cloud Computing in E-learning Systems. In International of Cloud Computing* (pp. 208–213). Technologies, Applications & Management.

Rădulescu, Ş. (2014). *A perspective on E-Learning and Cloud Computing*. Social and Behavioral Sciences.

Rajkumar, B., Christian, V., & Thamarai, S. (2013). *Mastering Cloud Computing: Foundations and Applications Programming*. Burlington, MA: Elsevier Science.

Rountree, D., & Castrillo, I. (2013). *The Basics of Cloud Computing: Understanding the Fundamentals of Cloud Computing in Theory and Practice*. Burlington, MA: Elsevier Science.

Samra, H. E., Li, A. S., Soh, B., & AlZain, M. A. (2017). A Cloud-Based Architecture for Interactive E-Training. *International Journal of Knowledge Society Research*, *8*(1), 67–78. doi:10.4018/IJKSR.2017010104

Sorensen, B., Ramachandran, S., & Salvendy, S. (2007). *Simulation-Based Automated Intelligent Tutoring*. Heidelberg, Germany: Human Interface, Part II Springer-Verlag Berlin. doi:10.1007/978-3-540-73354-6_51

Sosinsky, B. (2010). *Cloud Computing Bible*. Hoboken, NJ: Wiley. doi:10.1002/9781118255674

Tian, W., Sheng, S., & Guoming, L. (2010). *A framework for implementing and managing platform as a service in a virtual cloud computing lab*. Education Technology and Computer Science. doi:10.1109/ETCS.2010.126

Tshiamo, W. B., Kgositau, M., & Magowe, M. (2017). Use of Information and Communication Technology by Health Care Providers for Continuing Professional Development in Botswana. In K. H. Moshi & K. J. Bwalya (Eds.), *Health Information Systems and the Advancement of Medical Practice in Developing Countries* (pp. 181–191). IGI Global. doi:10.4018/978-1-5225-2262-1.ch011

Uden, L., Tao, Y.-H., Yang, H.-C., & Ting, I.-H. (2014). *The 2nd International Workshop on Learning Technology for Education in Cloud.* Dordrecht: Springer.

Younge, A., Von, L., & Gregor, W. (2010). Efficient resource management for Cloud computing environments. *International Conference on Green Computing*, 357-364. 10.1109/GREENCOMP.2010.5598294

Zaigham, M. (2013). *Cloud Computing: Methods and Practical Approaches*. Dordrecht: Springer.

Chapter 16
Academicians Quality of Work Life and Its Influence on Value of Education

Chandra Sekhar Patro
Gayatri Vidya Parishad College of Engineering (Autonomous), India

ABSTRACT

The value of education is a dynamic and multi-dimensional concept that refers not only to the educational model but also to the institutional mission and its goals, as well as to the specific standards of the system, facility, program, or event. In today's competitive scenario, the academic institutions need to focus explicitly on providing effective and quality education to the students with the help of experienced academicians. The education quality would increase when the faculty members are having a better quality of work life, and this can be possible by providing better welfare conveniences to them by the academic institutions. Welfare amenities enable the staff members to live a quality and more satisfactory life. These facilities also help to keep their motivation levels high. The chapter articulates the existence of quality of work life programs in academic institutions in order to retain valuable faculty members who influence the value of education.

INTRODUCTION

Quality of Work Life (QWL) is probably the most powerful type of reward that an organization can offer to its employees in today's competitive business environment (Dargahi & Yazdi, 2007). The quality of work life has become an imperative issue to achieve the goals of an organization in every sector whether it is education, service sector, banking sector, tourism, manufacturing, or any other sector. The employees' commitment and productivity depends upon the dimensions of quality of work life such as job satisfaction, organizational commitment, reward and recognition, welfare facilities (Patro, 2017), participative management, work life balance, proper grievances handling, work environment etc (Yadav and Khanna, 2014). When an organization provides better quality of work, then it develops a healthy work environment as well as satisfied employee (Patro, 2013). High quality of work life results in better

DOI: 10.4018/978-1-5225-7347-0.ch016

organizational performance, effectiveness, innovativeness, etc. Quality of work life is the corroboration between employees and their organization as it improves the family life as well as work life of the individual (Patro, 2015).

Muftah (2011) mentioned that QWL was one of the key areas of human resource management that is attracting attention and research focus. It was a philosophy that considers people as the most important recourses in the organization and views them as an asset' to the organization rather than as costs. Patro (2012a) argued that welfare facilities play an important role in managing Human Resource in any organization. These facilities increase the job satisfaction level of employees and QWL, which in turn leads to overall development of organizations. By the result of improved quality of work life among employees, their involvement in job gets increased and results in increased productivity of the organization. Ahmad (2013) stated that the core pillar of QWL was to create a work environment that employee can work cooperatively with each other in order to achieve to organization objectives. Thus, QWL provides healthier, more satisfied and more productive employees, which in turn increases the efficiency, productivity and profitability of the organization (Sadique, 2003).

Most organizations today view QWL as an important mechanism, but do not formally link it to any of their strategic or business plans (Periman, 2006), which affects the employee job satisfaction and retention (Newaz, Ali, & Akhter, 2007). This scenario has created an urge for the private academic institute policy makers to identify and evaluate the underlying situations and reasons and has brought them to the consideration of the QWL issue. Faculty members play the key role in manipulating their services through providing better education and building the nation, thus faculty turnover has a crucial effect on the ultimate education system of any country (Hasan, Chowdhury, & Alam, 2008). To improve the value of education in the academic sector skilled, knowledgeable, experienced faculty are very important resource. Because of the influence of present socio economical factors, the retention rate of faculties are reducing day by day. In order to maintain this resource QWL interventions are more important. It is essential for the academic institutions to develop quality relation between its employees and working environment (Patro, 2014). In order to attract and retain employees, an organization has to develop a high quality of work life. Organizations by adopting QWL programmes ensure to create excellent work condition and job for its employees. Hence, QWL seeks to create such a work environment where the employees work co-operatively and make positive contribution in achieving organizational objectives.

In the present scenario the academic institutions are facing the challenges of increased competitions. In such environment, faculty retention is very important factor for competition and survival. This chapter aims at exploring the perception of employees towards quality of work life in academic institutions. The main objective is to know the impact of Quality of Work Life (QWL) on Value of Education (VOE) in academic institutions.

BACKGROUND

Many researchers identified that Quality Work Life of the faculty has an impact on the Value of Education that is provided to the students in academic institutions. Serey (2006) argued that QWL was associated with meaningful and satisfying work. It includes an opportunity to utilize one's skills and capacities, to confront challenges and situations that require self-initiative and self-direction; an activity should be practiced by the individuals in organization. Mariappan (2002) outlined the education sector that produces the human resource has a pivotal role in the quality movement that demands total quality approach in the education system to live up to the requirements of the industry. Thakkar *et al.* (2006) pointed towards the growing interest in applying TQM in education and for a wide variety of reasons. Some of the reasons include: pressures from industry for continuous upgrading of academic standards with changing technology; government schemes with allocation of funds which encourage research and teaching in the field of quality; increasing competition between various private and government academic institutions; and a reduction in the pool of funds for research and teaching (Temponi, 2005). Patro (2012b) in a study on total quality management in engineering education with a case study undertaken in various engineering colleges identifies and analyses the quality of engineering education at the educational institutions which adopt the total quality management system and recommends useful ingredients to improve the quality.

Shahbaji, *et al.* (2011) identified the relationship between the quality of work life and performance of Esfahan University and Esfahan medical University employees. According to this study, performance was directly related to adequate and fair compensation, safe and healthy work environment, development of human capacities, growth and security, social integration and work environment, constitutionalism (rule of law), work life space, and social relevance of work life. Out of these constructs developments of human capacities, social integration, constitutionalism, and work and life space were more effectively related to performance. They concluded that level of quality of work life was different from university to university. Hong, Tan and Bujang (2010) explained that respondents only had moderate work life quality. There were also moderate relationships between work life quality and work commitment, stress, and satisfaction but there were no differences in the quality of work life based on demographic variables. On studying the work commitment based on gender the researcher found the difference in the level of commitment. Research also revealed that there were also no differences in work stress and satisfaction based on the same demographic variables. Similarly, the findings suggested that work commitment, work stress and satisfaction were not significantly correlated. Improvement in QWL of the teachers will ensure that students gain benefits from teaching. Nimalathasan (2010) identified four factors of QWL practices such as, job benefits for family, physically safe, payment for work, and creativity of outside. Policy implications may be useful for overall improvement of QWL of academic professions. The study found that the Universities should provide job security, conducive working environment, research facilities, and overall career advancement opportunities for their academic professionals (Kumar and Deo, 2011). These factors will help the faculty to live quality work life and also help in providing improved quality of education to the students.

Tabassum, Rahman and Jahan (2010) found in their study that male employees perceived higher QWL than their female colleagues. In another study conducted by the same authors, it was found that there is a significant difference among the QWL of the employees of local private and foreign commercial banks of Bangladesh (Tabassum, Rahman, and Jahan, 2011). Hoque and Rahman (1999) found that QWL is important for job performance, job satisfaction, labor turnover, labor management relations which play a crucial role in determining the overall well-being of any industrial organization. They found that workers in private sector textile mills perceived significantly higher QWL than the workers in public sector. Faculty members indicated positive job satisfaction and would continue to stay in the same job only if they have opportunity for growth and development along with organizational prestige, financial factors. In this direction the major cause of disgruntlement was found to be advancement opportunity, organizational prestige and financial factors. So, the college administration must give due wait age to these factors as respondents have held these factors responsible for retaining them in their present jobs (Abbas *et al.*, 2010).

Sadique (2003) conducted a study on the employees of sugar mills and explored a significant difference between the white collar and blue collar employees' QWL. Hossain and Islam (1999) found a positive relationship between QWL and job satisfaction among government hospital nurses in Bangladesh. Uddin *et al.* (2006) also found a positive relationship between QWL and job satisfaction. Elias and Saha (1995) found in their research that female workers' quality of working life was significantly lower than that of their male counterparts in the tobacco industry. Wadud (1996) found that QWL was notably higher among private sector women employees than their counterparts in the public sector. The four major determinants of QWL, i.e. decision-making authority, growth and development, recognition and appreciation, and promotional avenues were missing and the teachers would like an environment that included these determinants (Subash and Parampal, 1993). If the QWL of teachers is below average then its resultant impact will be on teaching and research work and these are the basis for the progress of any society. QWL and Quality of life has a significant association in teaching environment (Umaselvi and Kumar, 2011).

QWL programmes provide opportunity for growth and development by facilitating training to the employees which consequently increases job satisfaction. QWL is concerned with creating work environment which is conducive and congenial. There is a significant relation between job satisfaction, personal growth, and team effectiveness even in the academic sector (Reena and Jayan, 2012). A high QWL is required for the growth of both the employees and the institutions. Bindu and Yashika (2014) found that QWL in Indian academic sector is of low level. A planned change in the working environment is required to improve QWL in academic sector. Training, redesign of work, workshops for knowledge enhancement and personal growth, valuable participation in decision making, modification in promotion scheme etc. are some of the ways through which we can improve QWL. QWL has direct bearing with productivity, as improved QWL will ultimately lead to higher productivity and job satisfaction (Ghasemizad *et al.*, 2012). A study on the university employees revealed that there is a positive relationship between job satisfaction and QWL dimensions. QWL significantly contributes towards increasing the job satisfaction or dissatisfaction depending upon the employee's negative or positive perception of QWL dimensions (Ganguly, 2010).

QWL of academicians, particularly in the Private Technical Institute, is not in a better condition. Factors such as salary and wages biasness between same qualified employees, advancement opportunity for growth is low, salary and job security issues are badly affecting the relationship with administration and academicians, dissatisfaction regarding leave flexibility etc. are responsible for low QWL of respondents (Vishwakarma *et al.*, 2013). Tabassum (2012) used Walton's proposed eight components of quality of work life to measure the relationship between the components of QWL and job satisfaction in faculty members of private universities in Bangladesh, the study revealed that all the components are positively associated with the job satisfaction of faculty. Islam *et al* (2009) investigated the relation between QWL dimensions and organization performance. The study found that QWL has positive significant relationship with job satisfaction. Variable Union policy has positive relation but no significant with job satisfaction and QWL is related to Job Satisfaction. Further, QWL has no significant relationship with Organizational Performance.

RESEARCH METHODOLOGY

The chapter focuses on the performance measurement of the academicians working in private academic institutions. Analysis of the efficiency of academic institutions is valuable not only from the point of view of faculty QWL but also the students, who are concerned with getting quality education from these institutions. A study was undertaken in fifteen private technical institutions at Visakhapatnam city to measure the QWL of the faculty and their performance, which impact on the quality of teaching in those institutions.

This study attempts to explain the various characteristics related to QWL in teaching environment and to find out association between the selected variables and dimensions of QWL. In this study, the primary data is collected from 15 academic institutions to measure the performance of the academic institutions through QWL questionnaire distributed to the faculty members of various academic institutions. The QWL parameters are taken from 7 (seven) academicians each from 15 private academic institutions who are having more than seven years of experience to find out their satisfaction level on their Quality of Work Life (QWL). Also the VOE questionnaire was distrusted to the students of various private academic institutions. The VOE parameters are taken from 10 students each from 15 technical institutions who are in their final semester to know about the Value of Education (VOE) provided by these academic institutions.

Questionnaire for QWL

The QWL Questionnaire is grouped into six categories with different parameters in each category (Patro, 2015):

1. **Compensation Benefits (CB):** It includes Fair and adequate salaries, Appreciation and Recognition to efforts, Reward and Increment system.
2. **Work Environment (WE):** It includes Work load and job stress, Safety and healthy working conditions, Organization culture and climate, Social relevance of work life.
3. **Growth Opportunity (GO):** It includes Career advancement opportunities, Personal growth and autonomy, Training and development activities.
4. **Welfare Measures (WM):** It includes Medical facilities, Housing facilities, Transport facilities, Canteen facilities, Recreation facilities.
5. **Employee-Employer Relationship (EE):** It includes Relation with Superiors/Management, Cooperation by the Management, Trust on Employee capabilities, and Participation in decision making.
6. **Job Satisfaction (JS):** Job security, Retention Policies, Balance between work and family, Self Esteem or Respect from Superiors, Equity and Grievance Handling.

Questionnaire for VOE

The VOE Questionnaire is grouped into five categories with different parameters in each category (Patro, 2015):

1. **Quality of Teaching (QT):** Knowledge on related subject, Communicate effectively in class room, Ability to stimulate interest and curiosity, Usage of modern learning process.
2. **Attitude and Behavior (AB):** Punctuality to class, Teachers' attitudes and behavior, Personal bearing, dignity and maturity, Class room discipline and control.
3. **Initiative and Responsibility (IR):** Organizing co-curricular activities, Initiative and resourcefulness, Focus given to performance standards, Concentration on weak students.
4. Problem Solving Skills (PS): Continue support for student-centered learning, Capacity to motivate the students, Train how to deal with the exams, Spare time for interaction with students outside the class room.
5. **Support and Cooperation (SC):** Communication with students and parents, Teachers' support and leadership, Emphasis on improvement of results, Appreciation and reward for good results.

STATISTICAL DATA ANALYSIS

The study analyzes the satisfaction level of the academicians on the identified factors of quality of work life parameters and the satisfaction of the students on the value of education parameters. The analysis shows the mean values of all the parameters related to Academicians QWL and Students VOE.

The descriptive statistics related to the satisfaction level of the academicians towards their QWL is shown in table 1. The analysis shows that Job Satisfaction (JS) and Employee-Employer Relationship (EE) are the highest rated parameters with a mean value of 4.08. The parameters Growth Opportunity (GO) and Compensation Benefits (CB) and second and third highest rated with mean values of 3.87 and 3.70 respectively. The parameters Welfare Measures (WM) and Work Environment (WE) are lowest rated with the mean value of 3.32 and 3.53. It can be observed that all the parameters got positive ratings by the respondents.

With regard to the compensation benefits, all the variables fair and adequate salaries, appreciation and recognition to efforts, and reward and increment system secured same ratings with a mean value of 3.7. In the case of work environment, the variable work load and job stress secured highest rating with a mean value of 4.40 followed by the variables social relevance of work life (μ=3.70) and organization culture and climate (μ=3.30). The variable safety and healthy working conditions got lowest rating of 2.70. For the parameter growth opportunity, the variable personal growth and autonomy got highest rating with a mean value of 4.30 followed by training and development activities (μ=3.70) and career advancement opportunities (μ=3.60). In the case of welfare measures, the variables medical facilities and housing facilities secured highest mean value of 4.70 followed by recreation facilities (μ=3.60). The variables, transport facilities and canteen facilities got lowest rating of 2.00 and 1.60 respectively. Regarding the employee-employer relationship, the variable participation in decision making got highest rating with a mean value of 4.70 followed by the variables trust on employee capabilities (μ=4.30) and cooperation by the management (μ=4.00). The variable Relation with Superiors/Management (μ=3.30) got lowest rating by the respondents. With reference to the parameter job satisfaction, Job security secured highest rating of 4.70 followed by retention policies (μ=4.30) and balance between work and family (μ=4.00). The variables self esteem or respect from superiors (μ=3.70) and equity and grievance handling (μ=3.70) also got positive rating from the respondents. Therefore, it can be observed that the welfare measures taken by the academic institutions can be improved so as to enhance the performance of the academicians.

Table 1. Descriptive Statistics of Academicians QWL (n=105)

Parameters		Mean
Compensation Benefits (CB):		**3.70**
	Fair and adequate salaries	3.70
	Appreciation and Recognition to efforts	3.70
	Reward and Increment system	3.70
Work Environment (WE):		**3.53**
	Work load and job stress	4.40
	Safety and healthy working conditions	2.70
	Organization culture and climate	3.30
	Social relevance of work life	3.70
Growth Opportunity (GO):		**3.87**
	Career advancement opportunities	3.60
	Personal growth and autonomy	4.30
	Training and development activities	3.70
Welfare Measures (WM):		**3.32**
	Medical facilities	4.70
	Housing facilities	4.70
	Transport facilities	2.00
	Canteen facilities	1.60
	Recreation facilities	3.60
Employee-Employer Relationship (EE):		**4.08**
	Relation with Superiors/Management	3.30
	Cooperation by the Management	4.00
	Trust on Employee capabilities	4.30
	Participation in decision making	4.70
Job Satisfaction (JS):		**4.08**
	Job security	4.70
	Retention Policies	4.30
	Balance between work and family	4.00
	Self Esteem or Respect from Superiors	3.70
	Equity and Grievance Handling	3.70

The descriptive statistics related to the satisfaction level of the students towards the VOE is shown in table 2. The analysis shows that the parameter Initiative and Responsibility (IR) scored highest rating with a mean value of 4.35. The parameters Attitude and Behaviour (AB) and Problem Solving Skills (PS) secured second and third highest ratings with mean values of 3.77 and 3.60 respectively. The parameters Support and Cooperation (SC) and Quality of Teaching (QT) got least ratings by the respondents with mean value of 3.45 and 3.23. It can be observed that all the parameters got positive ratings by the respondents.

Table 2. Descriptive Statistics of Students VOE (n=150)

Parameters		Mean
Quality of Teaching (QT):		**3.23**
	Knowledge on related subject	2.30
	Communicate effectively in class room	2.60
	Ability to stimulate interest and curiosity	3.70
	Usage of modern learning process	4.30
Attitude and Behaviour (AB):		**3.77**
	Teachers' attitudes and behavior	3.90
	Personal bearing, dignity and maturity	3.70
	Class room discipline and control	3.70
Initiative and Responsibility (IR):		**4.35**
	Organizing co-curricular activities	4.40
	Initiative and resourcefulness	4.70
	Focus given to performance standards	4.85
	Concentration on weak students	3.30
Problem Solving Skills (PS):		**3.60**
	Continue support for student-centered learning	4.00
	Capacity to motivate the students	3.70
	Train how to deal with the exams	2.70
	Spare time for interaction with students outside the class room	4.00
Support and Cooperation (SC):		**3.45**
	Communication with students and parents	3.70
	Teachers' support and leadership	3.70
	Emphasis on improvement of results	3.00
	Appreciation and reward for good results	3.40

With reference to the parameter quality of teaching, the variables usage of modern learning process secured highest rating with a mean value of 4.30 followed by the variable Ability to stimulate interest and curiosity (μ=3.70). The variables communicate effectively in class room (μ=2.60) and knowledge on related subject (μ=2.30) got lowest ratings from the respondents. In the case of Attitude and Behaviour, the variable Teachers' attitudes and behavior secured highest rating with a mean value of 3.90 followed by the variables personal bearing, dignity and maturity (μ=3.70) and classroom discipline and control (μ=3.70). The parameter initiative and responsibility reveals that the variable focus given to performance standards secured highest rating with a mean value of 4.85 followed by the variables initiative and resourcefulness (μ=4.70) and organizing co-curricular activities (μ=4.40). The variable concentration on weak students (μ=3.30) got lowest mean value by the respondents. With regard to problem solving skills, the variables continue support for student-centered learning and spare time for interaction with students outside the class room secured highest mean value of 4.00 followed by capacity to motivate the students (μ=3.70). The variable train how to deal with the exams (μ=2.70) secured lowest rating. In the case of support and cooperation, the variables communication with students and parents and teachers' support and leadership secured highest ratings with a mean value of 3.70 followed by the variable appreciation and reward for good results with a value of 3.40. The variable emphasis on improvement of results (μ=3.00) got lowest rating by the respondents. Thus, it can be observed that the academic institutions need to give more emphasis on the quality of teaching and support and cooperation from the academicians.

RESULTS AND DISCUSSION

Based on the analysis of collected data from the academicians and students of private academic institutions, the following are the QWL factors that are influencing the value of education:

1. It is identified that in some institutions which are having poor working environments, due to that the faculty are not able to concentrate on the quality of teaching.
2. In the low performance institutions due to heavy workload, job stress and unhealthy organizational culture and climate, the faculty is not able to deliver quality of knowledge to the students.
3. It is found that due to lack of trust on the employee capabilities and poor relationships with management, principal and head of the departments, the faculty is not involved in decision making.
4. The low performed academic institutions are lacking in providing an opportunity to learn new skills. The management is not concentrating on the training and development activities for the faculty in the institutions to develop their skills.
5. It is observed that in low performing institutions fair and adequate compensation benefits are not provided, which is affecting the quality of teaching provided to the students.
6. Due to lack of recognition to the efforts, proper of reward and increment system also the faculties are continuously changing the institutions.
7. The poor performed academic institutions also lacks in providing housing, transport, medical and other welfare facilities to the employees due to which they are not satisfied.
8. It is identified that respect towards the faculty and solving their grievances are not considered in some educational institutions.
9. Faculty retention is another important element that gives security to the faculty to work for the institutions in better manner is also neglected in some academic institutions.
10. The teachers who are not well-motivated and duly rewarded or recognized are the causes of quality of education.

It is also identified that due to the above factors the faculty members are not able to concentrate on the following elements:

- Provide effective knowledge on related subject;
- Communicate effectively in class room with the students;
- Ability to create interest and curiosity among the students;
- Change in teachers' attitudes and behavior;
- Class room discipline and control the students;
- Lack of initiative in organizing co-curricular activities;

- Focus given to performance standards;
- Concentrate and support weak students;
- Capacity to motivate the students;
- Spare time for interaction with students outside the class room;
- Lack of time to communicate with students and parents;
- Emphasis on improvement of results.

IMPACT OF FACULTY QWL ON STUDENT VOE

The literature review have shown that QWL is a multidimensional construct; these constructs have to be considering during the job design process which helps in providing VOE to the students and enhance their knowledge. The literature would also reveal the contribution of the ten important components of QWL which associated with the faculty satisfaction and their opinion on QWL in different sectors and their impact on the VOE.

1. **Work Environment:** Work environment is a place in which an individual works. It is a social and professional environment in which a faculty is supposed to interact with a number of students, and have to work in coordination with other staff members. If the organization is providing safety and healthy working conditions, a faculty can concentrate on the student's value of education. His attitude and behavior with the student also changes with a change in work environment.
2. **Organization Culture and Climate:** Organizational features such as policies and procedures, leadership style, operations, and general contextual factors of setting, all have a profound effect on how staff views the quality of work life. When an academic institution has good organizational culture and climate, it has an impact on the faculty and the student also.
3. **Relation and Co-Operation:** Relation and co-operation is a communication between management and faculty, concerning workplace decision, resolving conflicts and problems. QWL is associated with career development and career is evolving from such interaction of individuals within the organizations. When the management extends support and cooperation to the faculty, they can put more efforts in improving the quality of education provided to the students.
4. **Training and Development:** Training and development is an organizational activity aimed at bettering the performance of individuals and groups. To improve the performance efficiency of faculty the management can conduct workshops, seminars, faculty development programmes, etc. These programs help in improving the quality of working life of faculty and also the quality of teaching.
5. **Compensation and Rewards:** Compensation and rewards are motivational factors, the best performer is given the rewards, and this builds competition among the staff to work hard and achieve both organizational and individual goals. QWL is associated with satisfaction with wages, working hours and working conditions, equitable wages, equal employment opportunities and opportunities for advancement. This gives satisfaction to the faculty and helps in taking initiative and responsibility towards the students by providing them with quality of education.

6. **Welfare Facilities:** Facilities like food service, transportation, security, etc, play major role in actualization of the goals and objectives by satisfying both the physical and emotional needs of the faculty. This is one way to improve faculty productivity and morale. These facilities help in maintaining good communication with students and parents, support and put emphasis on improvement of students' performance standards.
7. **Job Satisfaction and Job Security:** Job satisfaction is the favorableness or un-favorableness with which employees view their work. Different aspects of the job, such as pay, promotions, fringe benefits, co-workers support, etc., are associated with levels of satisfaction. Permanent employment provides job security to the employees and improves their QWL. This makes the faculty to help stimulate the students' knowledge.
8. **Autonomy of Work:** In autonomous work groups, employees are given the freedom of decision making, workers themselves plan, coordinate and control work related activities. If the academic institutions provide appropriate authority to design work activities to the individual faculty, then it is highly possible that the work activities can match their needs that contribute to the quality of teaching.
9. **Adequacy of Resources:** Resources should match with stated objectives otherwise, faculty will not be able to attain the end result. Adequacy of resources has to do with availability of sufficient time and equipment, adequate information and help to complete assignments. Provision of adequate resources by the management helps the faculty to contribute more towards the quality of education to the students.
10. **Growth Opportunity:** It includes career advancement opportunities, personal growth and autonomy. When the faculty is satisfied with the opportunities provided to develop his knowledge and his career, it influences the same on the student's development.

The above ten components will motivate the faculty members towards their satisfaction on the QWL provided by the management's of various academic institutions. These components will help the faculty members to concentrate on the teaching abilities and give quality of education to the students.

MANAGERIAL IMPLICATIONS

The managerial implications are based on the personal discussions with the faculty members, students, experts and managements and also through distributed questionnaires of private academic institutions to improve the QWL of the faculty members and provide value of education (VOE) to the students:

1. The management of academic institutions has to put emphasis on the improvement of quality of work life of the faculty members, as it will give satisfaction to them and concentrate on quality of teaching to the students.
2. The teachers have to be well motivated and recognize their efforts by the management in order to improve their teaching ability.
3. The faculty has to pay more attention to increase the performance standards of the students and improvement of their practical skills.

4. The management and faculty have to communicate with the parents regularly and monitor the students on their performance.
5. They need to spare time for interaction with students outside the class room and motivate them to achieve their goals.
6. The staff members have to concentrate on the students' examination skills and give appropriate examination material and suggestions to improve their results.
7. The faculty must have the ability to create interest and curiosity among the students on the subject and provide effective knowledge on related subject.
8. The management has to focus on retaining the faculty for a long period of time, so that they can have good rapport with the students, colleagues and management.
9. Quality of education for the students must be given upmost preference in the academic institutions and this is possible only when the faculty members are provided with good organizational environment and culture.
10. It is the mutual responsibility of the management and the staff members to improve the quality of education to the students and improve their competitive spirit.

CONCLUSION

The chapter provides valuable implications for the private academic institutions that have a growing interest in attracting and retaining quality faculty members to distinguish them in providing value education. The research revealed the most positively and negatively perceived QWL dimensions by the faculty members of private academic institutions. Thus, the management should consider the policy implications based on the concerned issues of QWL improvement. It can be undoubtedly said that an improved QWL leads to a higher level of job satisfaction, which in turn reduces the employee turnover rate. That is why it will help to minimize the faculty member turnover rate that is currently prevailing in academic institutions.

Faculty members of private institutions play a significant role for economic growth by contributing their knowledge, skills and effort. In order to attain these goals the teacher should not only be committed and devoted but also competent and creative and for that matter they should be provided a better quality of wok life. QWL of academicians, particularly in the private academic institutions, is not in a better condition. Factors such as salary and wages biasness between same qualified faculty, advancement opportunity for growth is low, salary and job security issues are badly affecting the relationship with the management and academicians, etc., are responsible for low QWL of faculty members. These factors show the impact on the quality of education provided to the students of these private academic institutions.

A planned change in the working environment is required to improve QWL in academic sector. Training, redesign of work, workshops and seminars for knowledge enhancement and personal growth, valuable participation in decision making, modification in promotion scheme etc. are some of the ways through which we can improve QWL. Improved QWL is beneficial for both the faculty and the institute in providing QOE, so it's the mutual responsibility of the two. It is essential for the private academic institutions to develop quality relation between its staff and working environment. In order to attract and retain employees, an organization has to develop a high quality of work life. Hence, QWL seeks to create such a work environment where the employees work co-operatively and make positive contribution towards providing quality of education.

LIMITATIONS AND FUTURE RESEARCH DIRECTIONS

Although there are notable contributions in this area, especially for employee retention by ensuring QWL, the results of this study need to be viewed and acknowledged in light of its limitations. In this study only a few academic institutions were included and the sample is also limited to only one district. Thus, the findings cannot be generalized. Therefore, future research should be conducted on a larger scale by considering more private academic institutions to authenticate the faculty members' perceptions about QWL. Furthermore, to enhance the development of QWL initiatives in the HRM arena, the current QWL issues should be refined in terms of modern HR practices.

REFERENCES

Abbas, S. M., Vandana, P., & Anat, J. (2010). Job satisfaction in management faculties of a metropolitan and proximate area: A study on private colleges. *International Journal of Research in Commerce and Management*, *3*(4), 11–22.

Ahmad, S. (2013). Paradigms of Quality of Work Life. *Journal of Human Values*, *19*(1), 73–82. doi:10.1177/0971685812470345

Bindu, J., & Yashika, S. (2014). Quality of Work Life with Special Reference to Academic Sector. *Research Journal of Management Sciences*, *3*(1), 14–17.

Dargahi, H., & Yazdi, M. K. S. (2007). Quality of work life in Tehran University of Medical Sciences Hospitals' clinical laboratories employees. *Pakistan Journal of Medical Sciences*, *23*(4), 630–633.

Elias, M. S., & Saha, N. K. (2005). Environmental Pollution and Quality of Working Life in Tobacco Industries. *Journal of Life Earth Science*, *1*(1), 21–24.

Ganguly, R. (2010). Quality of work life and job satisfaction of a group of university employees. *Asian Journal of Management Research*, *1*(1), 209–216.

Ghasemizad, A., Amirian Zadeh, M., & Bagheri, S. (2012). A study of the relationship between teachers and principals' spiritual leadership, quality of work life, job satisfaction and productivity. *American Journal of Scientific Research*, *49*, 11–20.

Hasan, K. M. S., Chowdhury, M., & Alam, M. A. (2008). Faculty Turnover in Private Universities of Bangladesh-A Critical Evaluation. *Business and Economic Review*, *1*(2), 99–113.

Hong, K., Tan, K., & Bujang, S. (2010). Relationships Between Work Life Quality Of Teachers With Work Commitment, Stress And Satisfaction: A Study In Kuching, Sarawak, Malaysia. *Jurnal Teknologi*, *52*(1), 1–15. doi:10.11113/jt.v52.133

Hoque, M. E., & Rahman, A. (1999). Quality of working life & Job Behaviour of Workers in Bangladesh: A comparative study of private and public sectors. *Indian Journal of Industrial Relations*, *35*(2), 175–184.

Hossain, M. M., & Islam, M. T. (1991). QWL & Job Satisfaction of GVT Hospital Nurses in Bangladesh. *Indian Journal of Industrial Relations*, *35*(3).

Islam, M. Z., & Siengthai, S. (2009). Quality of work life and organizational performance: Empirical evidence from Dhaka Export Processing Zone. In *Proceeding of ILO Conference on Regulating for Decent Work*. Geneva: International Labour Office.

Kumar, D., & Deo, J. M. (2011). Stress and Work Life of College Teachers. *Journal of Academy of Indian Academy of Applied Psychology*, *37*, 78–85.

Muftah, H., & Lafi, H. (n.d.). Impact of QWL on employee satisfaction case of oil and gas industry in Qatar. *International Scientific Press, 1*(2), 107-134.

Nanjundeswaraswamy, T. S. (2013). Quality of Work life of Employees in Private Technical Institution. *International Journal of Qualitative Research*, *7*(3), 3–14.

Newaz, M. K., Ali, T., & Akhter, I. (2007). Employee Perception Regarding Turnover Decision- In Context of Bangladesh Banking Sector. *BRAC University Journal*, *4*(2), 67–74.

Nimalathasan, B. (2010). Quality of Work Life Practices of Academic Professionals in Bangladesh: A Factor Analytical Approach. *International Journal of Research in Commerce & Management*, *1*(7), 6–11.

Patro, C. S. (2012a). Employee Welfare Activities in Private Sector and Their Impact on Quality of Work Life. *International Journal of Productivity Management and Assessment Technologies*, *1*(2), 18–29. doi:10.4018/ijpmat.2012040102

Patro, C. S. (2012b). A Study on Adaptability of Total Quality Management in Engineering Education Sector. *International Journal of Quality Assurance in Engineering and Technology Education*, *2*(4), 25–37. doi:10.4018/ijqaete.2012100103

Patro, C. S. (2013). Work Life Balance Policies for Employees in Private Sector. *Global Research Thoughts, 2,* 9-13.

Patro, C. S. (2014). A Study on the Impact of Employee Retention Policies on Organisation Productivity in Private Sector. *International Journal of Asian Business and Information Management*, *5*(3), 48–63. doi:10.4018/ijabim.2014070104

Patro, C. S. (2015). A Study on the Impact of Faculty QWL on Quality of Education in Academic Institutions. *International Journal of Knowledge Society Research*, *6*(1), 1–16. doi:10.4018/IJKSR.2015010101

Patro, C. S. (2017). Employee Welfare Measures in Public and Private Sectors: A Comparative Analysis. In Public Health and Welfare: Concepts, Methodologies, Tools, and Applications (pp. 1026-1042). IGI Global. doi:10.4018/978-1-5225-1674-3.ch047

Periman, S. L. (2006). *Human Resource Innovators, integrating quality of work life into organizational performance*. Retrieved May 15, 2018, from http://www.quaiindia.com

Reena, G., & Jayan, C. (2012). Role Of Quality of Work Life on The Job Attitude and Personal Effectiveness of Engineering College Teachers. *South Asian Academic Research Journals, 2*(6).

Saad, H. A., Samah, A. J. A., & Juhdi, N. (2008). Employees' perception on quality work life and job satisfaction in a private higher learning institution. *International Review of Business Research Papers*, *4*(3), 23–34.

Sadique, Z. (2003). Quality of Work Life among White Collar and Blue Collar Employees. *Journal of the Institute of Bangladesh Studies*, *26*, 169–174.

Serey, T. T. (2006). Choosing a Robust Quality of Work Life. *Business Forum, 27*(2), 7-10.

Shahbazi, B., Shokrzadeh, S., Bejani, H., Malekinia, E., & Ghoroneh, D. (2011). A Survey of Relationship between the Quality of Work Life and Performance of Department Chairpersons of Esfahan University and Esfahan Medical Science University. *Procedia: Social and Behavioral Sciences*, *30*(1), 1555–1560. doi:10.1016/j.sbspro.2011.10.302

Tabassum, A. (2012). Interrelations between Quality of Work Life Dimensions and Faculty Member Job Satisfaction in the Private Universities of Bangladesh. *European Journal of Business and Management*, *4*(2), 78–89.

Tabassum, A., Rahman, T., & Jahan, K. (2010). Quality of Work Life among the Male and Female Employees of Private Commercial Banks in Bangladesh. In *Proceedings of International Conference on Business Competencies in a Changing Global Environment*. South East University.

Tabassum, A., Rahman, T., & Jahan, K. (2011). A Comparative Analysis of Quality of Work Life among the Employees of Local and Foreign Banks in Bangladesh. *World Journal of Social Science*, *1*(1), 17–33.

Tabassum, A., Rahman, T., & Jahan, K. (2012). An Evaluation of the Quality of Work Life: A Study of the Faculty Members of Private Universities in Bangladesh. *ABAC Journal*, *32*(3), 36–57.

Temponi, C. (2005). Continuous improvement framework: Implications for academia. *Quality Assurance in Education*, *13*(1), 17–36. doi:10.1108/09684880510578632

Thakkar, J., Deshmukh, S. G., & Shastree, A. (2006). Total quality management (TQM) in self-financed technical institutions: A quality function deployment (QFD) and force field analysis approach. *Quality Assurance in Education*, *14*(1), 54–74. doi:10.1108/09684880610643610

Uddin, M. T., Islam, M. T., & Ullah, M. O. (2006). A Study on the Quality of Nurses of Government Hospitals in Bangladesh. *Proc. Pakistan Acad. Science*, *43*(2), 121–129.

Umaselvi, B. P., & Kumar, S. (2011). *Quality of Work Life: Perception of College Teachers*. Hallamark Business School, Trichy, Tamilnadu, India, Paper (27868).

Vishwakarma A. C., Lakhawat P. S., & Poonam. (2013). Study on quality of work life among academicians of private technical educational institutions: Allahabad city, Uttar Pradesh. *International Research journal of Management Science and Technology, 4*(3), 51-60.

Wadud, N. (1996). Job Stress & QWL among Working Women. *Bangladesh Psychological Studies*, *6*, 31–37.

Yadav, R., & Khanna, A. (2014). Literature Review on Quality of Work Life and Their Dimensions. *IOSR Journal of Humanities and Social Science*, *19*(9), 71–80. doi:10.9790/0837-19957180

KEY TERMS AND DEFINITIONS

Behavior: It refers to the way in which one acts or conducts oneself, especially towards others.

Education: Education is the process of facilitating learning, or the acquisition of knowledge, skills, values, beliefs, and habits.

Effectiveness: It is referred as the capability of producing a desired result or the ability to produce desired output.

Opportunity: An opportunity is a situation in which it is possible for you to do something that you want to do.

Perception: It is referred as the ability to see, hear, or become aware of something through the senses.

Productivity: Productivity is an economic measure of output per unit of input. Inputs include labor and capital, while output is typically measured in revenues and other gross domestic product (GDP) components such as business inventories.

Quality of Work Life (QWL): QWL is referred as the overall quality of an individual's working life. QWL is considered as a sub-concept of the broad concept of quality of life, which refers to the overall quality of an individual's life.

Satisfaction: Satisfaction is referred as the customer level of approval when comparing a product's perceived performance with his or her expectations.

Compilation of References

Abasiattai, M. B. (1987). *Akwa Ibom and Cross River State: The land, the people and the culture, Calabar*. Wusen Press.

Abbas, S. M., Vandana, P., & Anat, J. (2010). Job satisfaction in management faculties of a metropolitan and proximate area: A study on private colleges. *International Journal of Research in Commerce and Management*, *3*(4), 11–22.

Abdelaziz, M., Kamel, S., Karam, O., & Abdelrahman, A. (2011). Evaluation of E-learning program versus traditional lecture instruction for undergraduate nursing students in a faculty of nursing. *Teaching and Learning in Nursing*, *6*(2), 50–58. doi:10.1016/j.teln.2010.10.003

Abduraman, A. E., Berrani, S. A., & Mérialdo B. (2011). TV program structuring techniques: A review. In *TV Content Analysis: Techniques and Applications*. Academic Press.

Abhichandani, T. (2008). *Evaluation of E-Government Initiatives for Citizen-Centric Delivery: Analysis of Online Public Transit Information Services*. VDM Verlag Publishing.

Abu Dhabi. (2008). *The Abu Dhabi Economic Vision 2030*. Abu Dhabi: The Government of Abu Dhabi.

Abu-Libdeh, H., & Princehouse, L. (2010). RACS: a case for cloud storage diversity. In *Proceedings of the 1st ACM symposium on Cloud computing*. ACM. 10.1145/1807128.1807165

Academic research. (2018). *Protege Ontology Library*. Retrieved from http://protegewiki.stanford.edu/wiki/Protege_Ontology_Library

Adamko, M., Navrat, P., & Kovarova, A. (2015). Personalised Recommendation of Who to Follow Based on Fellowship of Followers. *Editorial Policy, 1*.

Adamson, L. B., & Chance, S. E. (1989). Coordinating Attention to People, Objects, and Language. In A. M. Wetherby, S. F. Warren, & J. Reichle (Eds.), *Transitions in Prelinguistic Communication* (pp. 15–38). Baltimore, MD: Paul H. Brookes.

Adamu, S. O. (1978). *The Nigerian Statistical System: Retrospect and Prospect*. Ibadan: University of Ibadan Press.

Adekola, J., Dale, V. M., & Gardiner, K. (2017). Development of an institutional framework to guide transitions into enhanced blended learning in higher education. *Research in Learning Technology*, 25.

Adinolfi, P., D'Avanzo, E., Lytras, M. D., Novo-Corti, I., & Picatoste, J. (2016). Sentiment analysis to evaluate teaching performance. *International Journal of Knowledge Society Research*, *7*(4), 86–107. doi:10.4018/IJKSR.2016100108

Agboola, T., & Mabawonku, A. O. (1996). Indigenous knowledge environmental education and sanitation: application to an African city. In D. M. Warren, L. Egunjobi, & B. Wahab (Eds.), Indigenous knowledge in education. Ibadan: Indigenous Knowledge Study Group.

Aggeliki, D., Dimitri, T., & Timos, S. (2015). *Top-k-size keyword search on tree structured data*. Elsevier Information Systems; doi:10.1016/j.is.2014.07.002

Agrawal, D., & El Abbadi, A. (2009). Database Management as a Service: Challenges and Opportunities. In *Proceedings of The 2009 25th International Conference on Data Engineering*. IEEE. 10.1109/ICDE.2009.151

Ahmad, S. (2013). Paradigms of Quality of Work Life. *Journal of Human Values*, *19*(1), 73–82. doi:10.1177/0971685812470345

Ahn, J. B.-K. (2003). Sexual norms and morality among the Agikuyu people of Kenya. In M. Katola (Ed.), MIASMU Research Integration Papers to Moral Teaching and Practices of African Religion. Academic Press.

Aity, A. (1994). Seclusion and nurse education. In Seclusion and mental health (pp. 110-126). London: Chapman & Hall.

Ajzen, I. (1991). The theory of planned behavior. *Organizational Behavior and Human Decision Processes*, *50*(2), 179–211. doi:10.1016/0749-5978(91)90020-T

Al-Almaee, S. M., & Khayyat, W. R. (2010). Strategic planning and supporting initiatives: Saudi e-government program. In *Proceedings of the10th European Conference on e-Government, National Centre for Taxation Studies*. University of Limerick.

Alba, F. (2015). International migration: Consolidation of emerging patterns. *Demos (Mexico City, Mexico)*, (13), 10–11.

Al-Huneiti, R., Hunaiti, Z., & Al-Masaeed, S. (2018). Nurses' Attitudes Towards E-Learning for E-Health Education. In I. R. Association (Ed.), *Nursing Education, Administration, and Informatics: Breakthroughs in Research and Practice* (pp. 67–74). IGI Global. doi:10.4018/978-1-5225-5490-5.ch005

Al-Jenaibi, B. (2010). Differences between gender treatments in the Work Force. *Cross-Cultural Communication*, *6*(2), 63–74.

Al-Jenaibi, B. (2011). The practice of Public Relations Departments in increasing social support in the diverse workplaces of the United Arab Emirates. *Cross-Cultural Communication*, *7*(3), 41–54.

Al-Jenaibi, B. (2012). The scope and impact of workplace diversity in the United Arab Emirates–A preliminary study. *Malaysia Journal of Society and Space*, *8*(1), 1–14.

Al-Jenaibi, B. (2017). Corporate social responsibility: Case study in UAE organizations. *International Journal of Knowledge Society Research*, *8*(4), 85–100. doi:10.4018/IJKSR.2017100105

Al-Jenaibi, B. (2017). E-governance: A Study of Public Agencies in the United Arab Emirates, South. *Asian Journal of Management*, *24*(1), 2017.

Al-khaleej Times. (2012). *UAE among emerging e-Government leaders*. Retrieved on June 8,2015, from http://www.khaleejtimes.com/mobile/inside.asp?xfile=/data/todayevent/2012/September/todayevent_September18.xml&section=todayevent

Al-Khouri, A. M (2012). E-Government Strategies The Case of the United Arab Emirates (UAE). *European Journal of ePractice,* 135-140.

Al-Khouri, A. M. (2013). *Technological and Mobility Trends in E-Government*. Retrieved from www.id.gov.ae/assets/zbOAvc6B.pdf.aspx

Al-Khouri, A. M., & Bal, J. (2007). Electronic Government in the GCC Countries. *The International Journal of Social Sciences (Islamabad)*, *1*(2), 83–98.

Allan, P. (1989). Nursing education - a luxury or necessity? Current Issues in Nursing.

Allan, K., & Kate, B. (2006). *Forbidden Words: Taboo and the Censoring of Language*. Cambridge, UK: Cambridge University Press; doi:10.1017/CBO9780511617881

Al-Ohali, Y., Al-Oraij, A. A., & Shahzad, B. (2011, July). KSU News Portal: A Case Study. In The 2011 international conference on Internet Computing (ICOMP'11) (pp. 18-21). Academic Press.

Al-Shafi, S. (2012). Free Wireless Internet park Services: An Investigation of Technology Adoption in Qatar from a Citizens Perspective. *Journal of Cases on Information Technology*, *10*(3), 21–34. doi:10.4018/jcit.2008070103

Al-Shafi, S., & Weerakkody, V. (2008). The Use Of Wireless Internet Parks To Facilitate Adoption And Diffusion Of E-Government Services: An Empirical Study In Qatar. *Proceedings of the 14th Americas Conference on Information Systems (AMCIS 2008).*

Alshouiby, A., & Bakry, S. (2017). A Framework of Key E-Services Issues. *International Journal of Knowledge Society Research*, *8*(4), 1–22. doi:10.4018/IJKSR.2017100101

Al-Sulaiym, H., Al-Muammar, A., & Bakry, S. (1997). An Overview of Public, Research and Private Data Networks. *International Journal of Network Management*, *7*(2), 103–113. doi:10.1002/(SICI)1099-1190(199703/04)7:2<103::AID-NEM230>3.0.CO;2-0

Alwagait, E., Shahzad, B., & Alim, S. (2014). Impact of social media usage on students' academic performance in Saudi Arabia. *Computers in Human Behavior.*

AlZain, M. A. (2014). *Data security, data management and performance evaluation in a multi-cloud computing model* (PhD thesis). La Trobe University.

AlZain, M. A., Li, A. S., Soh, B., & Masud, M. (2016). Byzantine fault-tolerant architecture in cloud data management. *International Journal of Knowledge Society Research*, *7*(3), 86–98. doi:10.4018/IJKSR.2016070106

AlZain, M. A., Li, A. S., Soh, B., & Pardede, E. (2015). Multi-Cloud Data Management using Shamir's Secret Sharing and Quantum Byzantine Agreement Schemes. *International Journal of Cloud Applications and Computing*, *5*(3), 35–52. doi:10.4018/IJCAC.2015070103

AlZain, M. A., & Pardede, E. (2012). Cloud Computing Security: From Single to Multi-clouds. In *Proceedings of The 2012 45th Hawaii International Conference on System Science (HICSS).* Maui, HI: IEEE. 10.1109/HICSS.2012.153

AlZain, M. A., & Soh, B. (2011). MCDB: Using Multi-clouds to Ensure Security in Cloud Computing. In *Proceedings of The 2011 Ninth International Conference on Dependable, Autonomic and Secure Computing (DASC).* Sydney, Australia: IEEE. 10.1109/DASC.2011.133

AlZain, M. A., & Soh, B. (2012). A New Approach Using Redundancy Technique to Improve Security in Cloud Computing. In *Proceedings of The 2012 International Conference on Cyber Security, Cyber Warfare and Digital Forensic (CyberSec12).* Kuala Lumpur, Malaysia: IEEE. 10.1109/CyberSec.2012.6246174

AlZain, M. A., & Soh, B. (2013). A Byzantine Fault Tolerance Model for a Multi-cloud Computing. In *Proceeding of The 2013 16th International Conference on Computational Science and Engineering CSE.* Sydney, Australia: IEEE. 10.1109/CSE.2013.30

AlZain, M. A., Soh, B., & Pardede, E. (2013). A Survey on Data Security Issues in Cloud Computing: From Single to Multi-Clouds. *Journal of Software*, *8*(5), 1068–1078. doi:10.4304/jsw.8.5.1068-1078

Amazon. (2006). *Amazon Web Services. Web services licensing agreement.* Author.

An, A., & Kargar, M. (2011). Keyword search in graphs: Finding r-cliques. *Proceedings of the VLDB Endowment*, *4*(10), 681-692. DOI: 10.14778/2021017.2021025

Andemariam, M. (2001). Place of norms in Gikuyu morality. In L. Magesa (Ed.), MIASMU Research Integration Papers to Moral Teaching and Practices of African Religion. Academic Press.

Apache Jena. (2018). *The Apache Jena project*. Retrieved from https://jena.apache.org/

Aranguiz, M. S., & Quintana, M. G. (2016). ICT resources to improve learning in higher education. *International Journal of Knowledge Society Research*, *7*(4), 1–11. doi:10.4018/IJKSR.2016100101

Arenas, E. (2015). Affordances of learning technologies in higher education multicultural environments. *Electronic Journal of E-Learning*, *13*(4), 217–227.

Arkorful, V., & Abaidoo, N. (2015). The role of e-learning, advantages and disadvantages of its adoption in higher education. *International Journal of Instructional Technology and Distance Learning*, *12*(1), 29–42.

Arvis, V., & Debain, C., Berducat, M., & Benassi, A. (2004). Generalization of the Co-occurence Matrix for Color Images: Application to Color Textures Classification. *Image Analysis & Stereology*.

Asante, K. W. (2000). *Zimbabwe Dance: Rhythmic Forces, Ancestral Voices-An Aesthetic Analysis*. Trenton: African World Press, Inc.

Asghar, M. N., Hussain, F., & Manton, R. (2014). Video indexing: A survey. *Int. J. of Computer and Information Technology*. doi:10.1007/978-3-319-24306-1_2

As-Saber, S., Hossain, K., & Srivastava, A. (2007). Technology, society and e-government: In search of an eclectic framework. *Electronic Government. International Journal (Toronto, Ont.)*, *4*(2), 156–178.

Athanasakos, K. C., Doulamis, A. D., & Karanikolas, N. N. (2007). A Signature Tree Content-based Image Retrieval System. *10th International Conference on Computer Graphics and Artificial Intelligence*.

Australia. (2003). *An Over Review of the E-government Benefits Study*. Retrieved April 20, 2015 from: www.finance.gov.au/agimo-archive/__data/assets/file/.../benefits.pdf

Awan, M. (2003). E-Government: Assessment of GCC (Gulf Co-operating Council) countries and service provided. *Lecture Notes in Computer Science*, *2739*, 500–503. doi:10.1007/10929179_94

Aydın, M. (2011). Values, functions and morals. *Eğitime Bakış*, *19*, 39–45.

Bacanlı, H. (2011). Value is valuable. *Eğitime Bakış*, *19*, 18–21.

Backus, M. (2001). *E-Governance and Developing Countries, Introduction and examples, Research published on April 2001*. Retrieved May 12, 2015 from: from www.iicd.org/files/report3. doc

Bahrain. (2008). *Our Vision. From Regional Pioneer to Global Contender. The Economic Vision 2030 for Bahrain.* Manama: Academic Press.

Bakerman, J., Pazdernik, K., Wilson, A., Fairchild, G., & Bahran, R. (2018). Twitter Geolocation: A Hybrid Approach. *ACM Transactions on Knowledge Discovery from Data*, *12*(3), 34.

Bakhsh, S. T., Tahir, S., Altalhi, A. H., Subhan, F., & Sheikh, M. A. (2016). Resource optimization routing and scatternet formation protocols for Bluetooth networks. *International Journal of Knowledge Society Research*, *7*(2), 84–123. doi:10.4018/IJKSR.2016040106

Bakry, A., & Bakry, S. (2005). Enterprise resource planning: A review and a STOPE view. *International Journal of Network Management*, *15*(5), 363–370. doi:10.1002/nem.584

Bakry, S. (2003). Development of security policies for private networks. *International Journal of Network Management*, *13*(3), 203–210. doi:10.1002/nem.472

Bakry, S. (2004). Development of e-government: A STOPE view. *International Journal of Network Management*, *14*(5), 339–350. doi:10.1002/nem.529

Bakry, S., & Bakry, F. (2001). A strategic view for the development of e-business. *International Journal of Network Management*, *11*(2), 103–112. doi:10.1002/nem.395

Bakry, S., Bakry, Z., & Muhaya, F. (2016). An Approach for the Assessment of E-Government Benefits to Governments, citizens and Businesses. *Journal of Socialomics*, *05*(02). doi:10.4172/2167-0358.1000155

Balkin, J. M. (2005). Digital speech and democratic culture: A theory of freedom of expression for the information society. In A. D. Moore (Ed.), *Information Ethics, Privacy, Property, and Power* (pp. 297–354). Seattle, WA: University of Washington Press.

Bandara, I., Ioras, F., & Maher, K. (2014). Cyber security concerns in e-learning education. *Proceedings of ICERI2014 Conference*.

Bao, Z., & Wu, H. (2011). Object-oriented XML keyword search. In *International Conference on Conceptual Modeling* (pp. 402-410). Academic Press. DOI:10.1007/978-3-642-24606-7_30

Bao, Z., Chen, Z., Guo, J., Ling, T. W., Lin, X., Wang, W., & Zhou, J. (2012). Fast SLCA and ELCA computation for XML keyword queries based on set intersection. In *IEEE 28th International Conference on Data Engineering* (pp. 905-916). IEEE. DOI: 10.1109/ICDE.2012.75

Bao, Z., Lee, M. L., Ling, T. W., & Zeng, Z. (2013). A semantic approach to keyword search over relational databases. In *International Conference on Conceptual Modeling* (pp. 241-254). Academic Press. DOI: 10.1007/978-3-642-41924-9_21

Bao, Z., Lu, J., Ling, T. W., Wu, H., & Xu, L. (2010). An effective object-level XML keyword search. In *International Conference on Database Systems for Advanced Applications* (pp. 93-109). Academic Press. DOI: 10.1007/978-3-642-12026-8_10

Barth, J., Fietkiewicz, K. J., Gremm, J., Hartmann, S., Henkel, M., Ilhan, A., … Stock, W. G. (2017b). Informationswissenschaft in der Urbanistik. Teil 1: Konzeptioneller Forschungsrahmen und Methoden. *Information – Wissenschaft & Praxis, 68*(5-6), 365-377.

Barth, J., Fietkiewicz, K. J., Gremm, J., Hartmann, S., Henkel, M., Ilhan, A., … Stock, W. G. (2018). Informationswissenschaft in der Urbanistik. Teil 2: Erste empirische Ergebnisse zu smarten Städten. *Information – Wissenschaft & Praxis, 69*(1), 31-46.

Barth, J., Fietkiewicz, K. J., Gremm, J., Hartmann, S., Ilhan, A., Mainka, A., ... Stock, W. G. (2017a). Informational urbanism. A conceptual framework of smart cities. In *Proceedings of the 50th Hawaii International Conference on System Science* (pp. 2814-2823). Washington, DC: IEEE Computer Society. 10.24251/HICSS.2017.340

Basavaraj, S., Anami, Ramesh, S., Wadawadagi, & Veerappa, B. (2014). Machine Learning Techniques in Web Content Mining: A Comparative Analysis. *Journal of Information & Knowledge Management, 13*(1).

Belkin, N. J. (1980). Anomalous states of knowledge as a basis for information retrieval. *Canadian Journal of Information Science*, *5*(1), 133–143.

Ben Kharrat, F., Elkhleifi, A., & Faiz, R. (2016). Improving collaborative filtering algorithms: Sentiment-based approach in social network. *International Journal of Knowledge Society Research*, *7*(3), 99–118. doi:10.4018/IJKSR.2016070107

Benbunan-Fich, R., & Hiltz, S. R. (1999). Impacts of Asynchronous Learning Networks on Individual and Group Problem Solving: A Field Experiment. *Group Decision and Negotiation*, *8*(5), 409–426. doi:10.1023/A:1008669710763

Bergstra, J., Yamins, D., & Cox, D. (2013). Making a science of model search: Hyperparameter optimization in hundreds of dimensions for vision architectures. In *Proceedings of The 30th International Conference on Machine Learning* (pp. 115-123). Academic Press.

Berns, A., Isla-Montes, J., Palomo-Duarte, M., & Dodero, J. (2016). Motivation, students' needs and learning outcomes: A hybrid game-based app for enhanced language learning. *SpringerPlus*, *5*(1), 1–23. doi:10.118640064-016-2971-1 PMID:27547679

Beyer, B. K. (1988). *Developing a thinking skills programme*. Boston: Allyn & Bacon.

Beynon-Davies, P. (2007). Models for e-government, Transforming Government: People. *Process Policy*, *1*(1), 7–28.

Bhalotia, G., Chakrabarti, S., Hulgeri, A., Nakhe, C., & Sudarshan, S. (2002). Keyword searching and browsing in databases using BANKS. In *18th International Conference on Data Engineering, 2002. Proceedings,* (pp. 431-440). Academic Press. 10.1109/ICDE.2002.994756

Bhattacharya, D., Gulla, U., & Gupta, M. P. (2012). E-service quality model for Indian government portals: Citizens' perspective. *Journal of Enterprise Information Management*, *25*(3), 246–271. doi:10.1108/17410391211224408

Bicchieri, C. (2006). *The Nature and Dynamics of Social Norms*. New York: Cambridge University Press.

Bin-Abbas, H., & Bakry, S. (2012). Knowledge Management. *International Journal of Knowledge Society Research*, *3*(3), 58–67. doi:10.4018/jksr.2012070105

Bindu, J., & Yashika, S. (2014). Quality of Work Life with Special Reference to Academic Sector. *Research Journal of Management Sciences*, *3*(1), 14–17.

Biricik, G., & Diri, B. (2009). Impact of a New Attribute Extraction Algorithm on Web Page Classification. In *The 2009 International Conference on Data Mining* (pp. 481-485). Las Vegas, NV: Academic Press.

Bishop, C. M. (2006). *Pattern recognition and machine learning*. Springer.

Black, M., Chapman, D., & Clark, A. (2018). The Enhanced Virtual Laboratory: Extending Cyber Security Awareness through a Web-based Laboratory. *Information Systems Education Journal*, *16*(6), 4.

Black, T. (2006). Helping Novice Programmers Succeed. *Journal of Computing Sciences in Colleges*, *22*(2), 109–114.

Blau, I., & Shamir-Inbal, T. (2017). Re-designed flipped learning model in an academic course: The role of co-creation and co-regulation. *Computers & Education*, *115*, 69–81. doi:10.1016/j.compedu.2017.07.014

Bloomfield, J., Cornish, J., Parry, A., Pegram, A., & Moore, J. (2013). Clinical skills education for graduate-entry nursing students: Enhancing learning using a multimodal approach. *Nurse Education Today*, *33*(3), 247–252. doi:10.1016/j.nedt.2011.11.009 PMID:22178595

Bloomfield, J., & Jones, A. (2013). Using e-learning to support clinical skills acquisition: Exploring the experiences and perceptions of graduate first-year pre-registration nursing students — A mixed method study. *Nurse Education Today*, *33*(12), 1605–1611. doi:10.1016/j.nedt.2013.01.024 PMID:23473860

Bogin, B., & Varela-Silva, M. I. (2010). Leg Length, Body Proportion, and Health: A Review with a Note on Beauty. *International Journal of Environmental Research and Public Health*, *7*(3), 1047–1075. doi:10.3390/ijerph7031047 PMID:20617018

Bojars, U., & Breslin, J. (2018). *Sioc core ontology specification*. Retrieved from http://rdfs.org/sioc/spec/

Bojars, U., Breslin, J., Peristeras, V., Tummarello, G., & Decker, S. (2008). Interlinking the Social Web with Semantic. *Journal IEEE Intelligent Systems*, *23*(3), 29–40. doi:10.1109/MIS.2008.50

Boss, J. A. (1999). *Analyzing Moral Issues*. Belmont: Mayfield Publishing Company.

Bourdillon, M. F. C. (1987). *The Shona Peoples: An Ethnography of the Contemporary Shona, with special Reference to their Religion*. Gweru: Mambo Press.

Bowers, K. D., & Juels, A. (2009). HAIL: A high-availability and integrity layer for cloud storage. In *Proceedings of the 16th ACM conference on Computer and communications security*. ACM. 10.1145/1653662.1653686

Brahimi, T., Sarirete, A., & Ibrahim, R. M. (2016). The impact of accreditation on student learning outcomes. *International Journal of Knowledge Society Research*, *7*(4), 51–62. doi:10.4018/IJKSR.2016100105

Braun, V., & Clarke, V. (2006). Using thematic analysis in psychology. *Qualitative Research in Psychology*, *3*(2), 77–101. doi:10.1191/1478088706qp063oa

Brickley, D., & Miller, L. (2015). *FOAF (2000-2015+*. Retrieved from http://www.foaf-project.org/

BRICS. (2017). *BRICS Information Sharing and Exchange Platform*. BRICS Business Council. Retrieved from http://www.brics-info.org/

Brookfield, S. (1987). *Developing critical thinkers*. Open University Press.

Browne, M. N., & Keeley-Vasudeva, M. L. (1992). Classroom controversy as an antidote for the spage model of learning. *College Student Journal*, *26*, 368–373.

Burn, J. & Robins, G. (2003). Moving towards e-government: a case study of organizational change processes. *Logistics Information Management, 3*(2), 25-35.

Button, D., Harrington, A., & Belan, I. (2014). E-learning & information communication technology (ICT) in nursing education: A review of the literature. *Nurse Education Today*, *34*(10), 1311–1323. doi:10.1016/j.nedt.2013.05.002 PMID:23786869

Bwalya, K. J., & Healy, M. (2010). Harnessing e-government adoption in the SADC region: a conceptual underpinning. *Electronic Journal of E-Government, 8*(1), 23-32.

Bwalya, K. J. (2012). *E-Government in Emerging Economies: Adoption, E-Participation, and Legal Frameworks*. Hershey, PA: IGI Global Publishing.

Cachin, C., & Haas, R. (2010). Dependable storage in the Intercloud. *IBM Research*, *3783*, 1–6.

Cadorin, L., Rei, A., Dante, A., Bulfone, T., Viera, G., & Palese, A. (2015). Enhancing self-directed learning among Italian nursing students: A pre- and post-intervention study. *Nurse Education Today*, *35*(6), 746–753. doi:10.1016/j.nedt.2015.02.004 PMID:25735910

Camera. (2018). *Protege Ontology Library*. Retrieved from http://protegewiki.stanford.edu/wiki/Protege_Ontology_Library

Cao, J., & Nguyen, K. (2010). Exploit keyword query semantics and structure of data for effective xml keyword search. In *Proceedings of the Twenty-First Australasian Conference on Database Technologies* (*vol. 104*, pp. 133-140). Academic Press.

Carlos Filipe,, F., António,, M., Lino,, F., & João,, B., António, S., & Carlos, R. (2015). Context aware middleware in ambient intelligent environments. *International Journal on Computer Science and Engineering*, *10*(4), 347–358.

Carrillo, F. J. (2006). *Knowledge Cities: Approaches, Experiences, and Perspectives*. New York, NY: Butterworth Heinemann.

Carrillo, F. J., Yigitcanlar, T., García, B., & Lönnqvist, A. (2014). *Knowledge and the City: Concepts, Applications and Trends of Knowledge-based Urban Development*. New York, NY: Routledge.

Carroll, A. B. (2004). Managing ethically with global stakeholders: A present and future challenge. *Academy of Management Review*, *18*(2), 114–120.

Cascio, P. D. (2017). Unsettling trends in academia: A review of the predicament of minorities and members of the LGBTQ community in higher educational institutions in America. *International Journal of Knowledge Society Research*, *8*(2), 36–43. doi:10.4018/IJKSR.2017040103

Castells, M. (1989). *The Informational City: Information Technology, Economic Restructuring, and the Urban-regional Process*. Oxford, UK: Basil Blackwell.

Castells, M. (2009). *The Rise of the Network Society: The Information Age - Economy, Society, and Culture* (2nd ed.; Vol. 1). Chichester, UK: John Wiley & Sons. doi:10.1002/9781444319514

Castro, M., & Liskov, B. (1999). Practical Byzantine Fault Tolerance. *Proceedings of the 1999 3rd Symposium on Operating Systems Design and Implementation.*

Çengel, Y. (2011). Human values for a safety future. *Eğitime Bakış*, *19*, 13–17.

Chaffee, J. (1992). Critical thinking skills: The cornerstone of development education. *Journal of Development Education, 15*(3), 2–8.

Chaisorn, L., & Chua, T. S. (2006). *Story boundary detection in news video using global rule induction technique. Procs of the IEEE Inte. Conf. on Multimedia and Expo.*

Champagne, D. (2007). *Social Change and Cultural Continuity among Native Nations*. New York: Altamira Press.

Chan, C. Y., Goenka, A. K., & Chan, C. (2007). Multiwayslca-based keyword search in xml data. In *Proceedings of the 16th international conference on World Wide Web* (pp. 1043-1052). Academic Press. DOI: 10.1145/1242572.1242713

Chan, K. (2017). *What Is A 'Smart City'? Expatriate Lifestyle*. Retrieved on March, 17, 2018 from https://www.expatriatelifestyle.com/life-and-style/What-Is-A-Smart-City

Chaoa, K., Jamesa, A., Nanosa, A., Chena, J., Stan, S., Munteanb, I., ... Capelle, J. (2015). Cloud E-learning for Mechatronics: CLEM. *Future Generation Computer Systems*, *48*, 46–59. doi:10.1016/j.future.2014.10.033

Chatterton, P. (2002). Will the real creative city please stand up? *City*, *4*(3), 390–397. doi:10.1080/713657028

Chatur, P. N., & Lambole, P. R. (2017). Effective XML Keyword Query Processing. *International Conference on Electronics, Communication and Aerospace Technology ICECA 2017*, 523-528.

Chen, L., Liu, X., & Wan, C. (2011). Returning clustered results for keyword search on XML documents. *IEEE Transactions on Knowledge and Data Engineering*, *23*(12), 1811-1825. Doi:10.1109/TKDE.2011.183

Chen, X., & Wang, Y. (2004). Use synchronous chat to improve online learning experience. *AMCIS 2004 Proceedings*, 365.

Cherichi, S., & Faiz, R. (2017). Leveraging temporal markers to detect event from microblogs. *International Journal of Knowledge Society Research, 8*(3), 54–67. doi:10.4018/IJKSR.2017070104

Chianese, A., Marulli, F., & Piccialli, F. (2016). Sensitivity mining in social pulses to address cultural heritage competitive intelligence. *International Journal of Knowledge Society Research, 7*(2), 40–56. doi:10.4018/IJKSR.2016040103

Chigidi, W. L. (2009). Shona Taboos and Norms: The Language of Manufacturing Fears for Sustainable Development. *The Journal of Pan African Studies, 1*(3), 174–188.

Chizari, A., Jamali, M.-V., Abdollahramezani, S., & Jawad, A. (2017). Visible light for communication, indoor positioning, and dimmable illumination: A system design based on overlapping pulse position modulation. *Optik International Journal for Light and Electron Optics, 151*, 110–122. doi:10.1016/j.ijleo.2017.08.003

Choroś, K. (2015a). Automatic fast detection of anchorperson shots in temporally aggregated TV news videos. In *ACIIDS 2015. LNCS*. Heidelberg, Germany: Springer. doi:10.1007/978-3-319-15705-4_33

Choroś, K. (2015b). Automatic Categorization of Shots in News Videos Based on the Temporal Relations. *Computational Collective Intelligence, 7th International Conference, ICCCI 2015.*

Chourabi, H., Nam, T., Walker, S., Gil-Garcia, J. R., Mellouli, S., Nahon, K., & Scholl, H. J. (2012). Understanding smart cities: An integrative framework. In *Proceedings of the 45th Annual Hawaii International Conference on System Sciences* (pp. 2289-2297). Washington, DC: IEEE Computer Society.

Cinnamon, J., & Schuurman, N. (2013). Confronting the data-divide in a time of spatial turns and volunteered geographic information. *GeoJournal, 78*(4), 657–674. doi:10.100710708-012-9458-6

Čiuladienė, G., & Katkonienė, A. (2017). Framing the conflict: How students see it. *International Journal of Knowledge Society Research, 8*(4), 75–84. doi:10.4018/IJKSR.2017100104

Clarke, P. (2009). Understanding the experience of stroke: A mixed-method research agenda. *The Gerontologist*. PMID:19386828

Clement, A., & Wong, E. (2009). *Making Byzantine fault tolerant systems tolerate Byzantine faults*. USENIX Association.

Cleveland, M., & Cleveland, S. (2018b). Cybercrime Post-incident Leadership Model. *Proceedings of the Thirteenth Midwest Association for Information Systems Conference.*

Cleveland, S. (2013). On-the-job learning practices for IS students. *Proceedings of the 16th Southern Association for Information Systems Conference (SAIS)*, 16-21.

Cleveland, S. (2016). Social Media Systems in the Workplace: Toward Understanding Employee Knowledge Creation via Microblogging within Shared Knowledge Domains. *Informatics, 3*(3), 1–19. doi:10.3390/informatics3030011

Cleveland, S., & Block, G. (2017). Toward knowledge technology synchronicity framework for asynchronous environment. *International Journal of Knowledge Society Research, 8*(4), 23–33. doi:10.4018/IJKSR.2017100102

Cleveland, S., & Cleveland, M. (2018a). Toward Cybersecurity Leadership Framework. *Proceedings of the Thirteenth Midwest Association for Information Systems Conference.*

Cleveland, S., Ellis, T., & Hinojosa, C. (2015). Knowledge exchange via unified communication services: A grounded theory approach. *Proceedings of the 21st Americas Conference on Information Systems.*

Coffin Murray, M., & Pérez, J. (2015). Informing and performing: A study comparing adaptive learning to traditional learning. *Informing Science*, *18*, 111–125. doi:10.28945/2165

Cohen, S., Kanza, Y., Mamou, J., & Sagiv, Y. (2003). XSEarch: A semantic search engine for XML. In *Proceedings of the 29th international conference on Very large data bases* (*vol. 29*, pp. 45-56). Academic Press. 10.1016/B978-012722442-8/50013-6

Colesca, S. E. (2009). Understanding Trust in e-Government, Inzinerine Ekonomika –. *The Engineering Economist*, (3), 7–15.

Combi, C., & Masini, A., Olibon, B. I. B., & Zorzi, M. (2017). A Hybrid Logic for XML Reference Constraints. *Data & Knowledge Engineering*, 1–20. doi:10.1016/j.datak.2018.02.005

Coskun, Y. D. (2015). Promoting digital change in higher education: Evaluating the curriculum digitalisation. *Journal of International Education Research*, *11*(3), 197–204.

Cournède, B., Goujard, A., & Pina, Á. (2013). *How to Achieve Growth-and Equity-friendly Fiscal Consolidation?* Academic Press.

Crémilleux, B., Boizumault, P., Loudni, S., & Rojas, W.-U. (2014). *Modélisation et extraction de motifs optimaux*. Reconnaissance de Formes et Intelligence Artificielle.

Crooks, A., Croitoru, A., Stefanidis, A., & Radzikowski, J. (2013). # Earthquake: Twitter as a distributed sensor system. *Transactions in GIS*, *17*(1), 124–147. doi:10.1111/j.1467-9671.2012.01359.x

Cross, R., & Sproull, L. (2004). More than an answer: Information relationships for actionable knowledge. *Organization Science*, *15*(4), 446–462. doi:10.1287/orsc.1040.0075

D'Avanzo, E., & Pilato, G. (2016). The good, the ugly and the bad situation awareness in the big data: A cognitive architecture for social forecasting. *International Journal of Knowledge Society Research*, *7*(2), 25–39. doi:10.4018/IJKSR.2016040102

Dammak, S. M., Jedidi, A., & Bouaziz, R. (2017). Filtering method for the annotated and non-annotated web pages. *International Journal of Knowledge Society Research*, *8*(1), 1–22. doi:10.4018/IJKSR.2017010101

Daniela, L., Kalniņa, D., & Strods, R. (2017). An overview on effectiveness of technology enhanced learning (TEL). *International Journal of Knowledge Society Research*, *8*(1), 79–91. doi:10.4018/IJKSR.2017010105

Dargahi, H., & Yazdi, M. K. S. (2007). Quality of work life in Tehran University of Medical Sciences Hospitals' clinical laboratories employees. *Pakistan Journal of Medical Sciences*, *23*(4), 630–633.

David, C. (1991). *Practice of the video*. Edition LAROUSSE.

Dayananda, P., & Sowmyarani, C. N. (2016). Comprehensive study on keyword search on semi structured data. *International Journal of Knowledge Society Research*, *7*(3), 62–85. doi:10.4018/IJKSR.2016070105

De Jong, T., & Ferguson-Hessler, M. G. (1996). Types and qualities of knowledge. *Educational Psychologist*, *31*(2), 105–113. doi:10.120715326985ep3102_2

Dennis, A. R., Fuller, R. M., & Valacich, J. S. (2008). Media, tasks, and communication processes: A theory of media synchronicity. *Management Information Systems Quarterly*, *32*(3), 575–600. doi:10.2307/25148857

Dewey, J. (1933). *How we think*. Boston: Heath.

Dhiba, S. F., & Prastiti, D. C. (2017). Social education institution a way to achieve sustainable education in remote areas: A case study of 'Harapan' Bajulmati, Indonesia. *International Journal of Knowledge Society Research*, *8*(3), 21–33. doi:10.4018/IJKSR.2017070102

Di Serio, A. D., Ibanez, M. B., & Kloos, C. D. (2013). Impact of an augmented reality system on students' motivation for a visual art course. *Computers & Education*, *68*, 586–596. doi:10.1016/j.compedu.2012.03.002

Digital Government. (2018). *Tech and data: can 'digital government' be radically smarter?* Retrieved on March, 17, 2018 from https://apolitical.co/flagship/big-data-and-digital-government/

Ding, B., Lin, X., Qin, L., Wang, S., Yu, J. X., & Zhang, X. (2007). Finding top-k min-cost connected trees in databases. In *2007 IEEE 23rd International Conference on Data Engineering* (pp. 836-845). Academic Press. 10.1109/ICDE.2007.367929

Dou, W., Yu, L., Wang, X., Ma, Z., & Ribarsky, W. (2013). Hierarchicaltopics: Visually exploring large text collections using topic hierarchies. Visualization and Computer Graphics. *IEEE Transactions on*, *19*(12), 2002–2011. PMID:24051766

Dowling, C., Godfrey, J. M., & Gyles, N. (2003). Do hybrid flexible delivery teaching methods improve accounting students' learning outcomes? *Accounting Education*, *12*(4), 373–391. doi:10.1080/0963928032000154512

Dredze, M., Paul, M. J., Bergsma, S., & Tran, H. (2013, June). Carmen: A twitter geolocation system with applications to public health. In AAAI workshop on expanding the boundaries of health informatics using AI (HIAI) (pp. 20-24). AAAI.

Drewett, M. (1995). There must be some way out there: A case study in the teaching of critical thinking. *Suid Afrikaanse Tydskrif vir Wysbegeerte*, *14*(2), 72–76.

Dubai eGovernment. (2013). *eGovernment to 24/7 smart government*. Retrieved on May, 22, 2015, from the website: http://www.dsg.gov.ae/SiteCollectionImages/Content/DeG%20Documents/June-2013-en.pdf

Dubai Strategic Plan. (2015). Highlights: Dubai Strategic Plan. Dubai, UAE: Academic Press.

Dumont, E., & Quénot, G. (2012). Automatic Story Segmentation for TV News Video Using Multiple Modalities. *International Journal of Digital Multimedia Broadcasting*, *2012*, 1–11. doi:10.1155/2012/732514

Dunker, P., Gruhne, M., & Sturtz, S. (2008). Personal Television: A Cross modal Analysis Approach. *IEEE International Symposium on Consumer Electronics*.

Durkheim, E. (1963). *Incest: The nature and the origin of the taboo*. New York: Lyle Stuart.

Durlauf, S. N., & Blume, L. E. (2008). *New Palgrave Dictionary of Economics* (2nd ed.). London: Macmillan.

Dutta, S., & Lavin, B. (Eds.). (2016). The Global Innovation Index (GII) 2016: The Human Factor in Innovation. Cornell University, INSEAD and WIPO.

Dwivedi, O. P. (1987). Ethics, the Public Service and Public Policy. *International Journal of Public Administration*, *10*(1), 21–50. doi:10.1080/01900698708524556

Ekong, E. (1983). *Sociology of Ibibio*. Calabar: Scholars Press Nig. Ltd.

El Khoury, E., Sénac, C., & Joly, P. (2014). Audiovisual diarization of people in video content. *Multimedia Tools and Applications Journal, 68*(3).

Elias, M. S., & Saha, N. K. (2005). Environmental Pollution and Quality of Working Life in Tobacco Industries. *Journal of Life Earth Science*, *1*(1), 21–24.

Ellahi, A., Zaka, B., & Sultan, F. (2017). A Study of supplementing conventional business education with digital games. *Journal of Educational Technology & Society*, *20*(3), 195–206.

Epstein, R. L. (2006). *Critical thinking*. Belmont, CA: Wadsworth Thomas Learning.

Ergazakis, K., Metaxiotis, K., & Psarras, J. (2006). An emerging pattern of successful knowledge cities' main features. In F. J. Carrillo (Ed.), *Knowledge Cities: Approaches, Experiences, and Perspectives* (pp. 3–15). New York, NY: Butterworth Heinemann. doi:10.1016/B978-0-7506-7941-1.50004-X

Esema, J. D. (2002). *Norms, Culture, Customs and Traditions of Akwa Ibom people of Nigeria, Uyo.* Mocomp Publishing Co.

Essen, A. J. A. (1982). *Ibibio Profile, Calabar.* APCON Press Ltd.

Essien, D. E. (2013). The Hidden Dimension of Ibibio Taboos. *Uniuyo Journal of Humanities, 16-17*, 256-275.

Essien, D. E. (2012). Ethics in the Nigerian Public Service: Enhancing Role, Professionalism and Ethical Value. *Journal of Religious Studies*, *11*(2), 40–48.

Essien, D. E. (2017). Toward a Normative Ethical Claims of Ibibio Social Norms, Values, and Beliefs. *International Journal of Knowledge Society Research*, *8*(1), 51–66. doi:10.4018/IJKSR.2017010103

Ewers, M. C., & Malecki, E. J. (2010). Leapfrogging into the knowledge economy: Assessing the economic development strategies of the Arab Gulf states. *Tijdschrift voor Economische en Sociale Geografie*, *101*(5), 494–508. doi:10.1111/j.1467-9663.2010.00628.x

Faithman, N. U. (1999). *Ibibio Jews of Nigeria.* Uyo: Menorah Publishing Coy.

Fan, W., & Yan, Z. (2010). Factors affecting response rates of the web survey: A systematic review. *Computers in Human Behavior*, *26*(2), 132–139. doi:10.1016/j.chb.2009.10.015

Fasihuddin, H., Skinner, G., & Athauda, R. (2012). A Holistic Review of Cloud-Based E-learning System. *IEEE International Conference on Teaching, Assessment, and Learning for Engineering (TALE).* 10.1109/TALE.2012.6360325

Fehr, E., Urs, F., & Simon, G. (2002). Strong reciprocity, human cooperation, and the enforcement of social norms. *Human Nature (Hawthorne, N.Y.)*, *13*(1), 1–25. doi:10.100712110-002-1012-7 PMID:26192593

Feng, J., Li, G., Ooi, B. C., Wang, J., & Zhou, L. (2008). EASE: an effective 3-in-1 keyword search method for unstructured, semi-structured and structured data. In *Proceedings of the 2008 ACM SIGMOD international conference on Management of data* (pp. 903-914). ACM. DOI: 10.1145/1376616.1376706

Feng, J., Li, G., Wang, J., & Zhou, L. (2007). Effective keyword search for valuable lcas over xml documents. In *Proceedings of the sixteenth ACM conference on Conference on information and knowledge management* (pp. 31-40). ACM. DOI: 10.1145/1321440.1321447

Fields, A. (2005). *Discovering statistics using SPSS*. Beverly Hills, CA: Sage Publications.

Fietkiewicz, K. J., & Stock, W. G. (2015). How "smart" are Japanese cities? An empirical investigation of infrastructures and governmental programs in Tokyo, Yokohama, Osaka and Kyoto. In *Proceedings of the 48th Hawaii International Conference on System Sciences* (pp. 2345-2354). Washington, DC: IEEE Computer Society.

Fietkiewicz, K. J., Mainka, M., & Stock, W. G. (2017). eGovernment in cities of the knowledge society. An empirical investigation of Smart Cities' governmental websites. *Government Information Quarterly*, *34*(1), 75–83. doi:10.1016/j.giq.2016.08.003

Fietkiewicz, K. J., & Pyka, S. (2014). Development of informational cities in Japan: A regional comparison. *International Journal of Knowledge Society Research*, *5*(1), 69–82. doi:10.4018/ijksr.2014010106

Fisher, R. B. (1997). West African Religious Traditions. New York: Maryknoll.

Flak, L., Nordheim, S., & Munkvoid, B. (2008). Analyzing Stakeholder Diversity in G2G Efforts: Combining Descriptive Stakeholder Theory and Dialectic Process Theory. *E-Service Journal, 6*(2), 6.

Flanagan, B. (2011). *UAE is most high-tech country in Mena region.* Retrieved on June, 17, 2015 from: http://www.thenational.ae/business/technology/uae-is-most-high-tech-country-in-mena-region

Floos, A. Y. (2016). Arabic rumours identification by measuring the credibility of Arabic tweet content. *International Journal of Knowledge Society Research*, *7*(2), 72–83. doi:10.4018/IJKSR.2016040105

Florida, R. (2002). The rise of the creative class: Cities without gays and rock bands are losing the economic development race. *The Washington Monthly*, *34*(5), 15–25.

Florida, R. (2005). *Cities and the Creative Class.* New York, NY: Routledge.

Florida, R. (2012). *The Rise of the Creative Class: Revisited.* New York, NY: Basic Books.

Fox, E. A. (1983). *Extending the Boolean and Vector Space Models of Information Retrieval with P-Norm Queries and Multiple Concept Types* (PhD thesis). Cornell University, Ithaca, NY.

Fraser, J. G. (1994). *The Golden Bough: A Study in Magic and Religion.* New York: Oxford University Press.

Frechtling, J. A. (1997). *User-friendly handbook for mixed method evaluations.* Diane Publishing.

Freely, A. J. (1993). *Argumentation and debate.* Woodsworth.

Freud, S. (1955). Totem and Taboo: Some Points of Agreement between the Mental Lives of Savages and Neurotics. In J. Strachey (Ed.), The Standard Edition of the Complete Psychological Works of Sigmund Freud (Vol. 13). London, UK: Hogarth Press. (Orig. pub. 1913)

Freud, S. (1985). *The Origins of Religion: Totem and Taboo, Moses and Monotheism and Other Works.* Harmondsworth, UK: Penguin Books Ltd.

Freud, S. (2000). *Totem and Taboo. Translated into Chinese by Yung-yi Yang*. Taipei: Chih-wen Publishing Co.

Friedmann, J. (1986). The world city hypothesis. *Development and Change*, *17*(1), 69–83. doi:10.1111/j.1467-7660.1986.tb00231.x

Friedmann, J. (1995). Where we stand: A decade of world city research. In P. Knox & P. Taylor (Eds.), *World Cities in a World-system* (pp. 21–47). Cambridge, UK: Cambridge University Press. doi:10.1017/CBO9780511522192.003

Ganguly, R. (2010). Quality of work life and job satisfaction of a group of university employees. *Asian Journal of Management Research*, *1*(1), 209–216.

García, E., Poudereux, P., Hernández, A., García, J.-J., & Ureña, J. (2013). DS- UWB indoor positioning system implementation based on FPGAs. *Sensors and Actuators. A, Physical*, *201*, 172–181. doi:10.1016/j.sna.2013.06.019

Garfinkel, S. L. (2007). *An evaluation of amazon's grid computing services: EC2, S3, and SQS.* Retrieved from http://simson.net/clips/academic/2007.Harvard.S3.pdf

Garfinkel, S. L. (2003). Email-based identification and authentication: An alternative to PKI? *IEEE Security and Privacy*, *1*(6), 20–26. doi:10.1109/MSECP.2003.1253564

Garrison, R. (2011). *E-Learning in the 21st Century: A Framework for Research and Practice*. Hoboken, NJ: Taylor and Francis.

Garson, D. G. (2006). *Public Information Technology and E-Governance*. Sudbury, MA: Jones and Bartlett Publishers.

GCC. (2017). *The Cooperation Council of the Arab States of the Gulf, Secretariat General*. Retrieved from http://www.gcc-sg.org/eng/

Gelfand, M. (1968). *African Crucible: An ethno-religious study with special reference to the Shona-speaking people*. Capetown: Juta & Company Limited.

Gelfand, M. (1979). *Growing up in a Shona Society: From Birth to Marriage*. Mambo Press.

Gezgin, D. M., Adnan, M., & Acar Guvendir, M. (2018). Mobile learning according to students of computer engineering and computer education: A comparison of attitudes. *Turkish Online Journal of Distance Education, 19*(1), 4–17. doi:10.17718/tojde.382653

Ghasemizad, A., Amirian Zadeh, M., & Bagheri, S. (2012). A study of the relationship between teachers and principals' spiritual leadership, quality of work life, job satisfaction and productivity. *American Journal of Scientific Research, 49*, 11–20.

Ghazizadeh, A. (2012). Cloud Computing Benefits And Architecture In E-Learning. *IEEE International Conference on Wireless, Mobile and Ubiquitous Technology in Education*. 10.1109/WMUTE.2012.46

Giffinger, R., Fertner, C., Kramar, H., Kalasek, R., Pichler-Milanović, N., & Meijers, E. (2007). *Smart Cities: Ranking of European Medium-sized Cities*. Vienna, Austria: Centre of Regional Science.

Goburdhun, S., & Ramsaha, J. (2017). The relevance of social sciences teacher education programme in preparing effective secondary school educators. *International Journal of Knowledge Society Research, 8*(2), 1–12. doi:10.4018/IJKSR.2017040101

Goffin, K., & Koners, U. (2011). Tacit knowledge, lessons learnt, and new product development. *Journal of Product Innovation Management, 28*(2), 300–318. doi:10.1111/j.1540-5885.2010.00798.x

Goldkuhl, G. (2006, October). What does it mean to serve the citizen?–Towards a practical theory on public e-services founded in socio-instrumental pragmatism. *Proceedings of the International Workshop on E-services in Public Administration*, 27-47.

Goldkuhl, G. (2007). What does it mean to serve the citizen in e-services?-Towards a practical theory founded in socio-instrumental pragmatism. *International Journal of Public Information Systems, 3*(3), 135–159.

Golenberg, K., Kimelfeld, B., & Sagiv, Y. (2008). Keyword proximity search in complex data graphs. In *Proceedings of the 2008 ACM SIGMOD international conference on Management of data* (pp. 927-940). ACM. 10.1145/1376616.1376708

Gomez, N. (2007). Healing Hidden Wounds. *Cultural Survival Quarterly, 31*(3).

Gonzalez-Martínez, J., Bote-Lorenzo, M., Gomez-Sanchez, E., & Cano-Parra, R. (2015). Cloud computing and education: A state-of-the-art survey. *Computers & Education, 80*, 132–151. doi:10.1016/j.compedu.2014.08.017

Goodchild, M. F. (2015). Twenty years of progress: GIScience in 2010. *Journal of Spatial Information Science,* (1), 3-20.

Goodyear, P., & Ellis, R. (2008). University students' approaches to learning: Rethinking the place of technology. *Distance Education, 29*(2), 141–152. doi:10.1080/01587910802154947

Goos, M., & Manning, A. (2007). Lousy and lovely jobs: The rising polarization of work in Britain. *The Review of Economics and Statistics*, *89*(1), 118–133. doi:10.1162/rest.89.1.118

Gorer, G. (1965). *Death, grief and mourning in contemporary Britain*. London: Cresset Press.

Goswami, A., Das, A.-K., & Sahoo, J. (2015). An efficient approach for mining association rules from high utility itemsets. *Expert Systems with Applications*, *42*(13), 5754–5778. doi:10.1016/j.eswa.2015.02.051

Gould, D., & Cleveland, S. (2018). Evolutionary Systems: Applications to Cybersecurity. *Proceedings of the Thirteenth Midwest Association for Information Systems Conference*.

Gowrishankar, S., Madhu, N., & Basavaraju, T. G. (2015). Role of BLE in proximity based automat- -ion of IoT: A practical approach. *Proceedings of the 2015 IEEE Recent Advances in Intelligent Computational Systems (RAICS)*, 400–405. 10.1109/RAICS.2015.7488449

Goyal, A., Punitha, P., Hopfgartner, F., & Jose, J. M. (2009). Split and merge based story segmentation in news videos. *Proc. of the 31th European Conference on IR Research on Advances in Information Retrieval*. 10.1007/978-3-642-00958-7_82

Gremm, J., Barth, J., Fietkiewicz, K. J., & Stock, W. G. (2018). *Transitioning Towards a Knowledge Society. Qatar as a Case Study*. Cham, Switzerland: Springer Nature. doi:10.1007/978-3-319-71195-9

Gremm, J., Barth, J., & Stock, W. G. (2015). Kuwait is the past, Dubai is the present, Doha is the future: Informational cities on the Arabian Gulf. *International Journal of Knowledge Society Research*, *6*(2), 51–64. doi:10.4018/IJKSR.2015040103

GSDP. (2008). *Qatar National Vision 2030*. Doha, Qatar: Qatar General Secretariat for Development Planning.

GSDP. (2011). *Qatar National Development Strategy 2011~2016*. Doha, Qatar: Qatar General Secretariat for Development Planning.

Guo, L., Shao, F., & Shanmugasundaram, J. (2003). XRANK: Ranked keyword search over XML documents. In *Proceedings of the ACM SIGMOD International conference on Management of data* (pp. 16-27). ACM. 10.1145/872757.872762

Hall, B. (2014). Designing Collaborative Activities to Promote Understanding and Problem-Solving. *International Journal of e-Collaboration*, *10*(2), 55–71. doi:10.4018/ijec.2014040104

Hall, P. (1997a). Modelling the post-industrial city. *Futures*, *29*(4/5), 311–322. doi:10.1016/S0016-3287(97)00013-X

Hall, P. (1997b). *The First Megacity Lecture: Megacities, World Cities and Global Cities*. Rotterdam, The Netherlands: Stichting Megacities.

Hall, R. E., Bowerman, B., Braverman, J., Taylor, J., Todosow, H., & von Wimmersperg, U. (2000). *The vision of a smart city*. In *2nd International Life Extension Technology Workshop*, Paris, France.

Han, Y., Hong, B., Lee, H., & Kim, K. (2017). How do we Tweet? The Comparative Analysis of Twitter Usage by Message Types, Devices, and Sources. *The Journal of Social Media in Society*, *6*(1), 189–219.

Hasan, K. M. S., Chowdhury, M., & Alam, M. A. (2008). Faculty Turnover in Private Universities of Bangladesh-A Critical Evaluation. *Business and Economic Review*, *1*(2), 99–113.

Havlik, D., Egly, M., Huber, H., Kutschera, P., Falgenhauer, M., & Cizek, M. (2013). Robust and trusted crowd-sourcing and crowd-tasking in the Future Internet. In Environmental Software Systems. Fostering Information Sharing (pp. 164-176). Springer Berlin Heidelberg. doi:10.1007/978-3-642-41151-9_16

Hechter, M., & Karl-Dieter, O. (Eds.). (2001). *Social Norms*. New York: Russell Sage Foundation.

Heeks, R. (2005). *Implementing and Managing E-Government: An International Text*. Sage Publishers.

He, H., Wang, H., Yang, J., & Yu, P. S. (2007). BLINKS: ranked keyword searches on graphs. In *Proceedings of the 2007 ACM SIGMOD international conference on Management of data*,(pp. 305-316). ACM. 10.1145/1247480.1247516

Henkel, M., Barth, J., Gremm, J., & Stock, W. G. (2018). Qatar National Library as part of a countrywide knowledge infrastructure. In *Proceedings of the 2018 International Conference on Library and Information Science*. Taipei, Taiwan: International Business Academics Consortium.

Heo, J., Kim, M., & Chan-Olmsted, S.-M. (2010). Perceived effectiveness and business structure among advertising agencies: A case study of mobile advertising in south Korea. *Journal of Media Business Studies*, *7*(2), 1–20. doi:10.1080/16522354.2010.11073504

Hesse-Biber, S. N. (2010). *Mixed methods research: Merging theory with practice*. Guilford Press.

Hiltz, S. R., & Wellman, B. (1997). Asynchronous learning networks as a virtual classroom. *Communications of the ACM*, *40*(9), 44–49. doi:10.1145/260750.260764

Hirschfeld, B. (2012). *Global Thesis Update: Technology and the Arab Spring*. Retrieved on April 11, 2015 from http://worldperspectivesprogram.org/2012/04/12/global-thesis-update-technology-and-the-arab-spring/

Ho, C.-L., Chang, H.-J., & Hung, L.-P. (2007). An anticipation model of potential customers' purchasing behavior based on clustering analysis and association rules analysis. *Expert Systems with Applications*, *32*(3), 753–764. doi:10.1016/j.eswa.2006.01.049

Höfer, C., & Karagiannis, G. (2011). Cloud computing services: Taxonomy and comparison. *Journal of Internet Services and Applications*, *2*(2), 81–94. doi:10.100713174-011-0027-x

Hoffman, L., Burley, D., & Toregas, C. (2012). Holistically building the cybersecurity workforce. *IEEE Security and Privacy*, *10*(2), 33–39. doi:10.1109/MSP.2011.181

Holden, L. (2000). *Encyclopedia of Taboos*. Oxford, UK: ABC CLIO Ltd.

Holgersson, J., Söderström, E., Karlsson, F., & Hedström, K. (2010). Towards a roadmap for user involvement in e-government service development. In Electronic Government (pp. 251-262). Springer Berlin Heidelberg. doi:10.1007/978-3-642-14799-9_22

Hong, K.-S., Lai, K.-W., & Holton, D. (2003). Students' satisfaction and perceived learning with a web-based course. *Journal of Educational Technology & Society*, *6*(1), 116–124.

Hong, K., Tan, K., & Bujang, S. (2010). Relationships Between Work Life Quality Of Teachers With Work Commitment, Stress And Satisfaction: A Study In Kuching, Sarawak, Malaysia. *Jurnal Teknologi*, *52*(1), 1–15. doi:10.11113/jt.v52.133

Hoque, M. E., & Rahman, A. (1999). Quality of working life & Job Behaviour of Workers in Bangladesh: A comparative study of private and public sectors. *Indian Journal of Industrial Relations*, *35*(2), 175–184.

Hore, B., & Mehrotra, S. (2004). A privacy-preserving index for range queries. In *Proceedings of the Thirtieth international conference on Very large data bases*. VLDB Endowment. 10.1016/B978-012088469-8.50064-4

Höselbarth, F. (2010). *The Education Revolution in the Gulf.* Hildesheim, Germany: Georg Olms.

Hossain, M. M., & Islam, M. T. (1991). QWL & Job Satisfaction of GVT Hospital Nurses in Bangladesh. *Indian Journal of Industrial Relations*, *35*(3).

Hrastinski, S. (2008). Asynchronous and synchronous e-learning. *EDUCAUSE Quarterly*, *31*(4), 51–55.

Huang, B., & Carley, K. M. (2017). *On Predicting Geolocation of Tweets using Convolutional Neural Networks.* arXiv preprint arXiv:1704.05146

Hua, S., Chen, G., Wei, H., & Jiang, Q. (2012). Similarity measure for image resizing using SIFT feature. *EURASIP Journal on Image and Video Processing*.

Huber, G. P. (1991). Organizational learning: The contributing processes and the literatures. *Organization Science*, *2*(1), 88–115. doi:10.1287/orsc.2.1.88

Huircan, J. I., Munoz, C., Young, H., Von Dossow, L., Bustos, J., Vivallo, G., & Toneatti, M. (2010). Zigbee-based wireless sensor network localization for cattle monitoring in grazing fields. *Computers and Electronics in Agriculture*, *74*(2), 258–264. doi:10.1016/j.compag.2010.08.014

Human Development Report. (2004). *Globalization and Cultural Choice*. Washington, DC: UNDP.

Human Development Report. (2014). Retrieved on March 25, 2015, from:www.pnud.org.br/arquivos/rdh2014.pdf

Hung, S.-Y., Chang, C.-M., & Yu, T.-Y. (2006). *Determinants of user acceptance of the ... agent: a research note*. European and Mediterranean Conference on Information Systems (EMCIS), Alicante, Spain.

Hung, Y. (2012). The Effect of Teaching Methods and Learning Style on Learning Program Design in Web-Based Education Systems. *Educational Computing Research*, *47*(4), 409–427. doi:10.2190/EC.47.4.d

Hussein, R., Mohamed, N., Rahman Ahlan, A., & Mahmud, M. (2011). E-government application: an integrated model on G2C adoption of online tax. *Transforming Government: People, Process and Policy*, *5*(3), 225–248.

Hu, W., Xie, N., Li, L., & Zeng, X. (2011). A Survey on Visual Content-Based Video Indexing and Retrieval. *IEEE Transactions on Systems, Man and Cybernetics. Part C, Applications and Reviews*, *41*(6).

Hwang, W., Shadiev, R., Wang, C., & Huang, Z. (2012). A Pilot Study of Cooperative Programming Learning Behavior and its Relationship with Students' Learning Performance. *Computers & Education*, *58*(4), 1267–1281. doi:10.1016/j.compedu.2011.12.009

Hwang, W., Wang, C., Hwang, G., Huang, Y., & Huang, S. (2008). A Web-Based Programming Learning Environment to Support Cognitive Development. *Interacting with Computers*, *20*(6), 524–534. doi:10.1016/j.intcom.2008.07.002

Ifemesia, C. C. (1978). Comments on the South Province of Nigeria for the year 1938. *Southern Eastern State: Heritage Magazine*, *3*(2), 1–16.

Ilika, A. L. (2005). Women's perception of partner violence in a rural Igbo community. *African Journal of Reproductive Health*, *9*(3), 77–88. doi:10.2307/3583414 PMID:16623192

Islam, M. Z., & Siengthai, S. (2009). Quality of work life and organizational performance: Empirical evidence from Dhaka Export Processing Zone. In *Proceeding of ILO Conference on Regulating for Decent Work*. Geneva: International Labour Office.

ITU. (2016). Measuring the Information Society Report: 2016. International Telecommunication Union.

Izquierdo-Yusta, A., Olarte-Pascual, C., & Reinares-Lara, E. (2015). Attitudes toward mobile advertising among users versus non-users of the mobile internet. *Telematics and Informatics*, *32*(2), 355–366. doi:10.1016/j.tele.2014.10.001

Jaatun, M., Zhao, G., & Rong, C. (2009). *Cloud Computing, First International CloudCom 2009*. Berlin: Springer-Verlag.

Jacobs, A., Miene, A., Ioannidis, G. T., & Herzog, O. (2004). Automatic shot boundary detection combining color, edge, and motion features of adjacent frames. *TRECVID Workshop Notebook Papers*. doi:10.1186/1687-5281-2012-6

Jagadish, H. V., Li, Y., & Yu, C. (2004). Schema-Free XQuery. In *Proc. of VLDB* (*vol. 30*, pp. 72-83). Academic Press.

Jdidou, Y., & Khaldi, M. (2016). Increasing the profitability of students in MOOCs using recommendation systems. *International Journal of Knowledge Society Research*, *7*(4), 75–85. doi:10.4018/IJKSR.2016100107

Jenhani El Jed, O. (2006). Classification des réponses d'un moteur de recherche et évaluation de leur pertinence. In *Actes du XXIV*[ème] *Congrès INFORSID* (pp. 291-306). Hammamet, Tunisie: Academic Press.

Jewkes, R., Penn-Kekana, L., & Rose-Junius, H. (2005). "If they rape me, I can't blame them": Reflections on gender in the social context of child rape in South Africa and Namibia. *Social Science & Medicine*, *61*(8), 1809–1820. doi:10.1016/j.socscimed.2005.03.022 PMID:15913860

Jiang, B., & Thill, J. C. (2015). Volunteered Geographic Information: Towards the establishment of a new paradigm. *Computers, Environment and Urban Systems*, *53*, 1–3. doi:10.1016/j.compenvurbsys.2015.09.011

Jin, Z. (1994). *Explore the Secret of Norms*. Taipei: Shu-xin.

Jivani, M. (2014). *UAE government launches world's first store for smart government application*. Retrieved on June 8,2015, from http://techview.me/2014/03/uae-government-launches-worlds-first-store-smart-government-application/

Joanna Briggs Institute Reviewers' Manual . (2014). Joanna Briggs Institute.

Johnson, C., & Altowairiki, N. (2016). Developing Teaching Presence in Online Learning Through Shared Stakeholder Responsibility. Handbook of Research on Innovative Pedagogies and Technologies for Online Learning in Higher Education, 151.

Johnson, B. W. (1988). *Design & analysis of fault tolerant digital systems*. Addison-Wesley Longman Publishing Co., Inc.

Jonassen, D., Davidson, M., Collins, M., Campbell, J., & Haag, B. B. (1995). Constructivism and computer-mediated communication in distance education. *American Journal of Distance Education*, *9*(2), 7–26. doi:10.1080/08923649509526885

Jonassen, D., & Hernandez-Serrano, J. (2002). Case-based reasoning and instructional design: Using stories to support problem solving. *Educational Technology Research and Development*, *50*(2), 65–77. doi:10.1007/BF02504994

Kai, S., Amy, S., Suhang, W., Jiliang, T., & Huan, L. (2017). Fake News Detection on Social Media: A Data Mining Perspective. *ACM SIGKDD Explorations Newsletter*, *19*(1), 22–36. doi:10.1145/3137597.3137600

Kaiser, H. F. (1974). An index of factorial simplicity. *Psychometrika*, *39*(1), 31–36. doi:10.1007/BF02291575

Kantardzic, M. (2011). *Data mining: concepts, models, methods, and algorithms*. John Wiley & Sons. doi:10.1002/9781118029145

Karanasios, S. (2011). New and Emergent ICTs and Climate Change in Developing Countries. Manchester, UK: Centre for Development Informatics, Institute for Development Policy and Management, SED, International Development Research Centre. Retrieved from http://www.niccd.org/KaranasiosClimateChangeEmergentICTs.pdf

Karimi, S., Zobel, J., Pohl, S., & Scholer, F. (2009). The challenge of high recall in biomedical systematic search. In *The ACM 3*[rd] *International Workshop of Data and Text Mining Methods in Bioinformatics* (pp. 89-92). Hong Kong: ACM. 10.1145/1651318.1651338

Karlsson, F., Holgersson, J., Söderström, E., & Hedström, K. (2012). Exploring user participation approaches in public e-service development. *Government Information Quarterly*, *29*(2), 158–168. doi:10.1016/j.giq.2011.07.009

Kenrick, J. (2000). The Forest Peoples of Africa in the 21st Century. *Indigenous Affairs, Hunters and Gatherers, 2000*(2), 10-24. Retrieved from http://www.iwgia.org

Khatiban, M., & Sangestani, G. (2014). The effects of using problem-based learning in the clinical nursing education on the students' outcomes in Iran: A quasi-experimental study. *Nurse Education in Practice*, *14*(6), 698–703. doi:10.1016/j.nepr.2014.10.002 PMID:25458229

Kinoti, H. W. (1992). African morality: past and present. In J. N. K. Mugambi & A. Nasimiyu-Wasike (Eds.), *Moral and ethical issues in African Christianity* (pp. 73–82). Nairobi: Initiatives.

Kirkwood, A., & Price, L. (2014). Technology-enhanced learning and teaching in higher education: What is 'enhanced' and how do we know? A critical literature review. *Learning, Media and Technology*, *39*(1), 6–36. doi:10.1080/17439884.2013.770404

Kivunja, C. (2015). Innovative methodologies for 21st century learning, teaching and assessment: A convenience sampling investigation into the use of social media technologies in higher education. *International Journal of Higher Education*, *4*(2), 1–26. doi:10.5430/ijhe.v4n2p1

Knipfer, K., Kump, B., Wessel, D., & Cress, U. (2012). Reflection as a catalyst for organizational learning. *Studies in Continuing Education*. doi:10.1080/0158037X.2012.683780

Köpke, J. (2017). Annotation Paths for Matching XML-Schemas. *Data & Knowledge Engineering*, 1–30. doi:10.1016/j.datak.2017.12.002

Kortemeyer, G. (2006). An analysis of asynchronous online homework discussion in introductory physics courses. *American Journal of Physics*, *74*(6), 526–536. doi:10.1119/1.2186684

Kosior, A., Barth, J., Gremm, J., Mainka, A., & Stock, W. G. (2015). Imported expertise in world-class infrastuctures. The problematic development of knowledge cities in the Gulf region. *Journal of Information Science Theory and Practice*, *3*(3), 17–44. doi:10.1633/JISTaP.2015.3.3.2

Koswatte, S., McDougall, K., & Liu, X. (2018). VGI and crowdsourced data credibility analysis using spam email detection techniques. *International Journal of Digital Earth*, *11*(5), 520–532. doi:10.1080/17538947.2017.1341558

Kotla, R., Alvisi, L., Dahlin, M., Clement, A., & Wong, E. (2007). Zyzzyva: Speculative byzantine fault tolerance. *Operating Systems Review*, *41*(6), 45–58. doi:10.1145/1323293.1294267

Kraut, R., Olson, J., Banaji, M., Bruckman, A., Cohen, J., & Couper, M. (2004). Psychological research online: Report of Board of Scientific Affairs' Advisory Group on the Conduct of Research on the Internet. *The American Psychologist*, *59*(2), 105–117. doi:10.1037/0003-066X.59.2.105 PMID:14992637

Krawczyk & Bellare. (1997). *HMAC: Keyed-hashing for message authentication.* Academic Press.

Krishnamurthy, Y., Pham, K., & Santos, A. (2016). Interactive Exploration for Domain Discovery on the Web. In *ACM KDD Workshop on Interactive Data Exploration and Analytics (IDEA)* (pp. 64-71). San Francisco, CA: ACM.

Krötzsch, M., Simancik, F., & Horrocks, I. (2012). *A description Logic.* Primer Technical Report, arXiv.org, volume CoRR abs/1201.4089

Kumar, D., & Deo, J. M. (2011). Stress and Work Life of College Teachers. *Journal of Academy of Indian Academy of Applied Psychology*, *37*, 78–85.

Kumar, J. N., & Reddy, M. J. (2013). Effective efficient Boolean retrieval. *International Journal of Computer Science and Mobile Applications*, *1*(5), 38–43.

Kuwait. (2015). *Kuwait Mid-Range Development Plan 2015/2016 – 2019/2020.* Kuwait City: Academic Press.

Kuznetsov, P., & Rodrigues, R. (2009). BFTW 3: Why? when? where? workshop on the theory and practice of byzantine fault tolerance. *ACM SIGACT News*, *40*(4), 82–86. doi:10.1145/1711475.1711494

Lakshmi, D., Damodaram, A., Sreenivasa, M., & Lal, J. (2008). Content based image retrieval using signature based similarity search. *Indian Journal of Science and Technology*, *1*(5).

Lam, A. (2000). Tacit knowledge, organizational learning and societal institutions: An integrated framework. *Organization Studies*, *21*(3), 487–513. doi:10.1177/0170840600213001

Lamport, L., Shostak, R., & Pease, M. (1982). The Byzantine generals problem. *ACM Transactions on Programming Languages and Systems*, *4*(3), 382–401. doi:10.1145/357172.357176

Lance, G. N., & Williams, W. T. (1967). A General Theory of Classificatory Sorting Strategies: Hierarchical Systems. *The Computer Journal*, *9*(4), 373–380. doi:10.1093/comjnl/9.4.373

Landry, C. (2000). *The Creative City: A Toolkit for Urban Innovation*. London, UK: Earthscan.

Landry, C., & Bianchini, F. (1995). *The Creative City*. London, UK: Demos.

Lansari, A., Tubaishat, A., & Al-Rawi, A. (2010). Using a learning management system to foster independent learning in an outcome-based university: A gulf perspective. *Issues in Informing Science & Information Technology*, *7*, 73–87. doi:10.28945/1193

Lansley, G., & Longley, P. A. (2016). The geography of Twitter topics in London. *Computers, Environment and Urban Systems*, *58*, 85–96. doi:10.1016/j.compenvurbsys.2016.04.002

Lantz, B. (2013). *Machine learning with R*. Packt Publishing Ltd.

Le, N., Bredin, H., Sargent, G., India, M., Lopez-Otero, P., & Barras, C. (2017). Towards large scale multimedia indexing: A case study on person discovery in broadcast news. *International Workshop on Content-Based Multimedia Indexing - CBMI 2017*. 10.1145/3095713.3095732

Lee, D., Morrone, A. S., & Siering, G. (2018). From swimming pool to collaborative learning studio: Pedagogy, space, and technology in a large active learning classroom. *Educational Technology Research and Development*, *66*(1), 95–127. doi:10.100711423-017-9550-1

Leetaru, K., Wang, S., Cao, G., Padmanabhan, A., & Shook, E. (2013). Mapping the global Twitter heartbeat: The geography of Twitter. *First Monday*, *18*(5). doi:10.5210/fm.v18i5.4366

Leite Lucas de Azevedo, V., & Borges, M. (2015). More Collaboration, More Collective Intelligence. *International Journal of Knowledge Society Research*, *6*(3).

Li, J., Liu, C., Wang, W., & Zhou, R. (2010). Suggestion of promising result types for XML keyword search. In *Proceedings of the 13th International Conference on Extending Database Technology* (pp. 561-572). Academic Press. 10.1145/1739041.1739108

Li, W., & He, J. (2005). A framework to support survivable web services. In *Proceeding of the 19th International Parallel and Distributed Processing Symposium*. IEEE.

Li, X-J. (1994). *Norms and Laziness*. Beijing: International Culture Publishing Co.

Liang, H., Chen, L. H., & Chen, W. S. (2018). Video Caption Extraction Using Spatio-Temporal Slices. *International Journal of Image and Graphics*, *18*(2).

Likert, R. (1932). A technique for the measurement of attitudes. *Archives de Psychologie*, *22*(140), 3–55.

Limperos, A. M., Buckner, M. M., Kaufmann, R., & Frisby, B. N. (2015). Online teaching and technological affordances: An experimental investigation into the impact of modality and clarity on perceived and actual learning. *Computers & Education, 83*, 1–9. doi:10.1016/j.compedu.2014.12.015

Lin. (2006). Impact of organizational support on organizational intention to facilitate knowledge sharing. *Knowledge Management Research & Practice, 4*(1), 26-35.

Linnefell, W. (2014). *E-government Policy Formation*. Retrieved on February 22, 2014 from: www.ejeg.com/issue/download.html?idArticle=348

Lin, W. (2013). Volunteered geographic information and networked publics? Politics of everyday mapping and spatial narratives. *GeoJournal, 78*(6), 949–965. doi:10.100710708-013-9490-1

Li, Q., Song, D., Liao, L., & Liu, L. (2015). Personalized Mention Probabilistic Ranking–Recommendation on Mention Behavior of Heterogeneous Social media. In *Web-Age Information Management* (pp. 41–52). Springer International Publishing. doi:10.1007/978-3-319-23531-8_4

Liu, C., Li, J., & Zhou, R. (2010). Fast ELCA computation for keyword queries on XML data. In *Proceedings of the 13th International Conference on Extending Database Technology* (pp. 549-560). Academic Press. DOI: 10.1145/1739041.1739107

Lohmann, S., & Riechert, T. (2010). Adding semantics to social software engineering: (re-) using ontologies in a community-oriented requirements engineering environment. In The Social Software Engineering Workshops (pp. 485-494). Paderborn, Germany: Academic Press.

Loncar, M., Barrett, N. E., & Liu, G. Z. (2014). Towards the refinement of forum and asynchronous online discussion in educational contexts worldwide: Trends and investigative approaches within a dominant research paradigm. *Computers & Education, 73*, 93–110. doi:10.1016/j.compedu.2013.12.007

Loo, R. (2002). Journaling: A learning tool for project management training and team-building. *Project Management Journal, 33*(4), 61–66.

Lootah, R., & Geray, O. (2006). *Dubai eGovernment Case Study*. Retrieved on February 23, 2014 from: http://www.oecd.org/dataoecd/4/40/36986277.pdf

Lowe, D. (2004). Distinctive Image Features from Scale-Invariant Key points. *International Journal of Computer Vision, 60*(2), 91–110. doi:10.1023/B:VISI.0000029664.99615.94

Lucas de Azevedo, V. L., & Borges, M. (2015). More collaboration, more collective intelligence. *International Journal of Knowledge Society Research, 6*(3), 1–18. doi:10.4018/IJKSR.2015070101

Lwin, S., & Poungchompu, S. (2017). Determinant of food security on upland agriculture households in Paletwa Township, Chin State of Myanmar. *International Journal of Knowledge Society Research, 8*(2), 44–53. doi:10.4018/IJKSR.2017040104

Maâlej Dammak, S. (2016). *A framework for achieving semantic annotation and interrogation for Web pages* (Unpublished doctoral dissertation). Sfax University, Tunisie.

Maâlej Dammak, S., Jedidi, A., & Bouaziz, R. (2013a). Semantic Annotation Framework for Web Resources. In *The fifth International Conference on Internet Technologies & Applications* (pp. 106-113). Wrexham, UK: Academic Press.

Maâlej Dammak, S., Jedidi, A., & Bouaziz, R. (2013b). Automation and evaluation of the semantic annotation of Web resources. In *The 8th International Conference for Internet Technology and Secured Transactions* (pp. 448-453). London, UK: Academic Press.

Maâlej Dammak, S., Jedidi, A., & Bouaziz, R. (2014a). Fuzzy semantic annotation of Web resources. In *The World Symposium on Computer Applications & Research* (pp. 253-258). Sousse, Tunisie: Academic Press.

Maâlej Dammak, S., Jedidi, A., & Bouaziz, R. (2017). Filtering Method for the Annotated and Non-Annotated Web Pages. International Journal of Knowledge Society Research - IJKSR, 8 (1), 1-22.

Maâlej Dammak, S., Jedidi, A., & Bouaziz, R. (2014b). Automation of the semantic annotation of web resources. *International Journal of Internet Technology and Secured Transactions*, *5*(2), 133–148. doi:10.1504/IJITST.2014.064513

MacKenzie, S. (2012). *Human-Computer Interaction: An Empirical Research Perspective*. Burlington, MA: Elsevier Science.

MacQueen, J. B. (1967). Some methods for classification and analysis of multivariate observations. In *The fifth Berkeley Symposium on Mathematical Statistics and Probability* (pp. 281-297). University of California Press.

Madzar, S. (2001). Subordinates' information inquiry: Exploring the effect of perceived leadership style and individual differences. *Journal of Occupational and Organizational Psychology*, *74*(2), 221–232. doi:10.1348/096317901167325

Mahdi, W., & Zlitni, T. (2016). Navigating news programs in large TV-streams: A knowledge-based approach. *International Journal of Knowledge Society Research*, *7*(1), 80–98. doi:10.4018/IJKSR.2016010106

Mainka, A. (2018). 21st *Century Cities.* Berlin: De Gryuter.

Mainka, A., Hartmann, S., Orszullok, L., Peters, I., Stallmann, A., & Stock, W. G. (2013). Public libraries in the knowledge society. Core services of libraries in informational world cities. *Libri*, *63*(4), 295–319. doi:10.1515/libri-2013-0024

Majima, Y., Maekawa, Y., Soga, M., & Sakoda, M. (2013). *A Proposal of the New System Model for Nursing Skill Learning Based on Cognition and Technique*. Berlin: Human-Computer Interaction, Part II. doi:10.1007/978-3-642-39262-7_16

Malarvizhi, P., & Pujeri, R. V. (2012). Distributed approach to web page categorization using map-reduce programming model. *International Journal of Engineering and Technology*, *3*(6), 373-386.

Manning, C. D., Raghavan, P., & Schutze, H. (2008). *Introduction to Information Retrieval*. Cambridge University Press. doi:10.1017/CBO9780511809071

Manson, G., Naturel, X., & Berrani, S. A. (2009). Automatic Program Extraction From TV Streams. *European Interactive TV Conference, EuroITV'09.*

Martindale, D. (2005). Are sex and death taboos in advertising? *International Journal of Advertising*, *25*, 1, 9–33.

Masaka, D., & Chemhuru, C. (2011). Moral Dimensions of Some Shona Taboos and Norms (Zviera). *Journal of Sustainable Development in Africa*, *13*(3).

Matiatou, M. (2015). From internal branding to cultural transformation: A virtuous circle. *International Journal of Knowledge Society Research*, *6*(2), 1–19. doi:10.4018/IJKSR.2015040101

Matsumoto, H., & Yutaka, E. (2011). *Dynamic resource management in cloud environment*. Fujitsu Sci. Tech.

Matsumura, N., Miura, A., Komori, M., & Hiraishi, K. (2016). Media mediate sentiments: Exploratory analysis of tweets posted before, during, and after the Great East Japan earthquake. *International Journal of Knowledge Society Research*, *7*(2), 57–71. doi:10.4018/IJKSR.2016040104

Matturro, G., & Silva, A. (2010). Using reflective guides to capture software projects experience. *Proceedings of the International Conference on Information and Knowledge Engineering*, 202-207.

Mbiti, J. S. (1969). *African religions and philosophy*. Nairobi: West African Publishers.

McIntyre, S., Watson, K., & Larsen, S. (2009). Strategies for large scale blended learning initiatives: Training, teaching and management. In E. Terry, A. Jefferies, & A. Bracq (Eds.), *Proceedings of the 4th International Blended Learning Conference* (pp. 295–308). Hatfield: University of Hertfordshire. Retrieved from unsworks.unsw.edu.au

Meena, R. S. (2009). Environmental-informatics---A solution for long term environmental research. *Global Journal of Enterprise Information System, 1*.

Menkiti, I. A. (1984). Person and Community in African Traditional Thought. In R. A. Wright (Ed.), *African Philosophy: An Introduction* (pp. 171–181). New York: University Press of America.

Menkiti, I. A. (2006). On the Normative Conception of a Person. In K. Wiredu (Ed.), *A Companion to African Philosophy* (pp. 324–331). Oxford, UK: Blackwell Publishing.

Merideth, M. G., & Iyengar, A. (2005). Thema: Byzantine-fault-tolerant middleware for web-service applications. In *Proceeding of the 24th Symposium on Reliable Distributed Systems (SRDS)*. IEEE.

Merryfield, M. (2001). The paradoxes of teaching a multicultural education course online. *Journal of Teacher Education, 52*(4), 283–299. doi:10.1177/0022487101052004003

Mertens, D. M. (2014). *Research and evaluation in education and psychology: Integrating diversity with quantitative, qualitative, and mixed methods*. Sage Publications.

Migdley, C., & Urdan, T. (2001). Academic self-handicapping and performance goals: A further examination. *Contemporary Educational Psychology, 26*(1), 61–75. doi:10.1006/ceps.2000.1041 PMID:11161640

Miller, H. B. (1983). Platonists and Aristotelians. In H. B. Miller & W. H. Williams (Eds.), *Ethics and Animals* (pp. 1–14). Clifton, NJ: Humana Press. doi:10.1007/978-1-4612-5623-6_1

Miller, V. D., & Jablin, F. M. (1991). Information seeking during organizational entry: Influences, tactics, and a model of the process. *Academy of Management Review, 16*(1), 92–120. doi:10.5465/amr.1991.4278997

Miller, W., & Walling, J. (2013). *Government in the twenty-first century: New Avenues of Study. Taking Sides*. New York, NY: McGraw Hill.

Misra, H., Hopfgartner, F., Goyal, A., Punitha, P., & Jose, J. (2010). TV News Story Based Segmentation one Semantic Coherence and Content Similarity. *Proc. of 16th International Conference one Multimedia Modeling*.

Mitilineos, S. A., Kyriazanos, D. M., Segou, O. E., Goufas, J. N., & Thomopoulos, S. C. (2010). Indoor localisation with wireless sensor networks. *Progress in Electromagnetics Research, 109*, 441–474. doi:10.2528/PIER10062801

Mohadab, M., Bouikhalene, B., & Safi, S. (2018). *Predicting rank for scientific research papers using supervised learning*. Applied Computing and Informatics.

Moller, L. (1998). Designing communities of learners for asynchronous distance education. *Educational Technology Research and Development, 46*(4), 115–122. doi:10.1007/BF02299678

Mong, L., Thuy, N. L., TokWang, L., & Zhong, Z. (2016). ORA-Semantics based Keyword Search in XML and Relational Databases. In *International Conference on data engineering* (pp. 157-160). Academic Press. DOI: 10.1109/ICDEW.2016.7495638

Moon, M. J. (2002). The evolution of e-government among municipalities: Rhetoric or reality? *Public Administration Review, 62*(4), 424–433. doi:10.1111/0033-3352.00196

Mourtada, R., & Alkhatib, F. (2014). *2014 UAE Social Media Outlook: Increasing Connectivity Between Government and Criticizes GCAA strategic Plan*. Retrieved on February 22, 2015 from: https://www.gcaa.gov.ae/

Muftah, H., & Lafi, H. (n.d.). Impact of QWL on employee satisfaction case of oil and gas industry in Qatar. *International Scientific Press, 1*(2), 107-134.

Muhaya, F., Bakry, S., & AlAlmaee, S. (2015). E-Government Success Factors: Views of Saudi Professionals. *J Entrepren Organiz Manag*, *4*, 128. doi:10.4172/2169-026X.1000128

Murugadas, D., Vieten, S., Nikolic, J., Fietkiewicz, K. J., & Stock, W. G. (2015). Creativity and entrepreneurship in informational metropolitan regions. *Journal of Economic and Social Development*, *2*(1), 14–24.

Naidoo, S., & Raju, J. (2012). Impact of the digital divide on information literacy training in a higher education context. *South African Journal of Library and Information Science*, *78*(1), 34–44. doi:10.7553/78-1-46

Nam, T., & Pardo, T. A. (2011a). Conceptualizing smart city with dimensions of technology, people, and institutions. In *Proceedings of the 12th Annual International Conference on Digital Government Research* (pp. 282-291). New York, NY: ACM. 10.1145/2037556.2037602

Nam, T., & Pardo, T. A. (2011b). Smart city as urban innovation: Focusing on management, policy, and context. In E. Estevez, & M. Janssen (Eds.), *5th International Conference on Theory and Practice of Electronic Governance* (pp. 185-194). New York, NY: ACM. 10.1145/2072069.2072100

Nandi, D., Hamilton, M., & Harland, J. (2012). Evaluating the quality of interaction in asynchronous discussion forums in fully online courses. *Distance Education*, *33*(1), 5–30. doi:10.1080/01587919.2012.667957

Nanjundeswaraswamy, T. S. (2013). Quality of Work life of Employees in Private Technical Institution. *International Journal of Qualitative Research*, *7*(3), 3–14.

Naomi, K. (1993). Culture: Perspectives and Challenges of Indigenous People. *Cultural Survival Quarterly*, *31*(4), 52–81.

Naturel, X., & Gros, P. (2008). Detecting repeats for video structuring. *Multimedia Tools and Applications*, 38.

Nefiodow, L. A. (1991). *Der fünfte Kondratieff: Strategien zum Strukturwandel in Wirtschaft und Gesellschaft.* Wiesbaden, Germany: Gabler, FAZ.

Newaz, M. K., Ali, T., & Akhter, I. (2007). Employee Perception Regarding Turnover Decision- In Context of Bangladesh Banking Sector. *BRAC University Journal*, *4*(2), 67–74.

Newell, S. (2004). Enhancing cross-project knowledge. *Engineering Management Journal*, *16*(1), 12–20. doi:10.1080/10429247.2004.11415234

Niaros, V., Kostakis, V., & Drechsler, W. (2017). Making (in) the smart city: The emergence of maker spaces. *Telematics and Informatics*, *34*(7), 1143–1152. doi:10.1016/j.tele.2017.05.004

Nimalathasan, B. (2010). Quality of Work Life Practices of Academic Professionals in Bangladesh: A Factor Analytical Approach. *International Journal of Research in Commerce & Management*, *1*(7), 6–11.

Nixon, P. G., & Rawal, R. (2011). *Enabling Democracy? e-Government, Inclusion and Citizenship*. The Hague University.

Noah, M. E. (1980). *Ibibio Pioneers in Modern Nigerian History*. Scholars Press.

Nock, R., & Nielsen, F. (2004). Statistical Region Merging. *Trans. Pattern Anal. Mach. Intell.*

Nonaka, I. (1994). A dynamic theory of organizational knowledge creation. *Organization Science*, *5*(1), 14–37. doi:10.1287/orsc.5.1.14

Northrop, D. (1973). Pre-colonial references to the Annang Ibibio. *Ikenga, 2*(1).

Nouali, O., & Blache, P. (2004). A semantic vector space and features-based approach for automatic information filtering. *International Journal Expert Systems with Applications*, *26*(2), 171–179. doi:10.1016/S0957-4174(03)00118-0

Noura, A., Ismahène, H. K., & Sami, F. (2017). Process of personalizing the ubiquitous advertisements. *International Journal of Knowledge Society Research*, *8*(2), 13–35. doi:10.4018/IJKSR.2017040102

Nssien, F. U. (1991). *The People called Ibibios, An annotated Biography*. Calabar: De-Rio Press.

Obio-Offiong, U. E. E. (1958). *A first step to the study of Ibibio History*. Aman Press.

Oku, E. E. (1989). *The Kings and Chiefs of Old Calabar (1785-1925)*. Glad Tiding Press Ltd.

Oldenburg, R. (1999). *The Great Good Place*. New York, NY: Marlowe & Company.

Oldenburg, R., & Brissett, D. (1982). The third place. *Qualitative Sociology*, *5*(4), 265–284. doi:10.1007/BF00986754

Opdenakker, R. (2006, September). *Advantages and Disadvantages of Four Interview Techniques in Qualitative Research*. Retrieved June 9,2014, from Forum Qualitative Sozialforschung website: http://www.qualitative-research.net/index.php/fqs/article/view/175/391

Osei, J. (2006). The Value of African Taboos for Biodiversity and Sustainable Development. *Journal of Sustainable Development in Africa*, *8*(3), 42–61.

Osm. (n.d.). Retrieved from https://www.openstreetmap.org

Ouidade, F. (2010). *Social Norms and Taboos*. Oxford, UK: Clarendon Press of Oxford University Press.

Pallant, J. (2011). *SPSS Survival Manual* (4th ed.). Crows Nest, Australia: Allen & Unwin.

Palmer, M. M., Shaker, G., & Hoffmann-Longtin, K. (2014). Despite faculty skepticism: Lessons from a graduate-level seminar in a hybrid course environment. *College Teaching*, *62*(3), 100–106. doi:10.1080/87567555.2014.912608

Papakonstantinou, Y., & Xu, Y. (2005). Efficient keyword search for smallest LCAs in XML databases. In *Proceedings of the 2005 ACM SIGMOD. International conference on Management of data* (pp. 527-538). ACM. DOI: 10.1145/1066157.1066217

Papakonstantinou, Y., & Xu, Y. (2008). Efficient LCA based keyword search in XML data. In *Proceedings of the 11th International conference on extending database technology: Advances in database technology* (pp. 535-546). ACM. DOI: 10.1145/1353343.1353408

Parajuli, K. P. (2016). Mobile learning practice in higher education in Nepal. *Open Praxis*, *8*(1), 41–54. doi:10.5944/openpraxis.8.1.245

Paramvir & Padmanabhan. (2000). RADAR: An in-building RF based user location and tracking system. *Proceedings of the Nineteenth Annual Joint Conference of the IEEE Computer and Communications Societies (INFOCOM 2000)*, 775–784.

Parasuraman, A., Zeithaml, V. A., & Berry, L. L. (1988). SERVQUAL: A multiple-item scale for measuring consumer perceptions of service quality. *Journal of Retailing*, *64*(1), 12–40.

Paredes-Valverde, M. A., Alor-Hernández, G., Rodríguez-González, A., Valencia-García, R., & Jiménez-Domingo, E. (2015). A systematic review of tools, languages, and methodologies for mashup development. *Software, Practice & Experience*, *45*(3), 365–397. doi:10.1002pe.2233

Parrinder, G. (1969). *Religion in Africa*. New York: Praeger Publishers.

Pasha, M. (2017). Semantically enabled communication framework for autonomous systems. *International Journal of Knowledge Society Research*, *8*(2), 54–68. doi:10.4018/IJKSR.2017040105

Patel, K., & Kanani, H. (2013). Dynamic Error Handling Tool. *International Journal of Engineering and Innovative Technology*, *2*(11), 34–36.

Patro, C. S. (2013). Work Life Balance Policies for Employees in Private Sector. *Global Research Thoughts, 2,* 9-13.

Patro, C. S. (2017). Employee Welfare Measures in Public and Private Sectors: A Comparative Analysis. In Public Health and Welfare: Concepts, Methodologies, Tools, and Applications (pp. 1026-1042). IGI Global. doi:10.4018/978-1-5225-1674-3.ch047

Patro, C. S. (2012a). Employee Welfare Activities in Private Sector and Their Impact on Quality of Work Life. *International Journal of Productivity Management and Assessment Technologies*, *1*(2), 18–29. doi:10.4018/ijpmat.2012040102

Patro, C. S. (2012b). A Study on Adaptability of Total Quality Management in Engineering Education Sector. *International Journal of Quality Assurance in Engineering and Technology Education*, *2*(4), 25–37. doi:10.4018/ijqaete.2012100103

Patro, C. S. (2014). A Study on the Impact of Employee Retention Policies on Organisation Productivity in Private Sector. *International Journal of Asian Business and Information Management*, *5*(3), 48–63. doi:10.4018/ijabim.2014070104

Patro, C. S. (2015). A Study on the Impact of Faculty QWL on Quality of Education in Academic Institutions. *International Journal of Knowledge Society Research*, *6*(1), 1–16. doi:10.4018/IJKSR.2015010101

Paul, R. W. (1988). Critical thinking in the classroom. *Teaching*, *K-8*, 49–51.

Paul, R. W. (1993). The logic of creative and critical thinking. *The American Behavioral Scientist*, *37*(1), 21–40. doi:10.1177/0002764293037001004

Pease, M., Shostak, R., & Lamport, L. (1980). Reaching agreement in the presence of faults. *Journal of the Association for Computing Machinery*, *27*(2), 228–234. doi:10.1145/322186.322188

Peng, Y., Fan, W., Dong, X., & Zhang, X. (2016). An Iterative Weighted KNN (IW-KNN) Based Indoor Localization Method in Bluetooth Low Energy (BLE) Environment. *Proceedings of the* 2016 *Intl IEEE Conferences on Ubiquitous Intelligence Computing, Advanced and Trusted Computing, Scalable Computing and Communications, Cloud and Big Data Computing, Internet of People, and Smart World Congress (UIC/ATC/ScalCom/CBDCom/IoP/SmartWorld),* 794–800.

Periman, S. L. (2006). *Human Resource Innovators, integrating quality of work life into organizational performance*. Retrieved May 15, 2018, from http://www.quaiindia.com

Perkins, D., Hancock, C., Hobbs, R., Martin, F., & Simmons, R. (1986). Conditions of learning in novice programmers. *Journal of Educational Computing Research*, *2*(1), 37–55. doi:10.2190/GUJT-JCBJ-Q6QU-Q9PL

Pernici, B., Francalanci, C., Scalia, G., Corsi, M., Grandoni, D., & Biscardi, M. A. (2018). *Geolocating social media posts for emergency mapping*. arXiv preprint arXiv:1801.06861

Pham, K. T., Sattigeri, P., Dhurandhar, A., Jacob, A. C., Vukovic, M., Chataigner, P., … Varshney, K. R. (2017). Real-time understanding of humanitarian crises via targeted information retrieval. *IBM Journal of Research and Development, 61*(6), 7.1-7.12.

Pipitone, A., & Pirrone, R. (2016). Aligning relational schema and OWL ontologies with hidden Markov model. *International Journal of Knowledge Society Research*, *7*(2), 1–24. doi:10.4018/IJKSR.2016040101

Pirolli, P., & Card, S. (1999). Information foraging. *Psychological Review*, *106*(4), 643–675. doi:10.1037/0033-295X.106.4.643

Pithers, R. T., & Soden, R. (2000). Critical thinking in education. *Review of Educational Research, 42*(3), 237–249.

Piwowarski, B., Gauthier, L.-A., & Gallinari, P. (2014). *Filtrage collaboratif et prise en compte de la polarité des jugements*. CORIA.

Pohl, S., Zobel, J., & Moffat, A. (2010). Extended Boolean Retrieval for Systematic Biomedical Reviews. In *The 33rd Australasian Computer Science Conference* (pp. 117-126). Brisbane, Australia: Academic Press.

Pohl, S., Moffat, A., & Zobel, J. (2012). Efficient Extended Boolean Retrieval. *IEEE Transactions on Knowledge and Data Engineering, 24*(6), 1014–1024. doi:10.1109/TKDE.2011.63

Polanyi, M. (1966). *The tacit dimension*. Gloucester, MA: University of Chicago Press.

Portnoy, M. (2012). *Virtualization Essentials*. Hoboken, NJ: Wiley.

Posey, G., Burgess, T., Eason, M., & Jones, Y. (2010). The Advantages and Disadvantages of the Virtual Classroom and the Role of the Teacher. *Southwest Decision Sciences Institute Conference*, 2-6.

Posner, E. (2000). *Law and Social Norms*. Cambridge, MA: Harvard University Press.

Quinn, B. (2016). Computational Thinking Guiding Change in Online Education. *Journal of Medical Biomedical and Applied Sciences, 4*(12), 8–17.

Radan, S. (2012). UAE among emerging e-Government leaders. *Khaleej Times*. Retrieved on June, 15, 2015 from http://www.khaleejtimes.com/

Radhakrishnan, N., Poorna Chelvan, N., & Ramkumar, D. (2012). *Utilization of Cloud Computing in E-learning Systems. In International of Cloud Computing* (pp. 208–213). Technologies, Applications & Management.

Rădulescu, Ş. (2014). *A perspective on E-Learning and Cloud Computing*. Social and Behavioral Sciences.

Rahmani, A. M., Hossaini, Z., & Setayeshi, S. (2009). Link Processing for Fuzzy Web Pages Clustering and Classification. *European Journal of Scientific Research, 27*(4), 620-627.

Rajkumar, B., Christian, V., & Thamarai, S. (2013). *Mastering Cloud Computing: Foundations and Applications Programming*. Burlington, MA: Elsevier Science.

Ramon, G. (2012). *Enacting electronic government success: An integrative study of government-wide websites, organizational capabilities, and institutions.* Springer.

Rashid. (2013). *HH launches Mobile Government, ICT Fund offers AED 200m in support of the initiative: Emirates News Agency*. Retrieved on May 22, 2014, from http://www.wam.org.ae/servlet/Satellite?c=WamLocEnews&cid=1290004835681&pagename=WAM%2FWAM_E_Layout&parent=Collection&parentid=1135099399983&rendermode=preview-admin-1135099398363

Ravani, R., Mirali, R. M., & Baniasadi, M. (2010). Parallel CBIR System Based on Color Coherence Vector. *Proc. of International Conference on Systems, Signals and Image.*

Reddick, C. G. (2010a). *Politics, Democracy and eGovernment: Participation and Service Delivery*. Hershey, PA: Information Science Reference. doi:10.4018/978-1-61520-933-0

Reddick, C. G. (2010b). *Citizens and E-Government: Evaluating Policy and Management*. Hershey, PA: Information Science Reference. doi:10.4018/978-1-61520-931-6

Reena, G., & Jayan, C. (2012). Role Of Quality of Work Life on The Job Attitude and Personal Effectiveness of Engineering College Teachers. *South Asian Academic Research Journals, 2*(6).

Resch, B. (2013). People as sensors and collective sensing-contextual observations complementing geo-sensor network measurements. In *Progress in Location-Based Services* (pp. 391-406). Springer Berlin Heidelberg.

Resch, B., Summa, A., Sagl, G., Zeile, P., & Exner, J. P. (2015). Urban emotions—geo-semantic emotion extraction from technical sensors, human sensors and crowdsourced data. In Progress in Location-Based Services 2014 (pp. 199-212). Springer International Publishing.

Ricker, B., Daniel, S., & Hedley, N. (2014). Fuzzy Boundaries: Hybridizing Location-based Services, Volunteered Geographic Information and Geovisualization Literature. *Geography Compass*, *8*(7), 490–504. doi:10.1111/gec3.12138

Rizzo, A. (2017). Why knowledge megaprojects will fail to transform Gulf countries in post-carbon economies: The case of Qatar. *Journal of Urban Technology*, *24*(3), 85–98. doi:10.1080/10630732.2017.1311569

Rockinson-Szapkiw, A. J. (2012). Should online doctoral instructors adopt audio feedback as an instructional strategy? Preliminary evidence. *International Journal of Doctoral Studies*, *7*, 7245–7258. doi:10.28945/1595

Rohr, J. (1978). *Ethics for Bureaucrats*. New York: Marcel Dekker.

Rokhman, A. (2011). e-Government adoption in developing countries; the case of Indonesia. *Journal of Emerging Trends in Computing and Information Sciences*, *2*(5), 228–236.

Rorissa, A., Demissie, D., & Pardo, T. (2011). Benchmarking e-government: A comparison of frameworks for computing e-government index and ranking. *Government Information Quarterly*, *28*(3), 354–362. doi:10.1016/j.giq.2010.09.006

Rossetto, L., Giangreco, I., Tănase, C., & Schuldt, H. (2017). Multimodal video retrieval with the 2017 IMOTION system. *Procs of the 2017 ACM International Conference on Multimedia Retrieval (ICMR 2017),* 457–460.

Rossi, F., & Villa-Vialaneix, N. (2011). Représentation d'un grand réseau à partir d'une classification hiérarchique de ses sommets. *Journal de la Société Française de Statistique SFDS*, *152*(3).

Rossi, I. (Ed.). (1980). *People in Culture: A Survey of Cultural Anthropology*. New York: Praeger.

Rothman, R. (2007). Building 'smart education systems'. *Education Week*, *26*(44), 25–27.

Rountree, D., & Castrillo, I. (2013). *The Basics of Cloud Computing: Understanding the Fundamentals of Cloud Computing in Theory and Practice*. Burlington, MA: Elsevier Science.

Rydberg, C. (2015). *Time Efficiency of Information Retrieval with Geographic Filtering. Digitala Vetenskapliga Arkivet.* Stockholm, Sweden: Digital Scientific Archive.

Saad, H. A., Samah, A. J. A., & Juhdi, N. (2008). Employees' perception on quality work life and job satisfaction in a private higher learning institution. *International Review of Business Research Papers*, *4*(3), 23–34.

Sadique, Z. (2003). Quality of Work Life among White Collar and Blue Collar Employees. *Journal of the Institute of Bangladesh Studies*, *26*, 169–174.

Salazar, J. C., Torres, M., & Moreno, M. (2016). *Monitoreo urbano de entidades y eventos geográficos basado en censado social* (Unpublished master dissertation). Centro de Investigación en Computación, Instituto Politécnico Nacional, Mexico.

Saldana-Perez, A. M. M., Moreno-Ibarra, M. A., & Torres-Ruiz, M. J. (2018). Classification of Traffic Events Notified in Social Networks' Texts. In Encyclopedia of Information Science and Technology, Fourth Edition (pp. 6973-6984). IGI Global.

Saldana-Perez, A. M. M., & Moreno-Ibarra, M. (2016). Traffic analysis based on short texts from social media. *International Journal of Knowledge Society Research*, *7*(1), 63–79. doi:10.4018/IJKSR.2016010105

Saleh, M., Alrabiah, A., & Bakry, S. (2007). A STOPE model for the investigation of compliance with ISO 17799-2005. *Information Management & Computer Security*, *15*(4), 283–294. doi:10.1108/09685220710817806

Salton, G., Fox, E. A., & Wu, H. (1983). Extended Boolean Information Retrieval. *Magazine Communications of the ACM*, *26*(11), 1022–1036. doi:10.1145/182.358466

Salton, G., Wong, A., & Yang, C. S. (1975). A vector space model for automatic indexing. *Magazine Communications of the ACM*, *18*(11), 613–620. doi:10.1145/361219.361220

Samer, A., & Siobhán, N. (2017). The Emergence Computation of Overflow in Dynamic X,L Tree Based on Prefix and Interval Labelling Schemes. *The International Conference on Engineering & Technology*, 21-23.

Samra, H. E., Li, A. S., Soh, B., & AlZain, M. A. (2017). A cloud-based architecture for interactive e-training. *International Journal of Knowledge Society Research*, *8*(1), 67–78. doi:10.4018/IJKSR.2017010104

Samwald, M., & Adlassnig, K. (2008). The bio-zen plus ontology. *Journal Applied Ontology*, *3*(4), 213–217.

Sassen, S. (2001). *The Global City: New York, London, Tokyo* (2nd ed.). Princeton, NJ: Princeton University Press. doi:10.1515/9781400847488

Saunders, F. C., & Gale, A. W. (2012). Digital or didactic: Using learning technology to confront the challenge of large cohort teaching. *British Journal of Educational Technology*, *43*(6), 847–858. doi:10.1111/j.1467-8535.2011.01250.x

Savolainen, R. (2006). Time as a context of information seeking. *Library & Information Science Research*, *28*(1), 110–127. doi:10.1016/j.lisr.2005.11.001

Scanlan, R. (2003). The concept of taboos in African society with reference to the Kikuyu of Kenya. In M. Getui (Ed.), MIASMU Research Integration Papers to African Culture: An overview. Academic Press.

Schmidt, A., Shirazi, A.-S., Zurmaar, B., Bial, D., Zaidan, F., Alt, F., . . . Lewen, T. (2011), Digifieds: Evaluating suitable interaction techniques for shared public notice areas. In *Adjunct Proceedings of the 9th International Conference on Pervasive Computing*. Springer.

Schneider, F. B. (1990). Implementing fault-tolerant services using the state machine approach: A tutorial. *ACM Computing Surveys*, *22*(4), 299–319. doi:10.1145/98163.98167

Schudt, K. (2000). Taming the Corporate Monster: An Aristotelian approach to corporate virtue. *Business Ethics Quarterly*, *10*(3), 711–723. doi:10.2307/3857900

Schwab, K., & Sala-i-Martin, X. (Eds.). (2016). *The Global Competitiveness Report*. World Economic Forum.

Seol, S., Lee, E.-K., & Kim, W. (2017). Indoor mobile object tracking using RFID. *Future Generation Computer Systems*, *76*, 443–451. doi:10.1016/j.future.2016.08.005

Serageldin, I. (2011). *Science and the Arab spring. Issues in Science and Technology*. Retrieved on May 16, 2015, from http://www.issues.org/27.4/p_serageldin.html

Seres, A. (2010). Three Database Management Systems (DBMS) compared. *Open Source Science Journal*, *2*, 65–82.

Serey, T. T. (2006). Choosing a Robust Quality of Work Life. *Business Forum*, *27*(2), 7-10.

Shadiev, R., Hwang, W., Yeh, S., Yang, S., Han, L., & Hsu, G. (2014). Effectts of Unidirectional vs. Reciprocal Teaching on Web-Based Computer Programming Learning. *Journal of Educational Computing Research*, *50*(1), 67–95. doi:10.2190/EC.50.1.d

Shahbazi, B., Shokrzadeh, S., Bejani, H., Malekinia, E., & Ghoroneh, D. (2011). A Survey of Relationship between the Quality of Work Life and Performance of Department Chairpersons of Esfahan University and Esfahan Medical Science University. *Procedia: Social and Behavioral Sciences*, *30*(1), 1555–1560. doi:10.1016/j.sbspro.2011.10.302

Shahzad, B., & Alwagait, E. (2014). Best and the Worst Times to Tweet: An Experimental Study. *WSEAS, 15th International Conference on Mathematics and Computers in Business and Economics (MCBE'14), Proceedings of 15th International Conference on Mathematics and Computers in Business and Economics.*

Shamir, A. (1979). How to share a secret. *Communications of the ACM*, *22*(11), 612–613. doi:10.1145/359168.359176

Shareef, M. A., Kumar, V., Kumar, U., & Dwivedi, Y. K. (2011). e-Government Adoption Model (GAM): Differing service maturity levels. *Government Information Quarterly*, *28*(1), 17–35. doi:10.1016/j.giq.2010.05.006

Sharjah. (2018). *Sharjah Economy*. Retrieved from https://www.sharjah.com/v/economy/

Sherif, J., & Sumner, O. (2010). Beyond Rules, Laws and Social Norms. *Journal of Personality and Social Psychology*, *34*(3), 599–607.

Shulman, L. S. (1986). Those who understand: Knowledge growth in teaching. *Educational Researcher*, *15*(2), 4–14. doi:10.3102/0013189X015002004

Sidorov, G., Velasquez, F., Stamatatos, E., Gelbukh, A., & Chanona-Hernández, L. (2014). Syntactic n-grams as machine learning features for natural language processing. *Expert Systems with Applications*, *41*(3), 853–860. doi:10.1016/j.eswa.2013.08.015

Silverman, K., Naik, D., Lenzo, K., & Henton, C. (2013). *U.S. Patent No. 8,583,418*. Washington, DC: U.S. Patent and Trademark Office.

Simaanya, M. (2014). *Employee Management System* (Doctoral dissertation). University of Zambia.

Sims, R., Dobbs, G., & Hand, T. (2002). Enhancing quality in online learning: Scaffolding planning and design through proactive evaluation. *Distance Education*, *23*(2), 135–148. doi:10.1080/0158791022000009169

Sipior, J. C., Ward, B. T., & Connolly, R. (2011). The digital divide and t-government in the United States: Using the technology acceptance model to understand usage. *European Journal of Information Systems*, *20*(3), 308–328. doi:10.1057/ejis.2010.64

Smart Dubai Initiative. (2017). *Dubai Initiative*. Retrieved on March, 17, 2018 from http://www.smartdubai.ae

Smeaton, A. F., Kraaij, W., & Over, P. (2003). TRECVID 2003 - an overview. *TRECVID 2003 - Text REtrieval Conference TRECVID Workshop.*

Sorensen, B., Ramachandran, S., & Salvendy, S. (2007). *Simulation-Based Automated Intelligent Tutoring*. Heidelberg, Germany: Human Interface, Part II Springer-Verlag Berlin. doi:10.1007/978-3-540-73354-6_51

Sosinsky, B. (2010). *Cloud Computing Bible*. Hoboken, NJ: Wiley. doi:10.1002/9781118255674

Sotenga, P. Z., Djouani, K., Kurien, A. M., & Mwila, M. M. (2017). Indoor Localisation of Wireless Sensor Nodes Towards Internet of Things. *Procedia Computer Science*, *109*, 92–99. doi:10.1016/j.procs.2017.05.299

Souza, C. L., Pádua, F. L. C., Nunes, C. F. G., Assis, G. T., & Silva, G. D. (2014). *A unified approach to content-based indexing and retrieval of digital videos from television archives*. Artificial Intelligence Research.

Srebnaja, J. (2017). The role and implementation of new technologies in the ELT (English language teaching). *International Journal of Knowledge Society Research*, *8*(3), 68–83. doi:10.4018/IJKSR.2017070105

Stallmeyer, J. C. (2009). Landscapes of informational urbanism. *Journal of Landscape Architecture*, *6*(3), 34–39.

Stallmeyer, J. C. (2011). *Building Bangalore. Architecture and Urban Transformation in India's Silicon Valley*. London: Routledge.

Stefanou, V., & Kotsovoulou, M. (2016). Use of PowerPoint in the classroom: A participatory research project. *International Journal of Knowledge Society Research*, *7*(4), 38–50. doi:10.4018/IJKSR.2016100104

Stehman, S. V., Fonte, C. C., Foody, G. M., & See, L. (2018). Using volunteered geographic information (VGI) in design-based statistical inference for area estimation and accuracy assessment of land cover. *Remote Sensing of Environment*, *212*, 47–59. doi:10.1016/j.rse.2018.04.014

Stella, M., Shanshan, C., Mansoor, A., Brian, L., Paula, M., Nick, R., . . . Ying, D. (2011). The VIVO Ontology: Enabling Networking of Scientists. In The ACM WebSci'11 (pp. 1-2). Koblenz, Germany: Academic Press.

Stinson, D. R. (1992). An explication of secret sharing schemes. *Designs, Codes and Cryptography*, *2*(4), 357–390. doi:10.1007/BF00125203

Stock, W. G. (2011). Informational cities: Analysis and construction of cities in the knowledge society. *Journal of the American Society for Information Science and Technology*, *62*(5), 963–986. doi:10.1002/asi.21506

Stock, W. G. (2015). Informational urbanism. *Journal of Systemics, Cybernetics and Informatics*, *13*(6), 62–69.

Stock, W. G., & Stock, M. (2013). *Handbook of Information Science*. Berlin: De Gruyter Saur. doi:10.1515/9783110235005

Stroe, A., Spruit, M. R., Beltman, B., & Koelemeijer, S. J. (2016). PMOMM: The Project Management Office maturity model. *International Journal of Knowledge Society Research*, *7*(3), 47–61. doi:10.4018/IJKSR.2016070104

Stryjakiewicz, T. (2010). Location factors of the creative and knowledge-intensive industries in European metropolitan regions. *Geografický Časopis. The Geographical Journal*, *62*(1), 3–19.

Su, A., Yang, S., Hwang, W., Huang, C., & Tern, M. (2014). Investigating the Role of Computer-Supported Annotation in Problem-Solving-Based Teaching: An Empirical Study of a Scratch Programming Pedagogy. *British Journal of Educational Technology*, *45*(4), 647–664. doi:10.1111/bjet.12058

Sun. (n.d.). Retrieved from http://blogs.sun.com/gbrunett/entry/amazon_s3_silent_data_corruption

Swan, K. (2001). Virtual interaction: Design factors affecting student satisfaction and perceived learning in asynchronous online courses. *Distance Education*, *22*(2), 306–331. doi:10.1080/0158791010220208

Sykes, E. (2014). New methods of mobile computing: From smartphones to smart education. *TechTrends*, *58*(3), 26–37. doi:10.100711528-014-0749-2

Tabassum, A. (2012). Interrelations between Quality of Work Life Dimensions and Faculty Member Job Satisfaction in the Private Universities of Bangladesh. *European Journal of Business and Management*, *4*(2), 78–89.

Tabassum, A., Rahman, T., & Jahan, K. (2010). Quality of Work Life among the Male and Female Employees of Private Commercial Banks in Bangladesh. In *Proceedings of International Conference on Business Competencies in a Changing Global Environment*. South East University.

Tabassum, A., Rahman, T., & Jahan, K. (2011). A Comparative Analysis of Quality of Work Life among the Employees of Local and Foreign Banks in Bangladesh. *World Journal of Social Science*, *1*(1), 17–33.

Tabassum, A., Rahman, T., & Jahan, K. (2012). An Evaluation of the Quality of Work Life: A Study of the Faculty Members of Private Universities in Bangladesh. *ABAC Journal*, *32*(3), 36–57.

Talbort, P. A. (1969). *The People of Southern Nigeria*. London: Frank Cass and Co. Ltd.

Tanfeedh. (2017). *The National Program for Enhancing Economic Diversification. TANFEEDH Handbook*. Muscat, Oman: National Program for Enhancing Economic Diversification.

Tang, H., Liao, S.-S., & Sun, S.-X. (2013). A prediction framework based on contextual data to support mobile personalized marketing. *Decision Support Systems*, *56*, 234–246. doi:10.1016/j.dss.2013.06.004

Tassabehji, R., & Elliman, T. (2006). *Generating Citizen Trust in e-Government using a Trust Verification United Nations Survey (2014). E-government for the Future we Want*. Retrieved on July, 28, 2015, from http://www2.unpan.org/egovkb/documents/2010/E_Gov_2010_Complete.pdf

Tatira, L. (2000a). *Zviera ZvaVaShona*. Gweru: Mambo Press.

Tatira, L. (2000b). The Role of Zviera in Socialisation. In E. Chiwome, Z. Mguni, & M. Furusa (Eds.), *Indigenous Knowledge and Technology in African and Diasporan Communities* (pp. 146–151). Harare: University of Zimbabwe.

Taylor, J., Thackray, H., Hodge, S. E., & McAlaney, J. (2018). Introducing Psychological Concepts and Methods to Cybersecurity Students. In *Psychological and Behavioral Examinations in Cyber Security* (pp. 98–108). IGI Global. doi:10.4018/978-1-5225-4053-3.ch006

Taylor, P. J. (2004). *World City Network: A Global Urban Analysis*. London, UK: Routledge.

Taylor, P. J., Hoyler, M., & Verbruggen, R. (2010). External urban relational process: Introducing central flow theory to complement central place theory. *Urban Studies (Edinburgh, Scotland)*, *47*(13), 2803–2818. doi:10.1177/0042098010377367

Temponi, C. (2005). Continuous improvement framework: Implications for academia. *Quality Assurance in Education*, *13*(1), 17–36. doi:10.1108/09684880510578632

Thakkar, J., Deshmukh, S. G., & Shastree, A. (2006). Total quality management (TQM) in self-financed technical institutions: A quality function deployment (QFD) and force field analysis approach. *Quality Assurance in Education*, *14*(1), 54–74. doi:10.1108/09684880610643610

Thite, M. (2011). Smart cities: Implications of urban planning for human resource development. *Human Resource Development International*, *14*(5), 623–631. doi:10.1080/13678868.2011.618349

Tian, W., Sheng, S., & Guoming, L. (2010). *A framework for implementing and managing platform as a service in a virtual cloud computing lab*. Education Technology and Computer Science. doi:10.1109/ETCS.2010.126

Tiwari, P. K., & Joshi, S. (2018). A comprehensive report on security and privacy challenges in software as a service. In *Multidisciplinary Approaches to Service-Oriented Engineering* (pp. 143–167). IGI Global. doi:10.4018/978-1-5225-5951-1.ch007

Tobler, W. R. (1970). A computer movie simulating urban growth in the Detroit region. *Economic Geography*, *46*, 234–240. doi:10.2307/143141

Torres-Ruiz, M. J., & Lytras, M. D. (2016). Urban computing and smart cities applications for the knowledge society. *International Journal of Knowledge Society Research*, *7*(1), 113–119. doi:10.4018/IJKSR.2016010108

Truong, N., Roe, P., & Bancroft, P. (2005, January). Automated Feedback for "Fill in the Gap" Programming Exercises. In *Proceedings of the 7th Australasian conference on Computing education* (pp. 117-126). Australian Computer Society, Inc.

Tshiamo, W. B., Kgositau, M., & Magowe, M. (2017). Use of Information and Communication Technology by Health Care Providers for Continuing Professional Development in Botswana. In K. H. Moshi & K. J. Bwalya (Eds.), *Health Information Systems and the Advancement of Medical Practice in Developing Countries* (pp. 181–191). IGI Global. doi:10.4018/978-1-5225-2262-1.ch011

Tukel, O., Rom, W., & Kremic, T. (2008). Knowledge transfer among projects using a learn-forget model. *The Learning Organization*, *15*(2), 179–194. doi:10.1108/09696470810852339

Türkkahramana, M. (2013). Social values and value education, *Procedia - Social and Behavioral Sciences, 116*(2014), 633 – 638.

Twitterapi. (n.d.). Retrieved from https://developer.twitter.com/en/docs/developer-utilities/twitter-libraries

Tynan, B., Ryan, Y., Hinton, L., & Lamont Mills, A. (2012). *Out of hours: Final report of the project e-teaching leadership: Planning and implementing a benefits-oriented costs model for technology enhanced learning*. Retrieved from http://www.olt.gov.au

UAE government. (2014). *EGovernment in the United Arab Emirates 2014. Retrieved from* government.ae/.../eGov...E.../22b5703a-d8d5-4fd4-ad1f-5d2c2bcde8cc

UAE. (2014). *Vision 2021*. Abu Dhabi: United in Ambition and Determination.

Uddin, M. T., Islam, M. T., & Ullah, M. O. (2006). A Study on the Quality of Nurses of Government Hospitals in Bangladesh. *Proc. Pakistan Acad. Science*, *43*(2), 121–129.

Uden, L., Tao, Y.-H., Yang, H.-C., & Ting, I.-H. (2014). *The 2nd International Workshop on Learning Technology for Education in Cloud.* Dordrecht: Springer.

Udo, E. A. (1984). *Who are the Ibibio?* Onitsha: Africana FEP Publishers.

Ullah, A., & Shah, M. (2016). Perception of Social Values and Its Role in Controlling Corruption Geniein Peshawar Region. *Journal of Accounting & Marketing*, *5*, 177. doi:10.4172/2168-9601.1000177

Ullmann-Margalit, E. (1977). *The Emergence of Norms*. Oxford, UK: Oxford University Press.

Umaselvi, B. P., & Kumar, S. (2011). *Quality of Work Life: Perception of College Teachers.* Hallamark Business School, Trichy, Tamilnadu, India, Paper (27868).

Underdown, K., & Martin, J. (2016). Engaging the Online Student: Instructor-Created Video Content for the Online Classroom. *Journal of Institutional Research*, *5*(1), 8–12. doi:10.9743/JIR.2016.2

UNDESA (United Nations Department of Economic & Social Affairs). (2001). *Benchmarking E-Government 2001: Assessing Progress of UN Member States*. Author.

UNDESA (United Nations Department of Economic & Social Affairs). (2005). *UN Global E-Government Readiness Report 2005: From E-Government to E-Inclusion*. Author.

UNDESA (United Nations Department of Economic & Social Affairs). (2016). *E-Government Survey 2016: E-Government in Support of Sustainable Development*. Author.

UNESCO. (2005). *United Nations Education Scientific and Cultural Organization*. Paris: Towards Knowledge Societies.

UNESCO. (2010). United Nations Education Scientific and Cultural Organization, Toward Inclusive Knowledge Societies: A review of UNESCO's action in implementing the WSIS outcomes, Paris, 2010. UNESCO.

UNESCO. (2017). *UNESCO Institute of Statistics: UIS.* UNESCO.

Utomo, M. N. Y., Adji, T. B., & Ardiyanto, I. (2018). Geolocation Prediction in Social Media Data Using Text Analysis. *RE:view*.

Uya, O. E. (1983). *A History of Oron People of the lower Cross River Basin*. Manson Publishing Company.

Uysal, M. (2014). Improving First Computer Programming Experiences: The Case of Adapting a Web-Supported and Well-Structured ProblemSolving Method to a Traditional Course. *Contemporary Educational Technology*, *5*(3), 198–217.

Van den Berg, L., van der Meer, J., & Otgaar, A. H. J. (2007). The attractive city: Catalyst of sustainable urban development. In *XVI Congreso de Estudios Vascos* (pp. 485-491). Bilbao, Spain: Academic Press.

Van den Heever, G. (2001). On how to be or not to be: Theoretical reflection on religion and identity in Africa. *Religion and Theology*, *8*(1), 1–25. doi:10.1163/157430101X00017

Van Rijsbergen, C. J. (1979). Information Retrieval (2nd ed.). Butterworth-Heinemann.

Vandenbosch, B., & Huff, S. L. (1997). Searching and scanning: How executives obtain information from executive information systems. *Management Information Systems Quarterly*, *1997*(1), 81–107. doi:10.2307/249743

Van-Gennep, A. (1904). *Tabou et totémisme à Madagascar: Étude descriptive et théorique*. Paris: Ernest Leroux.

Vargas-Vera, M. (2016a). Data integration framework: A children and parents cohort case study. *International Journal of Knowledge Society Research*, *7*(1), 99–112. doi:10.4018/IJKSR.2016010107

Vargas-Vera, M. (2016b). Methodology for the elaboration of quizzes using propositional logic calculus in an e-learning environment. *International Journal of Knowledge Society Research*, *7*(4), 63–74. doi:10.4018/IJKSR.2016100106

Vargas-Vera, M. (2016c). Reflections on the Second Life platform used in the development of a virtual university campus. *International Journal of Knowledge Society Research*, *7*(4), 12–23. doi:10.4018/IJKSR.2016100102

Vargas-Vera, M. (2017). The implementation of DSSim: A multi-agent ontology mapping system. *International Journal of Knowledge Society Research*, *8*(3), 34–53. doi:10.4018/IJKSR.2017070103

Varian, H. R., Farrell, J., & Shapiro, C. (2005). *The Economics of Information Technology*. New York: Cambridge University Press.

Vidal-Filho, J. N., Lisboa-Filho, J., de Souza, W. D., & dos Santos, G. R. (2013). Qualitative Analysis of Volunteered Geographic Information in a Spatially Enabled Society Project. In *Computational Science and Its Applications–ICCSA 2013* (pp. 378–393). Springer Berlin Heidelberg. doi:10.1007/978-3-642-39646-5_28

Virtanen, M. A., Kääriäinen, M., Liikanen, E., & Haavisto, E. (2017). The Comparison of students' satisfaction between ubiquitous and web-based learning environments. *Education and Information Technologies*, *22*(5), 2565–2581. doi:10.100710639-016-9561-2

Vishwakarma A. C., Lakhawat P. S., & Poonam. (2013). Study on quality of work life among academicians of private technical educational institutions: Allahabad city, Uttar Pradesh. *International Research journal of Management Science and Technology*, *4*(3), 51-60.

Visvizi, A., Lytras, M. D., & Daniela, L. (2018). (Re)Defining smart education: Towards dynamic education and information systems for innovation networks. In M. D. Lytras, L. Daniela, & A. Visvizi (Eds.), *Enhancing knowledge discovery and innovation in the digital era* (pp. 1–12). Hershey, PA: IGI Global. doi:10.4018/978-1-5225-4191-2.ch001

Vukolic, M. (2010). The Byzantine empire in the intercloud. *ACM SIGACT News*, *41*(3), 105–111. doi:10.1145/1855118.1855137

Vygotsky, L. (1980). *Mind in Society: The Development of Higher Psychological Processes*. Cambridge, MA: Harvard University Press.

Wadud, N. (1996). Job Stress & QWL among Working Women. *Bangladesh Psychological Studies*, *6*, 31–37.

Waldo, D. (1994). Public Administration and Ethics. In F. S. Lane (Ed.), *Current Issues in Public Administration*. New York: St. Martin's Press.

Walker, R., Voce, J., & Ahmed, J. (2012). *2012 survey of technology enhanced learning for higher education in the UK*. Retrieved from Oxford Universities and Colleges Information Systems Association: http://www.ucisa.ac.uk/groups/ssg/surveys.aspx

Walker, R., Voce, J., Ahmed, J., Nicholls, J., Swift, E., Horrigan, S., & Vincent, P. (2014). *2014 survey of technology enhanced learning: Case studies*. Retrieved from Oxford: Universities and Colleges Information Systems Association: http://www.ucisa.ac.uk/groups/dsdg/asg/~/media/7BCB3F2FF0E141A79A66BC87DDB34A14.ashx

Watson, G., & Glaser, E. (2002). *Watson–Glaser critical thinking appraisal—UK edition. Test booklet*. The Psychological Corporation.

WebCrawler. (2018). Retrieved from www.webcrawler.com

Weinzierl, A. (2014). *Global Agenda Council Launches Guide to Good Government and Trust-Building*. Retrieved on August 25,2015, from www.oecd.org/governance/WEF_NR_FutureOfGovernmentSmartToolb

Whyte, G., & Classen, S. (2012). Using storytelling to elicit tacit knowledge from SMEs. *Journal of Knowledge Management*, *16*(6), 950–962. doi:10.1108/13673271211276218

Williams, K.Y.B., & Griffin, J.A.G. (2018). Better security and encryption within cloud computing systems. *International Journal of Public Administration in the Digital Age, 5*(2), 1-11.

Williams, E., Gray, J., & Dixon, B. (2017). Improving geolocation of social media posts. *Pervasive and Mobile Computing*, *36*, 68–79. doi:10.1016/j.pmcj.2016.09.015

Williams, T. (2008). How do organizations learn lessons from projects-and do they? *Engineering Management. IEEE Transactions on*, *55*(2), 248–266.

Wiseman, A. W., Alromi, N. H., & Alshumrani, S. (2014). Challenges to creating an Arabian Gulf knowledge economy. In A. W. Wiseman, N. H. Alromi, & S. Alshumrani (Eds.), *Education for a Knowledge Society in Arabian Gulf Countries* (pp. 1–33). Bingley, UK: Emerald. doi:10.1108/S1479-367920140000024002

Wong, C.-H., Tan, B.-I., Tan, G.-W.-H., & Ooi, K.-B. (2015). Mobile advertising: The changing landscape of the advertising industry. *Telematics and Informatics*, *32*(4), 720–734. doi:10.1016/j.tele.2015.03.003

World Bank. (2017). *World-Bank Data*. Retrieved from http://data.worldbank.org/indicator/NY.GDP.MKTP.KD.ZG

World Economic Forum. (2007). *United Arab Emirates and the World: Scenarios to 2025*. Retrieved on September, 8, 2015, from www.weforum.org/pdf/scenarios/UAE.pdf

World Telecommunication Development Report. (2005). Geneva: Access Indicators for the Information Society.

WSIS. (2003). Declaration of Principles: Building the Information Society: a global challenge in the new Millennium. *World Summit on the Information Society (WSIS), Geneva 2003-Tunis 2005*.

WSIS. (2005). *World Summit on the Information Society: Declaration of Principles: Building the Information Society: a global challenge in the new Millennium, Geneva 2003-Tunis 2005*. WSIS.

Wyse, S. V., Perry, G. L., O'Connell, D. M., Holland, P. S., Wright, M. J., Hosted, C. L., ... Curran, T. J. (2016). *A quantitative assessment of shoot flammability for 60 tree and shrub species supports rankings based on expert opinion.* Academic Press.

Xia, B. S., & Rekola, M. (2013). Learning outcomes and knowledge sharing using web-based technologies in Finnish forest education from an educational experience point of view. *E-Learning and Digital Media, 10*(1), 95–106. doi:10.2304/elea.2013.10.1.95

Xu, K., Kim, H.-W., & Kankanhalli, A. (2010). Task and social information seeking: Whom do we prefer and whom do we approach? *Journal of Management Information Systems, 27*(3), 211–240. doi:10.2753/MIS0742-1222270308

Xu, Y., Tan, C., & Yang, L. (2006). Who will you ask? An empirical study of interpersonal task information seeking. *Journal of the American Society for Information Science and Technology, 57*(12), 1666–1677. doi:10.1002/asi.20339

Yadav, R., & Khanna, A. (2014). Literature Review on Quality of Work Life and Their Dimensions. *IOSR Journal of Humanities and Social Science, 19*(9), 71–80. doi:10.9790/0837-19957180

Yaghi & Al-Jenaibi. (2017). Happiness, Morality, Rationality, and Challenges in Implementing Smart Government Policy. Public Integrity, 1-16.

Yang, L., Dong, L., & Bi, X. (2016). An improved location difference of multiple distances based nearest neighbors searching algorithm. *Optik-International Journal for Light and Electron Optics, 127*(22), 10838–10843. doi:10.1016/j.ijleo.2016.08.091

Ye, C., Swilley, E., & Bacile, T.-J. (2014). From firm controlled to consumer contributed: Consumer co-production of personal media marketing communication. *Journal of Interactive Marketing, 28*(2), 117–133. doi:10.1016/j.intmar.2013.12.001

Yesser. (n.d.). *E-Government Program.* Saudi Ministry of Communication and Information Technology. Retrieved October 2016 from https://www.yesser.gov.sa/en/Pages/default.aspx

Yılmaz, Y., & Hero, A. O. (2018). Multimodal Event Detection in Twitter Hashtag Networks. *Journal of Signal Processing Systems for Signal, Image, and Video Technology, 90*(2), 185–200. doi:10.100711265-016-1151-4

Yin, J., Martin, J. P., Venkataramani, A., Alvisi, L., & Dahlin, M. (2003). Separating agreement from execution for byzantine fault tolerant services. *Operating Systems Review, 37*(5), 253–267. doi:10.1145/1165389.945470

Yonazi, J., Sol, H., & Boonstra, A. (2010). Exploring issues underlying citizen adoption of e-government initiatives in developing countries: The case of Tanzania. In *Proceedings of the 10th European conference on e-government: National center for taxation studies University of Limerick, Ireland* (p. 425). Academic Press.

Younessian, E., & Rajan, D. (2015). Multi-modal fusion for associated news story retrieval. *Multimedia Tools and Applications Journal, 74.*

Younge, A., Von, L., & Gregor, W. (2010). Efficient resource management for Cloud computing environments. *International Conference on Green Computing*, 357-364. 10.1109/GREENCOMP.2010.5598294

Zaigham, M. (2013). *Cloud Computing: Methods and Practical Approaches.* Dordrecht: Springer.

Zhang, F., Yuan, N. J., Wilkie, D., Zheng, Y., & Xie, X. (2015). Sensing the pulse of urban refueling behavior: A perspective from taxi mobility. *ACM Transactions on Intelligent Systems and Technology, 6*(3), 37. doi:10.1145/2644828

Zhao, W. (2007). BFT-WS: A Byzantine fault tolerance framework for web services. In *Proceeding of The Eleventh International Conference of EDOC'07.* IEEE. 10.1109/EDOCW.2007.6

Zhao, Z., Cheng, Z., Hong, L., & Chi, E. H. (2015, May). Improving User Topic Interest Profiles by Behavior Factorization. In *Proceedings of the 24th International Conference on World Wide Web* (pp. 1406-1416). International World Wide Web Conferences Steering Committee. 10.1145/2736277.2741656

Zhuhadar, L., Coleman, P., & Marklin, S. (2016). Applicative personalized learning: How gamification is driving learning. *International Journal of Knowledge Society Research*, *7*(4), 24–37. doi:10.4018/IJKSR.2016100103

Zhu, N., Zhao, H., Feng, W., & Wang, Z. (2015). A novel particle filter approach for indoor positioning by fusing WiFi and inertial sensors. *Chinese Journal of Aeronautics*, *28*(6), 1725–1734. doi:10.1016/j.cja.2015.09.009

About the Contributors

Miltiadis D. Lytras is Associate Professor at Deree-ACG with a research focus on semantic web, knowledge management and e-learning, with more than 100 publications. He has co-edited more than 45 special issues in International Journals (*IEEE Transaction on Knowledge and Data Engineering, IEEE Internet Computing, IEEE Transactions on Education, Computers in Human Behaviour, Interactive Learning Environments, Journal of Knowledge Management, Journal of Computer Assisted Learning*) and has authored/(co-)edited 42 books (e.g., *Open Source for Knowledge and Learning Management, Ubiquitous and Pervasive Knowledge Management, Intelligent Learning Infrastructures for Knowledge Intensive Organizations, Semantic Web Based Information Systems, China Information Technology Handbook, Real World Applications of Semantic Web and Ontologies, Web 2.0: The Business Model*, etc.). He (has) serves(ed) as the (Co) Editor in Chief of 8 international journals (e.g., *International Journal on Semantic Web and Information Systems, International Journal of Knowledge Society Research, International Journal of Knowledge and Learning, International Journal of Technology Enhanced Learning*).

Linda Daniela, Dr.paed., is a professor and senior researcher at the University of Latvia and Chair of the Council for promotion in Pedagogy of the University of Latvia. She is Head of Scientific Institute of Pedagogy in University of Latvia, expert in education at the Council of Sciences of the Republic of Latvia. Her research interests include virtual education, smart education, educational technologies for learning, educational robotics, educational processes and solutions for reducing social exclusion from the educational processes. She is an author and co-author of publications about processes in all dimensions of education. She has been involved in several projects like "Printed audio-learning materials for early childhood education", "Robotics-based learning interventions for preventing school failure and Early School Leaving", "Young children (0-8) and Digital technology", "The creation and implementation of support programs for the youth at the risk of being subjected to social exclusion. The creation of a support system". She is editor and co-editor of several books and journals and she is a scientific adviser for doctoral and master thesis

Anna Visvizi is Associate Professor at Deree-The American College of Greece and Head of Research at the Institute of East-Central Europe (IESW), Poland. A political scientist and economist, editor and research consultant, with an extensive experience in academia and the think-tank sector, Dr. Visvizi presented her work in many forums across Europe and the US, including Capitol Hill. A practiced team-worker, researcher, analyst and lecturer, Dr. Visvizi's expertise spans the following interrelated areas of research: political economy; migration; smart cities; innovation, competitiveness and policy-making, with the geographical focus of her research defined by the EU, China, and the Arab World.

* * *

Noura Abdaoui is a researcher student at the Higher Institute of Multimedia Arts in Manouba University (ISAMM). She received her Master's degree in Multimedia and decision from the University of Manouba in 2017. Her research interests include interaction between smart devices in pervasive environment and ubiquitous computing.

Badreya Al-Jenaibi is full Professor in Mass Communication at the UAEU. Her research interests include Pedagogy, international communication, public relations, as well as the uses and effects of mass media, and new media. Her research has appeared in the Journal of Applied Journalism & Media Studies, International Journals of Information Systems & Social Change, the International Journal of E-adoption, Global Media Journal, Journal for Communication and Culture, and the Journal of Cross Cultural Communication. She served in the editorial board in 18 peer-reviewed international journals, she published more than 32 peer-reviewed articles. Also, she received Khlifah Award for the best distinguished teacher and professor in the Arab world, Arab Youth and Academic Publishing Award, Arab Women Creative Award-best creative teacher, His Highness Hamdan Bin Mubarak Award for the best distinguished employee in specialized jobs – Academic, Quality Assurance and Community Service Award, UAEU Academic and Publishing Award, Volunteering Award. She is the director of Mubadrah student community engagement center, her activities were engaged with more than 121 local organizations.

Abdulaziz Alshouiby is Former Director General of the Saudi e-Government Program (Yesser), and former board member of the General Authority for Statistics, Kingdom Saudi Arabia. He also works as an advisor to the College of Computer Engineering and Sciences, Prince Sattam bin Abdulaziz University. He holds an MBA and a Master of IT from The University of Newcastle, Australia. During his past 20 years' experience, he made key contributions to the development of the Saudi National e-Payments Platform (SADAD); and he established Tamkeen Technologies Ltd and became its first CEO. He was also advisor for the College of Computer and Information Sciences, King Saud University.

Saad Haj Bakry is Full Professor of Information Networks at King Saud University since 1989 PhD from Aston University, England, UK, 1980 CEng and FIET, UK Consultant to the Ministry of Higher Education, Saudi Arabia, Consultant to the Saudi Commission for Information and Communication Technology.

Julia Barth has been assistant lecturer for Smart City Research at Heinrich Heine University Düsseldorf. She conducted field researches on cities in the Gulf Region (especially, in Kuwait, Qatar, U.A.E. and Oman) and published articles on the "Arabian Way" into knowledge society.

Gregory Block received his doctorate in Computer information Systems from Nova Southeastern University in 2013. In 2017 he was hired as the chair of the Strategic Analytics program in the Graduate Professional Studies division of the Rabb School of Continuing Studies at Brandeis University. In addition, he is a professor of information systems at the City University of Seattle. He writes on issues related to asynchronous learning, software engineering and data analytics.

Rafik Bouaziz is full professor on computer science at the Faculty of Economic Sciences and Management of Sfax University, Tunisia. He was the president of this University during August 2014 – December 2017, and the director of its doctoral school of economy, management and computer science during December 2011 – July 2014. His PhD has dealt with temporal data management and historical record of data in Information Systems. The subject of his accreditation to supervise research was "A contribution for the control of versioning of data and schema in advanced information systems". Currently, his main research topics of interest are temporal databases, real-time databases, information systems engineering, ontologies, data warehousing and workflows. Between 1979 and 1986, he was a consulting Engineer in the organization and computer science and a head of the department of computer science at CEGOS-TUNISIA.

Sowmyarani C. N. is currently working as an Associate Professor in department of Computer Science and Eng. at RVCE, Bangalore. Has published research papers in the field of Data engineering, which includes security and privacy. Areas of interests are, Data security, Educational technology and Cryptography.

Simon Cleveland is a professor of information systems at the City University of Seattle. He has also taught at Georgetown University, Brandeis University, and Northeastern University.Dr. Cleveland has a Ph.D. in Information Systems from Nova Southeastern University's College of Engineering and Computing. He is certified as a project management professional (PMP), Six Sigma Black Belt (SSBB) expert, Certified ScrumMaster (CSM), and Information Technology Infrastructure Library (ITIL) foundations expert. Dr. Cleveland is the author of numerous peer-reviewed publications in journals and conferences, and is a three-time recipient of the prestigious Dr. Harold Kerzner Scholarship which is awarded by the Project Management Institute Educational Foundation. Professionally, Dr. Cleveland has worked for organizations such as MD Anderson Cancer Center, Department of Homeland Security, NASA, Accenture, and AOL.

Essien D. Essien holds a Ph.D in Social and Religious Ethics from University of Ibadan, Ibadan, Nigeria. He is a Senior lecturer and a researcher in the University of Uyo, Uyo, Nigeria where he currently teaches Social ethics, Comparative Religious and Christian Ethics in the Department of Religious and Cultural Studies of the University. He writes and presents papers widely on issues of Governance and ethics, Ethics of political system, environmental ethics, Peace ethics, Social media and ethics, comparative religious ethics and development studies. He has authored three books, eight book chapters, and many journal articles in peer reviewed scholarly journals.

Sami Faïz is a PhD in Computer Science from the Laboratoire de Recherche en Informatique of Orsay University (Paris 11). He is actually Full Professor in Computer Science at the Higher Institute of Multimedia Arts of Manouba and member of the Laboratory of Remote Sensing and Spatial Information Systems. He published more than 150 papers in specialized conferences and journals. He's also the author of three books in the framework of GIS. Dr Faïz is a scientific and organizing member of various international conferences. He's also the founder of many national and international projects in Geomatics.

Julia Gremm has been assistant lecturer for Smart City Research at Heinrich Heine University Düsseldorf. She conducted field researches on cities in the Gulf Region (especially, in Kuwait, Qatar, U.A.E. and Oman) and published articles on the "Arabian Way" into knowledge society.

Ismahène Hadj Khalifa is an assistant professor. She was graduated from the Business School IHEC Carthage, Carthage, Tunisia in 2006. She received the master degree in Information System and Conception Engineering in 2007 and PhD in Computer Sciences in 2011 from Ecole Centrale de Lille, France. Her current research interests include itinerary optimization methods in real time and pervasive computing.

Anis Jedidi graduated in Computer Science applied in management at the University of Sfax (Tunisia, 1999) and received the PhD in Computer Science from Paul Sabatier University (France, 2005) on studies about Multimedia information system. He is currently a Lecturer and a Researcher within the MIRACL Laboratory at the University of Sfax (Tunisia), where he specified his research on Semantic Web and Multimedia. He has been teaching and disseminating XML technology since 2005 and is actively applying it to different domains, from multimedia system to intelligent and pervasif Web. His main interests and research areas concern the XML technology and new Web standards, the Semantic Web, multimedia systems and standards, knowledge representation and innovative information and communication technologies. He is the author of more than 50 international publications on journals, conferences and books in these areas.

Daiga Kalnina is assistant professor at the University of Latvia Author of more than 30 methodological materials, student's books and teacher's books about information literacy, science teaching and learning using inquiry approach and education for sustainable development. Works in the international projects: Motivate and Attract Students to Science, Lessons from Nature, UNESCO Associated School Project, Education for Changes, Research on Educational Treatment of Diversity in Inclusive contexts, Evaluation of Spanish cooperation and international teacher training programmes: the model for the development of future projects in the common European education space, Research of Methodology to Improve Learning Quality of the New Generation etc. Scientific interests: technologies in teaching-learning process, learning of young children, didactics, science teaching-learning process.

Alice Li is a Lecturer in the Department of Management at La Trobe University, Australia. Her research interests are in the areas of international business, knowledge management and innovation, international human resource management, business and economic history, cross-cultural management and organisational development.

Sahar Maâlej Dammak received the PhD in Computer Science from University of Sfax (Tunisia, 2016). She is currently a Researcher within the MIRACL Laboratory (Multimedia Information Systems and Advanced Computing Laboratory) at the University of Sfax (Tunisia), where she specified her search on Semantic Web. Her research focuses on the construction of a system to achieve semantic annotation and interrogation for Web pages. She has been teaching in the Higher Institute of Management of Gabes, Tunisia (http://www.isggb.rnu.tn/).

Walid Mahdi is currently Associate Professor in College of Computers and Information Technology, Taif University, Kingdom of Saudi Arabia. He received the Ph.D degree in Computer Science in 2001, from Ecole Centrale de Lyon (France) and the M.S. degree in Computer Science from Paris-Sorbonne University. He spent one year at Ecole Centrale-Lyon (France) as a PostDoc position before joining R&D Avivias SA that specializes in audio-video asset administration platforms for media and corporate markets. In 2004, he joined the Sfax University (US)-Tunisa, Higher Institute of Computer Science and Multimedia, when he finished his habilitation, the highest academic qualification, in 2010. During the last 10 years he has held many administrative positions including Director of Computer Engineering Departement, M.S. degree Departement and local expert at the European Investment Bank. He is also a member of the Multimedia, InfoRmation systems and Advanced Computing Laboratory (MIRACL). His research interests include Computer Vision and Image & video analysis. He was the Guest Editorial for the International Journal of Recent Contributions from Engineering, Science & IT (iJES), the chair, chairman and PC member in multiple conferences.

Mehedi Masud is an Associate Professor in the Department of Computer Science at the Taif University, Taif, KSA. Dr. Mehedi Masud received his Ph.D. in Computer Science from the University of Ottawa, Canada. His research interests include cloud computing, distributed algorithms, data security, data interoperability, formal methods, cloud and multimedia for healthcare. He has authored and co-authored around 50 publications including refereed IEEE/ACM/Springer/Elsevier journals, conference papers, books, and book chapters. He has served as a technical program committee member in different international conferences. He is a recipient of a number of awards including, the Research in Excellence Award from Taif University. He is on the Associate Editorial Board of IEEE Access, International Journal of Knowledge Society Research (IJKSR), and editorial board member of Journal of Software. He also served as a guest editor of ComSIS Journal and Journal of Universal Computer Science (JUCS). Dr. Mehedi is a Senior Member of IEEE, a member of ACM.

Marco Moreno-Ibarra is professor of the Laboratory of Intelligent Processing of Geospatial Information of the Centre for Computer Research (CIC) and CIO of the Instituto Politecnico Nacional (IPN) of Mexico. He received his PhD in Computer Science in 2007 and he is member of the National Research System level 2. Former Director of the Centro Nacional de Cálculo (CENAC) of the IPN, former Director of Systems, Informatics and Telecommunications of National Council of Research Science and Technology (CONACYT). He has over 100 articles in journals, national and international conferences as well as being a reviewer for publications. He teaches the courses of Foundations of Geographic Information Science and Tools for the design and implementation of GIS within the master and doctorate programs in computer science at CIC-IPN. His research areas include GIS design, automatic generalization, volunteered geographic information and geospatial semantic similarity.

Chandra Sekhar Patro is currently serving as Assistant Professor at Gayatri Vidya Parishad College of Engineering (Autonomous), India. He has a Ph.D in Faculty of Commerce and Management Studies from Andhra University, India. He has post-graduate degree in Master of Commerce (M.Com.) from Andhra University, Master in Financial Management (MFM) from Pondicherry University, and also MBA (HR & Finance) from JNT University. Mr. Patro has over 9 years of teaching experience in higher education. Mr. Patro has gained very good knowledge in Human Resource Management and

Accountancy/Finance subjects. He has published number of research papers in reputed National and International Journals and also presented papers in National and International Conferences.

Dayananda Pruthviraja is working as Associate professor in Information Science & Eng. in JS-SATE, Bangalore. He has published research papers in the field of Data science and Information retrieval.

Ana Maria Magdalena Saldana Perez was born in Mexico City, Mexico, in 1989. She received the Communications and Electronic engineer title from the Escuela Superior de Ingeniería Mecánica y Eléctrica, Intituto Politécnico Nacional (IPN), Mexico in 2011, and the Master in computing sciences degree from the IPN in 2014. Currently, she is studying a PhD. in computing sciences at Centro de Investigación en Computación (CIC), IPN. In 2013 she made a research stay in the Informatics Department from the Concepción University, in Concepción Chile. In 2017 she made a research stay in the Steamer team from the Laboratoire d'Informatique de Grenoble, Grenoble France. She has participated in different congresses and she is the author of more than five researching articles. Her research areas include social networks data analysis, spatiotemporal data representations, GIS and geospatial information managing, mobile devices programming, and machine learning study and implementations. M. Sc. Saldaña-Pérez has been member of the geospatial information-processing laboratory at CIC since 2012.

Halima E. Samra was received the bachelor degree in Computer Science from King Abdulaziz University, Jeddah, Saudi Arabia, in 1992. Master of Information Technology from La Trobe University, Melbourne Australia, in 2008. She also gets a second master degree in Computer Science from La Trobe University, in 2015. Currently, she is a PhD candidate at La Trobe University with a research interest in the area of Health Information Systems Management, Clinical Databases, and E-Training. In 1998-2010, she joined the Health Sciences College, in Saudi Arabia, as a Teaching Assistant in Computer Studies. In 2011, she joined the Department of Computer Science and Information Technology at King Abdulaziz University, Jeddah, Saudi Arabia as a lecturer. In 2015 she granted a scholarship from King Abdulaziz University to pursue her postgraduate study (PhD) in Computer Science.

Ben Soh is an Associate Professor in the Department of IT and Computer Science, La Trobe University, Australia. His areas of research are: Cloud Computing, IS Research, Educational Technology, Business Process Modeling and Privacy & Security.

Wolfgang G. Stock is professor for information science and head of the Information Science Department of Heinrich Heine University Düsseldorf, Germany. His research activities und lectures include information retrieval, knowledge representation, informetrics, smart cities, knowledge society, social media and the markets of digital information. He is author of more than 300 articles and of some basic text books in information science. Additionally, he is editor of the book series "Knowledge & Information. Studies in Information Science" (De Gruyter Saur).

Raimonds Strods is a Ph.D. candidate in pedagogy and a researcher at the University of Latvia, Faculty of Education, Psychology and Art. His research interests include digitalization of higher education, professional agency of academic staff and media pedagogy in general. He has been engaged in the OECD PISA as a testing assistant as well as in National research program on social telerehabilitation, European Commission research on digital literacy and Erasmus+ project on robotics-based learning.

Tarek Zlitni is an Assistant Professor at the Higher Institute of Computer Science and Multimedia, Sfax, Tunisia. He received the Ph.D degree in Computer Science in 2016, from Faculty of Economics and Management of Sfax (Tunisia). He is also a member of the Multimedia, InfoRmation systems and Advanced Computing Laboratory (MIRACL) at Sfax Uniersity, Tunisia. His research interests focus on video and image processing and analysis, multimedia indexing, and content-based video segmentation and structuring, especially the automatic structuring of digital TV streams.

Index

www.ingramcontent.com/pod-product-compliance
Lightning Source LLC
LaVergne TN
LVHW062141200926
848484LV00007B/1840